Algorithmic Trends in Computational Fluid Dynamics

ICASE/NASA LaRC Series

M.Y. Hussaini   A. Kumar   M.D. Salas
Editors

Algorithmic Trends in Computational Fluid Dynamics

With 121 illustrations

Springer-Verlag

New York Berlin Heidelberg London Paris
Tokyo Hong Kong Barcelona Budapest

M.Y. Hussaini
Institute for Computer Applications
  in Science and Engineering
Mail Stop 132-C
NASA Langley Research Center
Hampton, VA 23681-0001
USA

A. Kumar
M.D. Salas
Mail Stop 156
NASA Langley Research Center
Hampton, VA 23681-0001
USA

Library of Congress Cataloging-in-Publication Data
Algorithmic Trends in computational fluid dynamics /
    [edited by] M.Y. Hussaini, A. Kumar, M.D. Salas.
        p.   cm.
    Proceedings from a workshop held Sept. 15-17, 1991.
    Includes bibliographical references.
    ISBN 0-387-94014-6 (New York). -- ISBN 3-540-94014-6 (Berlin)
    1. Fluid dynamics--Data processing-Congresses. 2. Algorithms-
-Congresses. I. Hussaini, M. Yousuff. II. Kumar, Ajay.
    III. Salas, M. D.
QA911.A555 1993
    620.1'064'0285--dc20   92-44294

Printed on acid-free paper.

Production managed by Karen Phillips, manufacturing supervised by Vincent Scelta.
Camera-ready copy provided by the editors.
Printed and bound by Edwards Brothers, Inc., Ann Arbor, MI.
Printed in the United States of America.

9 8 7 6 5 4 3 2 1

ISBN 0-387-94014-6 Springer-Verlag New York Berlin Heidelberg
ISBN 3-540-94014-6 Springer-Verlag Berlin Heidelberg New York

DEDICATION

This book is dedicated to Professor Joseph L. Steger, a pioneer in computational fluid dynamics, who passed away on May 1, 1992. It is sadly noted that the first article in this volume is also his last contribution.

Joe, as his friends called him, graduated from Parks College, St. Louis University, in 1964. His capabilities, at the young age of 21, had not escaped the notice of the school faculty who that year recognized him as the St. Louis AIAA Outstanding Engineering Graduate of Parks College. After a brief stay at General Dynamics, Joe enrolled in Iowa State University where he obtained his Master's and Doctoral degrees in Aerospace Engineering. In 1969, with Doctoral degree in hand, Joe started his long and fruitful association with NASA Ames Research Center as a National Research Council Post Doctoral Fellow.

The years at Ames, 1970 to 1978 and later from 1983 to 1989, characterized by an outpouring of original ideas that forever changed the field of computational fluid dynamics. Joe developed new algorithms to solve transonic flow over airfoils and wings. He pioneered the flux vector splitting scheme for the solution of the Euler equations. He studied vortical flows, separated boundary layers, fluid-structures interactions, cascades, projectile base flows, helicopter rotor blades, and many other problems. Above all, Joe was a practical scientist. He looked for ways to maximize his resources so that he could study problems of ever greater complexity. This led him to zonal methods, where the local nature of the flow dictates the equations that are used. His interest in useful, practical work led him to a study of graphics and visualization techniques, and to many original ideas on grid generation, among them the development of the hyperbolic grid generation method and the Chimera scheme. It was the Chimera scheme, a set of structured grids that overlay one another and locally fit the body surface, that occupied most of his later years. With it, he was able to solve some of the most complex flow problems ever attempted, such as a simulation of the space shuttle in its ascent configuration. His many contributions gained him the respect and international recognition of his peers, and numerous awards, such as the NASA Exceptional Scientific Achievement Medal in 1986, the H. Julien Allen Award in 1987, a NASA Ames Research Center award given for the best scientific paper, and the Silver Snoopy Award in

1988, given to him by the crew of the STS-26 Space Shuttle for his contributions to make the shuttle safer. This period of his life was marked by another happy occasion, his marriage to Ellen Price on April 25, 1987.

Joe had a great interest in education. From 1980 to 1983 he was associate professor in the Department of Aeronautics and Astronautics at Stanford University, and from 1989 to 1992 he was a senior professor in the Department of Mechanical, Aeronautical and Materials Engineering at the University of California at Davis. Joe was an excellent teacher and he developed a very comprehensive set of notes from his classes in computational fluid dynamics. Joe was very involved in the school activities of his students. He was a strong advocate for improvements to the graduate program, he crusaded to make modern workstations available to the students, and having the students involved on practical engineering projects.

It is with great respect for his accomplishments as a great scientist, an excellent teacher and a gentle friend that we dedicate this book to Professor Joseph L. Steger.

M. Yousuff Hussaini, ICASE, NASA LaRC
Ajay Kumar, NASA LaRC
Manny Salas, NASA LaRC

DEDICATION

Joseph L. Steger
1943 - 1992

PREFACE

This volume contains the proceedings of the ICASE/LaRC Workshop on the "Algorithmic Trends for Computational Fluid Dynamics (CFD) in the 90's" conducted by the Institute for Computer Applications in Science and Engineering (ICASE) and the Fluid Mechanics Division of NASA Langley Research Center during September 15-17, 1991. The purpose of the workshop was to bring together numerical analysts and computational fluid dynamicists i) to assess the state of the art in the areas of numerical analysis particularly relevant to CFD, ii) to identify promising new developments in various areas of numerical analysis that will have impact on CFD, and iii) to establish a long-term perspective focusing on opportunities and needs.

This volume consists of five chapters – i) Overviews, ii) Acceleration Techniques, iii) Spectral and Higher-Order Methods, iv) Multi-Resolution/ Subcell Resolution Schemes (including adaptive methods), and v) Inherently Multidimensional Schemes. Each chapter covers a session of the Workshop. The chapter on overviews contains the articles by J. L. Steger, H.-O. Kreiss, R. W. MacCormack, O. Pironneau and M. H. Schultz. Steger was a pioneer in developing algorithms for complex configurations. His viewpoint on discretization schemes brings out the pros and cons of overset structured grids or a composite of prismatic and unstructured grid or a combination of both. Kreiss categorizes problems based on resolution requirements, and makes some remarks on how to deal with them. MacCormack who is at the forefront of computational fluid dynamics research discusses the current and future directions in this field. His article provides an excellent overview of issues associated with structured versus unstructured grids, upwind versus central differencing, and approximate factorization versus direct inversion. It also includes expert comments on implicit and multigrid procedures on parallel computers as well as turbulence and large eddy simulation. Pironneau's article reviews finite volume and finite element methods for compressible Navier-Stokes equations, and also sheds cursory light on mathematical questions such as the existence and uniqueness of the solutions of the relevant equations and convergence results on the discretized versions thereof. Shultz gives a brief overview of the challenges in a massively parallel computation.

The second chapter deals with acceleration techniques, and includes articles by Bram van Leer, R. Temam, Y. Saad and G. Golub and colleagues with the final comments on the session by M. Hafez. The third chapter is on spectral and higher order methods. It consists of four authoritative papers by C. Canuto, D. Gottlieb, S. Osher, and S. A. Orszag, respectively. The fourth chapter comprises of contributions by R. Glowinski and colleagues on wavelet methods, E. Harabetian on subcell-resolution methods, A. Harten on the multiresolution analysis of essentially nonoscillatory (ENO) schemes, and by K. Powell and colleagues on solution adaptive techniques. The fifth chapter treats the inherently multidimensional schemes in the form of four articles by P. L. Roe, H. Deconinck, and R. A. Nicolaides who are renowned researchers in the field. The last article by K. W. Morton provides some insightful comments on these contributions.

Finally, the editors would like to thank all these participants for their cooperation and contributions which made the Workshop a success. We also sincerely appreciate the assistance of Emily Todd who coordinated the collection of manuscripts and facilitated their editing. Thanks are also due to the staff of Springer-Verlag for their assistance in bringing out this volume.

MYH, AK and MDS

CONTENTS

SPECTRAL AND HIGHER-ORDER METHODS

MULTI-RESOLUTION AND SUBCELL RESOLUTION SCHEMES

CONTRIBUTORS

Claudio Canuto
Dipartimento di Matematica
Politecnico di Torino
10129 TORINO
ITALY

Herman Deconinck
Von Karman Institute for
    Fluid Dynamics
72 Chaussee de Waterloo
1640 Rhode-St-Genese
BELGIUM

Roland Glowinski
Department of Mathematics
University of Houston
4800 Calhoun Road
Houston, TX 77204-3476

Gene H. Golub
Computer Science Department
Stanford University
Stanford, CA 94305

David Gottlieb
Division of Applied Mathematics
Box F
Brown University
Providence, RI 02912

Mohamed Hafez
Department of Mechanical
    Engineering
University of California, Davis
Davis, CA 95616

Edward Harabetian
Department of Mathematics
University of Michigan
Ann Arbor, MI 48109-1003

Ami Harten
School of Mathematical Sciences
Tel-Aviv University
Ramat-Aviv, 69978
Tel-Aviv
ISRAEL

Heinz-Otto Kreiss
Department of Mathematics
University of California,
    Los Angeles
Los Angeles, CA

Robert MacCormick
Department of Aeronautics
    and Astronautics
Stanford University
Stanford, CA 94305

K.W. Morton
Oxford University Com-
    puting Lab.
Numerical Analysis Group
8-11 Keble Road
Oxford, OX1 3QD
ENGLAND

R.A. Nicolaides
Department of Mathematics
Carnegie-Mellon University
Pittsburgh, PA 15213

Steven A. Orszag
Program in Applied Mathematics
218 Fine Hall
Princeton University
Princeton, NJ 08544

S. Osher
Department of Mathematics
University of California
Los Angeles, CA 90024-1555

Olivier Pironneau
INRIA
Racquencourt
Le Chesnay 78153
FRANCE

Kenneth G. Powell
Department of Aerospace
    Engineering
The University of Michigan
Ann Arbor, MI 48109-2140

Philip L. Roe
Department of Aerospace
    Engineering
The University of Michigan
Ann Arbor, MI 48109-2140

Youcef Saad
Computer Science Department
University of Minnesota
Twin Cities
4-192 EE/CSci Building
200 Union Street S.E.
Minneapolis, MN 55455

Martin Schultz
Department of Computer Science
Yale University
New Haven, CT 06510

Joseph L. Steger
Department of Mechanical,
    Aeronautical and Materials Eng.
University of California at Davis
Davis, CA 95616

Roger Temam
Laboratoire d'Analyse Numérique
    d'Orsay
Universite de Paris-Sud
Batiment 425
91405 Orsay Cedex
FRANCE

Bram van Leer
Department of Aerospace
    Engineering
The University of Michigan
Ann Arbor, MI 48109-2140

OVERVIEWS

A VIEWPOINT ON DISCRETIZATION SCHEMES FOR APPLIED AERODYNAMIC ALGORITHMS FOR COMPLEX CONFIGURATIONS

Joseph L. Steger

University of California at Davis
Davis, California

ABSTRACT

In the 90's, nonlinear CFD methods must begin to routinely simulate viscous unsteady flow of complete aircraft configurations. While the best approach for such simulations has yet to be determined, it is argued that the discretization requirements for unsteady viscous flow simulation of complex configurations will best be satisfied using either composite overset structured grids or a composite of prismatic and unstructured meshes - or a hybrid of these two.

1. Introductory

The discretized field methods, by which is meant finite difference, finite volume, or finite element methods, have given CFD an ability to cope with nonlinear and multiscale flow phenomena. But the discretized field methods, unlike the linear panel methods that they are replacing, have yet to deal with complex geometries in a mundane way. Consequently, although significant effort is needed to develop considerably more accurate and efficient numerical algorithms, the content of this paper will deal not with numerical algorithm development, but with the discretization process that they require.

The crucial step for CFD in the 90's is the routine viscous flow simulation of complex configurations using the discretized field methods. For a typical aircraft flow simulation, loss of accuracy due to shoddy geometric representation will generally outweigh any loss of accuracy due to shabby treatment of nonlinear compressibility, viscous effects, or unsteadiness. Moreover, inadequate geometric representation is generally a much more significant source of error than that due to deficient turbulence models or not resolving shocks crisply. So the field methods, to fulfill their promise, must first reach and then surpass the level of geometry sophistication enjoyed by linear panel methods.

This paper expresses the author's current viewpoint on flow field discretization schemes amiable to complex configurations. It begins with a brief examination of discretization techniques, and then sets out what the author believes to be the critical technologies that must be incorporated into the next generation of unsteady viscous flow field discretization methods. From this background, it is conjectured that the discretization requirements for unsteady viscous flow simulation of complex configurations will best be satisfied with either composite overset structured grids or a composite of prismatic and unstructured meshes - or some combination of these two.

2. Background

Various ways of discretizing a flow field have been and are being applied. In the early developments of CFD, uniform and stretched rectangular grids with thin airfoil boundary conditions or special operators and flux balances at curvilinear boundary surfaces were employed. For inviscid problems without concentrated gradients, rectangular meshes can still be ideal for much of the flow field. Body fitted structured meshes have also been used from the earliest days of CFD, being well suited for 'blunt body problems' and viscous flow analysis about simple or clean configurations. For complex configurations, however, a simple body fitted structured grid becomes too skewed, will waste points, and is too hard to generate. These difficulties lead to the introduction of composite structured meshes of multi-block and overset form. Complex configurations and more powerful computers also brought on the development of unstructured meshes. More recently, various forms of hybrid structured and unstructured and semi-unstructured meshes have been proposed. A 'hierarchal' listing of discretization approaches, each having advantages and disadvantages, can be listed as:

- Rectangular Meshes - uniform and stretched
- Body Fitted Structured Meshes -without or with cuts
- Semi-Unstructured Meshes - prismatic
- Composite Structured Meshes - blocked and overset
- Unstructured Meshes -quadrilaterals, prisms,....
- Mixed Structured and Unstructured Meshes - patched and overset

The requirements of what constitutes a good discretization have been enumerated before (see, for example, [Thompson & Steger

1988], or a conference proceeding on grid generation such as [Arcilla et al 1991, or AGARD CP-464 1990]) and are briefly reviewed. A practical grid must be relatively easy to build in a short amount of time using readily available computer resources such as engineering workstations. The quality of the grid must be such that numerical accuracy and stability (robustness) does not degrade, and often this imposes constraints on grid smoothness and skewness. Numerically efficient and flexible algorithms must also be maintained, and may, for example, require choices between ordered data bases and efficient mesh adaption procedures. Insofar as computer processing is itself undergoing change, the discretization must efficiently adapt to constraints on storage, vectorization, massively parallel architecture, and distributed processing. Finally, what constitutes a good discretization depends on which application is being considered – inviscid flow, for example, has different constraints than viscous flow. Here it will be assumed that many practical problems of the next decade will dictate discretizations for unsteady viscous flow of complex configurations.

In general, the two mainstream CFD discretization approaches for flow about complex configurations are the composite structured grid approach and the unstructured grid approach. Compared to the unstructured grid method, the composite structured grid approach has more history and development in flow simulation. It currently employs more efficient numerical flow algorithms, requires less computer memory, has vectorized and ported to parallel computers more readily, and has employed special solvers in different grids. Composite structured grids have also dealt with unsteady viscous flow about complex configurations. Unstructured schemes have utilized points more efficiently using adaptive enrichment, and grid generation has been more automated for inviscid flow problems. (For viscous flow, the semi-structured prismatic grid approach appears the most automated [Nakahashi 1991].) With either unstructured or composite structured grids the capability to solve flow about complex configurations has been suitably demonstrated. Hybrid schemes which incorporate the best features of both approaches have already appeared and are certain to become more important.

3. Requirements for Suitable Discretizations in the 90's

Unsteady flow and body motion, viscous flow, and treatment of complex configurations will be assumed to be the critical properties that have to be in hand for CFD simulation codes of the 90's. Given these constraints, in my opinion suitable discretization schemes for the next decade should possess the following features:

Radiated Grids

To simplify grid generation time and effort, field (or volume) meshes should be generated with the user only having to define the boundary surface mesh. By generating the field mesh from only the boundary surface mesh, engineering set up time is reduced. (Ideally, the user should only have to define surface boundary curves to generate the surface mesh). For viscous flow the mesh should also radiate from the surface, that is, along **lines** that emanate from the body surface. Such ray structure allows one dimensional clustering to the body surface which is needed to efficiently resolve viscous boundary layer flow. The single prismatic mesh generation scheme of [Nakahashi 1991] is an excellent example of a radiated mesh. Figure 1 illustrates this scheme for the Boeing-747[1]. Here the unstructured triangular body surface distribution at $\zeta = 0$ is radiated outwards so that prisms are formed between each level surface of $\zeta = const$ and $\zeta = const + \Delta\zeta$. The chimera overset grid approach (c.f. [Benek, Buning, & Steger 1985, Parks, Buning, Steger,& Chan 1991]) generally uses radiated meshes, and Fig. 2 indicates [2] computed surface pressure for the integrated space shuttle along with segments of the grids which were hyperbolically generated outwards from surface grids. While chimera need not only use radiated meshes, future applications are likely to rely on meshes radiated from overset surface grids [Steger 1991].

Current multi-block structured meshes, while allowing one- dimensional clustering to a surface, require specification of outer block boundaries and therefore require more user input than only the surface grid distribution. Conventional unstructured meshes do not use one-dimensional clustering along rays and are therefore not satisfactory for resolving viscous boundary layers. However, prismatic grids blended to unstructured grids are quite apropos.

Adaptiveness

Adaptive griding, whether automatic or user specified, is needed to keep from under utilizing grid points, especially in three dimen-

---

[1]   Courtesy of Prof. K. Nakahashi, University of Osaka Prefecture, Japan

[2]   Unpublished, courtesy of the NASA Johnson Spacecraft Center Space Shuttle CFD Group.

sional applications. The overhead costs and complexity associated with mesh adaptation algorithms is sometimes unwarranted in two dimensions, but the percentage of this overhead is reduced in three dimensions, and the need to save points is more essential in three dimensions than in two dimensions.

Solution adaptation can be accomplished in a variety of ways including:

Migration. Migration or reclustering of existing mesh points to action areas has long been used in structured meshes. Migration can be difficult and costly to implement and has sometimes lead to overly skewed or warped grids. Nevertheless, a considerable amount of viable and cost effective software has been demonstrated for a variety of difficult problems, (c.f.[Benson & McRae 1991]).

Enrichment and Elimination. Enrichment, that is, simply adding points to where they are needed, has been most strikingly demonstrated with unstructured meshes as Fig. 3 demonstrates[3]. With an unstructured mesh, points can simply be added to where they are needed or eliminated from where they are not needed. Enrichment has also been practiced with composite structured meshes, although less aggressively than with unstructured meshes. Especially with the overset grid method, if a flow feature is identified, an additional refined grid can be overset to resolve it (c.f [Berger & Oliger 1987, Rizk, Schiff, & Gee 1990]). Meshes can also be allowed to follow an unsteady feature. Elimination has not been practiced with composite structured meshes because one usually starts with a coarse mesh and adds meshes as needed.

Zonal and Fitting. Besides migration and enrichment of grids points, the desired effects of adaption can often be achieved by using simplified equations (zonal approach) or fitting special discontinuities. Common examples include using a boundary layer approximation near a wall, using nonreflecting or asymptotic boundary conditions in the far field, and using potential equations in irrotational flow regions. Such special equation sets can be incorporated into the solution using connectivities at interface boundaries or via forcing functions. Although shock fitting procedures have only been successfully applied to relatively simple situations, they are nevertheless effective [Moretti 1975]. Moreover, as enrichment procedures

---

[3] Unpublished, courtesy of John Melton, Applied Aerodynamics Branch, NASA Ames Research Center, Moffett Field, CA.

have forcefully demonstrated in resolving shock waves, considerable expertise has now been obtained in treating complex data structures. Another effort in shock fitting may therefore be warranted using floating methods of imposing jump relations to individual cells rather than treating the discontinuity as a boundary surface. Such a procedure could enrich a shock captured solution were shocks are safely identified, leaving complex shock intersections to the capturing method.

Unsteady Body Motion

Most CFD codes are currently configured to steady state flow simulation. Yet, even if the body is stationary, viscous flow often has regions of unsteadiness beyond those that can be modeled as turbulent. There are also an increasing number of engineering problems in which one body can be moving with respect to another body. Some kind of CFD simulation has, for example, already been carried out for rocket staging, aircraft store separation, helicopter and engine rotors, various valve motions, trains moving into and out of tunnels, jets in ground effect, blast wave induced overturning, etc. Figure 4 illustrates[4] a simulation in which a small body is ejected from another. Such problems are difficult to carry out experimentally, and are therefore prime areas for CFD.

Distributed Computer Processing

A significant amount of computer research and financial investment is going into massively parallel computers, so CFD codes of the 90's must be able to efficiently port to massively parallel computers. In addition, workstations of 6 to 25 million floating point operations per second capability are becoming commonly available and are frequently idle at off hours. A ring of 6 such machines can give the performance of a single processor of current supercomputers, and at this particular time, such machines are thought to be increasing in performance and cost effectiveness more rapidly than supercomputers. Both composite structured and unstructured discretization schemes appear to be adapted to distributed processors by using a domain decomposition treatment of the data structure.

4. Discretizations for the 90's

If the preceding observations are correct, in the 90's multi-block

---

[4] Unpublished, courtesy of R. Meakin, University of California at Davis, but see [Meakin & Suhs 1989].

composite grid schemes will defer to the overset (chimera) approach, while unstructured griding will be joined to prismatic griding to resolve viscous body surfaces. Hybrids of both approaches will also continue to evolve.

In the chimera composite structured grid approach, discretization will be further automated by oversetting structured surface meshes on the body boundary surface so as to blanket the surface. Volume grids will then be radiated out from these surface meshes to either the far field or to a simple background grid depending on the problem. Intermediate meshes, perhaps with some capacity for adaptive migration, will be automatically inserted to isolate flow features and even shock fitting may reappear. Likewise, special grids will be inserted to impose efficient solvers. If a mixed structured-unstructured grid approach proves sometimes advantageous, the unstructured data arrays used to connect overset interface boundaries will be expanded to account for small isolated points or regions that will be solved for in an unstructured way.

In the unstructured mesh approach, unstructured surface meshes will be radiated out prismatically to resolve viscous boundary layers. This meshing would then blend into a fully unstructured mesh or would be continued to the far field using various kinds of elimination or enrichment of prisms to resolve gradients as needed. If a mixed structured-unstructured approach proves desirable, in time, regions of structured mesh will be inserted in some regions to enhance solution efficiency.

It is impossible to predict which approach will prove best, the composite overset grid approach (chimera) or a composite prismatic and unstructured mesh approach. Indeed, the most highly funded rather than the best scheme may emerge as the future champion. To me it is striking that the chimera approach (to which I have a vested interest) has remained so competitive, even in the lead, to unstructured meshes while having less organized development effort and support. Certainly this implies that oversetting (whether for structured or unstructured meshing) is a powerful feature not to be ignored.

5. Concluding Remarks

Nonlinear CFD methods must be able to routinely simulate viscous unsteady flow of complete aircraft configurations. While the best discretization approach for such simulations has yet to be determined, the competition is currently between unstructured meshes and composite structured (blocked and overset) meshes. It has been argued here that this competition will ultimately be between chimera schemes, unstructured meshes with radiated prismatic meshing at boundary surfaces, or hybrids of these two basic approaches. In either case, meshes for viscous flow will be generated by radiating the mesh distribution specified on body surfaces.

Overset mesh schemes do not receive the attention that unstructured (or blocked) grid methods have received. Nevertheless they have proved to be competitive, and, since oversetting adds an extra flexibility, they may prove to be the more powerful.

Acknowledgment

This work received support from the Air Force Wright Research and Development Center, Wright-Patterson Air Force Base.

References

Thompson, J. F. and Steger, J. L., 1988 "Three Dimensional Grid Generation for Complex Configurations - Recent Progress", AGAR-Dograph No. 309 .

Arcilla, A. S., Hauser, J., Eisemen, P. R., and Thompson, J. F., editors, 1991 "Numerical Grid Generation in Computational Fluid Dynamics and Related Fields", North Holland, Amsterdam.

Conference Proceedings of the AGARD Fluid Dynamics Panel Specialists' Meeting, 1990 "Application of Mesh Generation to Complex 3-D Configurations", AGARD CP-464.

Nakahashi, K., 1991 "Optimum Spacing Control of the Marching Grid Generation", AIAA Paper No. 91-0103.

Benek, J. A., Buning, P. G., and Steger, J. L., 1985 "A 3-D Chimera Grid Embedding Technique", AIAA Paper No. 85-1523.

Parks, S. J., Buning, P. G. , Steger, J. L., and Chan, W. M., 1991 " Grids for Intersecting Geometric Components within the Chimera Overlapped Grid Scheme", AIAA Paper 91-1587, AIAA 10th Computational Fluid Dynamics Conference, Honolulu, Hawaii.

Steger, J. L., 1991 "On Enhancements to the Chimera Method of Flow Simulation", Proceeding of Supercomputing Japan 91, Tokyo, Japan.

Benson, R. A. and McRae, D. S., 1991 " A Solution-Adaptive Mesh Algorithm for Dynamic/Static Refinement of Two and Three Dimensional Grids", in *Numerical Grid Generation in Computational Fluid Dynamics and Related Fields* Arcilla, A. S., Hauser, J., Eisemen, P. R., and Thompson, J. F., editors, North Holland, Amsterdam.

Berger, M.J. and Oliger, J.,1987 "Adaptive Mesh Refinements for Hyperbolic Partial Differential Equations", J. Comp. Physics, Vol. 53.

Rizk, Y. M., Schiff, L. B., and Gee, K., 1990 "Numerical Simulation of the Viscous Flow Around a Simplified f/a-18 at High Angle of Attack", AIAA Paper No. 90-2999.

Moretti, G., 1975 "A Circumspect Exploration of a Difficult Feature of Multidimensional Imbedded Shocks", Proceedings AIAA 2nd Computational Fluid Dynamics Conference, Hartford.

Meakin, R. L. and Suhs, N., 1989 "Unsteady Aerodynamic Simulation of Multiple Bodies in Relative Motion", AIAA-89-1996, AIAA 9th Computational Fluid Dynamics Conference, Buffalo, NY.

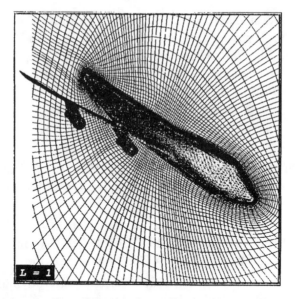

a. Unstructured surface mesh on B-747, $\zeta = 0$ plane.

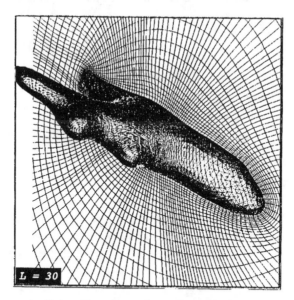

b. Prismatically radiated mesh at $\zeta = 30$ plane.

Figure 1. Two $\zeta = constant$ planes taken from the single prismatic mesh generated by Nakahashi for a Boeing 747.

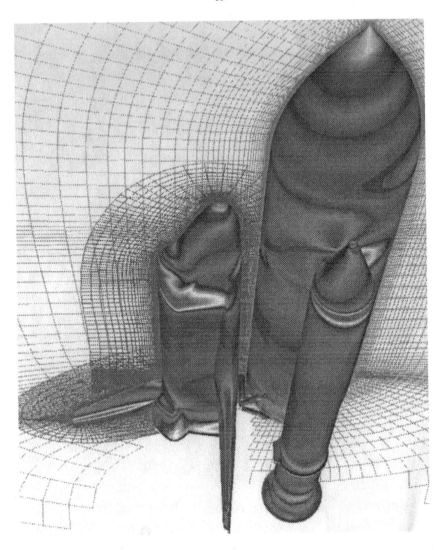

Figure 2. Pressure contours computed on the integrated space shuttle in its ascent configuration, $M_\infty = 0.9, \alpha = -4.5°, Re|_{length} = 6.25 \times 10^6$ (Courtesy of the NASA Johnson Spacecraft Center Space Shuttle CFD Group.)

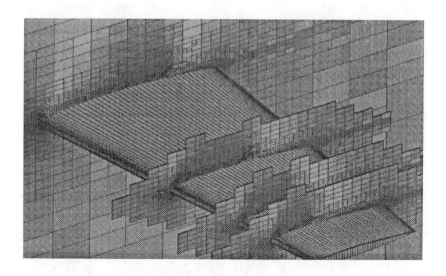

Figure 3. Mach distribution and mesh adaptation for an inviscid Euler equation solution of the ONERA M6 Wing using an unstructured Cartesian mesh, $M_\infty = 0.84, \alpha = 3.06°$. (Figure courtesy of John Melton, Applied Aerodynamics Branch, NASA Ames Research Center.)

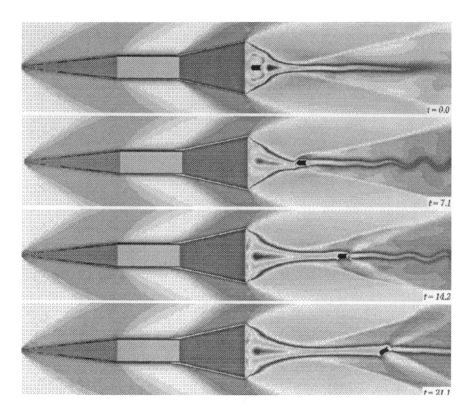

Figure 4. Pellet in free flight after ejection from awedge-flair combination. Wedge-flair at $M_\infty = 2.5, \alpha = 0°, Re|_{w-f\ length} = 9.5 \times 10^6$. (Nondimensional time shown with each frame is time in seconds scaled by $10 \times a_\infty/($wedge flair length$).)$

SOME REMARKS ABOUT COMPUTATIONAL FLUID DYNAMICS

Heinz-Otto Kreiss

University of California, Los Angeles

ABSTRACT

Recent results in computational fluid dynamics are discussed and open problems are indicated. It is stressed that we cannot hope for accurate results, if we do not resolve all the relevant scales. If the capacity of the computer is not adequate, then one has to modify the differential equations.

1. Introduction

During the past 30 years I have been involved in meteorological calculations. In 1955 I was a graduate student, when the first 24 hour weather predictions were performed in Stockholm. The computer, BESK, the largest in Europe at that time, had 1000 words of fast memory and two drums with 3600 words each. The one layer model was very rudimentary with gridpoints at a distance of 500 km. Thus, it had the problem, which plagues us even today in large scale calculations. The capacity of the machines is too small to resolve all the relevant scales. Enormous ingenuity has been developed to beat this restriction. However, satisfactory progress in accuracy was only achieved, when either the differential equations were changed or, due to increased machine capacity, the resolution became adequate. Typically, the accuracy of numerical weather prediction took a jump whenever a new generation of machines were introduced.

Roughly we can divide the problems we try to solve into three classes.

1) Problems with all scales resolved.

2) Problems with all scales resolved except at a few surfaces, where the solution becomes discontinuous (shocks).

3) Not all relevant scales are resolved (turbulence).

2. Problems with all scales resolved

For time dependent problems the theory is in very good shape. There is a complete stability and convergence theory, which is not just a mathematical game but is a rather accurate description of reality. This is not only true for the Cauchy problem but also for initial boundary value problems.

The theory tells us that higher order methods should be used. There are many difficulties with respect to accuracy and stability, if the mesh does not follow the boundary. Grids based on global mappings, overlapping and unstructured grids can overcome these problems. J. Steger gave a very interesting talk on the subject. There will be much development in this area in the next 5 – 10 years. For the moment it is not clear, which of the grid generating techniques will dominate the future.

These techniques can also be used to locally refine the meshes such that also the local scales are resolved, and we have heard excellent talks describing possible techniques. I am convinced that local refinement techniques will be perfected in the next decade.

We are often interested in steady state calculations. The situation here is far from satisfactory. Some of the problems are:

a) We do not know, whether the partial differential equations really converge to steady state.

b) How can we decide whether the steady state has been reached?

There are many examples, where the iteration converges almost to a steady state but then slowly drifts away. In practice, the process is often made quite unpenetrable by the addition of ad hoc tricks, which are meant to speed up the rate of convergence but which might also change the underlying differential equations. Research should be supported to clarify these questions.

3. Problems with discontinuity lines or surfaces

The solutions of many of our problems consist of patches, where all scales can be resolved, connected by discontinuities, for example, shocks. Numerically there are two ways to proceed.

1) **Regularization.** Instead of the inviscid (Euler) equations we use artificially viscous equations. We choose the viscosity such

that all scales are resolved, i.e., the viscosity depends on the meshsize we are willing (or better; forced by machine capacity) to use. Unfortunately, an efficient viscosity coefficient depends on the local "states". Although satisfactory ways to choose it have been found for particular problems, there is still no good theory to help in the choice. I am convinced that such a theory will be developed in the coming years.

2) **Riemann solvers, method of characteristics.** At the discontinuities the differential equations are replaced by algebraic conditions. A technique of this type known as the method of characteristics has been used for a long time. However, the difference with the modern versions is that the method of characteristics does not keep the equations in conservation form. The shockspeed and the position of the shock are calculated explicitly. Therefore, in more than one space dimension, the method requires that the shockfront is calculated explicitly, which can cause problems. The modern versions, based on Riemann solvers, keep the equations in conservation form and therefore, at least in one-dimensional problems, the shockspeed is automatically correct. In more space dimensions the theory is not satisfactory, because the methods in use are often based on dimensional splitting, which can only be justified for smooth solutions.

There has been an explosion of powerful methods. Without them many problems, especially with strong shocks, are difficult to solve. Still, the theory for systems in two or three space dimensions is not yet completely developed.

4. Not all scales can be resolved

Typical examples are turbulent flows, which are solutions of the incompressible Navier-Stokes (N-S) equations. We shall discuss some of the issues facing us.

a) There are very accurate and realistic estimates known for the smallest scale. Therefore we know very well what gridsize or how many modes we need to resolve all relevant scales.

b) In two-dimensional calculations today's computers can resolve these scales even for large Reynolds numbers. Thus we can solve the viscous N-S equations by direct simulation.

In three-dimensional calculations only moderate Reynolds numbers can be treated by direct simulation. It is also quite difficult to perform numerical experiments, because these experiments are performed at the limit of capacity. Practically no "playing around" is possible. There is little hope that the capacity of computers will increase so much in the next ten years that we can perform direct simulations for large Reynolds numbers. Therefore we have to modify the differential equations, i.e., use turbulence models. To improve the existing turbulence models is one of the most important tasks of the next decade.

c) If one can resolve all the relevant scales, then the numerical solution of the N-S equations poses no real difficulty. In the problems, that I have been involved in, it made not much of a difference whether one used a fourth order difference method or a spectral method. Perturbations at high frequencies did not change the solutions in an explosive manner. Instead, there was an inverse cascade, which slowly moved energy from high to low frequencies. However, in long time calculations the large scale is quite sensitive to a change of Reynolds number.

ALGORITHMIC TRENDS IN CFD IN THE 1990'S FOR AEROSPACE FLOW FIELD CALCULATIONS

Robert W. MacCormack

Department of Aeronautics and Astronautics
Stanford University
Stanford, California 94305

ABSTRACT

The development of computational fluid dynamics (CFD) procedures has progressed extremely rapidly during the past two decades. The parallel rapid development in computer hardware resources and architectures has not only matched the explosive algorithm development but has indeed provided and continues to provide its impetus. Together the resources are now available for the numerical simulation of the flow about complex three dimensional aerospace configurations. Yet, in many ways the discipline of CFD is still in its infancy. Major decisions concerning its future direction need to decided and many impeding obstacles must be overcome before its evolution into a mature discipline for solving the equations of compressible viscous flow.

While the battles of the past concerning CFD centered on shock fitting verses shock capturing algorithms or finite element verses finite difference procedures, present and future battles will be concerned with structured multi-block element grids verses unstructured single element grids, indirect relaxation or approximately factored procedures verses direct solution procedures, and turbulence modeling verses direct simulation of turbulent phenomena. These items will be discussed in the paper with regard to the development of CFD for applications to aerospace problems of current and future engineering interest. It may appear premature to consider the item of direct simulation of turbulence as an alternative today. Indeed it is with respect to the prediction of turbulent flow past full aerospace configurations at flight conditions for the foreseeable future; but there are flows at considerably lower Reynolds numbers past elements of such configurations that could be conceivably attacked using this approach during the 1990's and the CFD community should perhaps be gearing up for it now.

1. Introduction

Last year I stated [1] that although both numerical algorithms and computer resources had developed sufficiently during the past two decades for their routine application to the design of aerospace devices this transfer of computational fluid dynamics (CFD) technology to actual design had not occurred within US aerospace corporations to any large degree. I speculated on the reasons for this situation. Perhaps as a CFD radical I expected unrealistically too much too soon from industry, or

(1) CFD technology is too sophisticated for use by aerospace design engineers themselves, or

(2) CFD is too expensive an alternative for industry use, or

(3) industry has too little confidence in the reliability of computational results, or

(4) design management is too conservative to change from their "tried and true" design tools for a newer technology.

A fifth reason, unstated last year but quickly pointed out to me by industry representatives, is the sometimes long setup times required, namely mesh generation about the configuration under study, before CFD application can begin when answers are needed for design almost immediately. From my perspective the situation was different in both Europe and Japan, perhaps because the aerospace companies there have not only been rebuilt since the Second World War but have essentially been reborn again during the past two decades and new technologies consequently are more readily embraced. However, this year there are considerable signs of a significant increase in the use of CFD in design in industry in the US as well.

Before we get too far along, we should discuss what CFD is and what are the roles of universities, government research laboratories and industry in its development and aerospace application. In 1975 D. R. Chapman in a controversial paper [2] pitting computers vs. wind tunnels classified computational fluid dynamics into four stages. The numerical solution of

(1) linear equations describing fluid flow,

(2) non-linear equations for inviscid flow, i.e. the Euler equations,

(3) the Navier-Stokes equations with turbulence modeling for viscous flow,

and

(4) the Navier-Stokes equations on such a fine mesh scale that all significant turbulent eddies are resolved.

An essential ingredient in the above is complete flow field simulation, complete in that the entire flow field disturbed by the presence of the aerodynamic body is determined numerically, as are the forces and moments acting on the body itself, subject of course to the constraints of the chosen governing equations. The stage hierarchy given above is ordered by the increasing demand of computational resources required for their application. Roughly, the computer resources at US national aerospace laboratories (NASA) and large corporations were sufficient for the first stage by the early 1970's, the second by the mid 1970's and the third by the early 1980's. During the 1980's considerable emphasis shifted from large main frame computers toward scientific work stations capable of performing many of the smaller applications of the first three stages. The required resources for the fourth stage are not yet available but should begin to become available during the present decade. An analysis of the fourth stage by Chapman, seventeen years after his first discussion of it, is to appear early next year [3].

The development of CFD technology requires basic research on the elements of numerical algorithms, code development and validation, and aerospace application. In the US these tasks have been distributed with considerable overlap to universities, national laboratories, and industry. The ideal is to transfer the enabling technologies quickly to industrial use. This occurred for numerical procedures for solving linear potential flow equations (although it should be recognized that they were largely developed within industry itself) and for the techniques for solving the non-linear transonic small disturbance equation, the full potential equation, and the Euler equations. These techniques were in use by industry within a couple of years after their presentation. On the other hand, the transfer of technology to industry for methods for solving the Navier-Stokes equations has been exceedingly slow.

The main reasons for this hesitation by industry is the higher computer expense in running Navier-Stokes codes, the lack of confidence in the turbulence modeling, and the large investment by industry in inviscid flow codes plus boundary layer interaction programs. Until fairly recently, potential flow plus boundary layer equation solution techniques performed as well as or better than the more expensive Navier-Stokes plus turbulence modeling procedures for industrial use. A NASA Langley code, called TLNS3D (Thin Layer Navier-Stokes in 3 Dimensions), developed by Veer Vatsa and using a one equation Johnson-King turbulence model, has received high praise this year from the Boeing Commercial Airplane Group for the calculation of transonic flows past transport aircraft configurations [4].

Upper management at national laboratories and within industry want decisions made on which technical approach is the most appropriate to use for each class of problems of concern. For example, to solve for the flow about a complex aircraft configuration, should a multi-block structured mesh or a single unstructured mesh be used. Such choices could save both money and manpower by eliminating the support required for less appropriate computer code software. I have discussed this issue with CFD developers and first line managers at two NASA laboratories, Langley Research Center and Ames Research Center, and find, unfortunately, that there is little agreement in narrowing the field concerning future directions in CFD. Perhaps it is because CFD is still a young discipline far from maturity itself and is to be implemented on computer architectures that are themselves still evolving and presently far from reaching a steady state. The direction of computer architecture today is toward massively parallel hardware systems. Nevertheless, the question of the future direction of CFD is an important one and will be discussed herein.

2. Current and Future Directions of CFD

Two decades ago the questions asked concerning the future directions of CFD included:

(1) Which is better the finite difference or finite element approach in fluid dynamics?,

(2) Is shock fitting better than shock capturing?, and

(3) Can implicit methods be developed that are better than explicit methods for solving hyperbolic sets of equations, i.e. the Euler equations?

Two decades later these questions are still unresolved. The finite difference and finite element approaches are still both viable, have merged together with the finite volume approach, and have heavily borrowed key features from each other. New papers are presented each year exhibiting the benefits of shock fitting in sessions where new approximate Riemann solvers are presented in other papers demonstrating their shock capturing abilities. And, although implicit procedures have been developed during the past two decades that are highly efficient for viscous flow at high Reynolds numbers, the Euler equations are still predominantly solved today by explicit techniques.

The questions being asked today concern:

(1) structured vs. unstructured grids,

(2) upwind vs. central differencing,

(3) turbulence modeling vs. large eddy simulation (LES),

(4) approximate factorization and indirect relaxation procedures vs. direct inversion, and

(5) implicit and multi-grid procedures on massively parallel computers.

2.1. Structured vs. unstructured grids

The simple nodal point array-like structure of a grid in two or three dimensions is a natural choice for computer languages to express matrix like operations upon data associated with it. However, the array-like structure imposes undesirable constraints in flow fields where local mesh refinement is needed or body surface or flow structure topologies are illsuited to simple array orderings. Zonal or multi-block approaches avoid these constraints by partitioning the flow field into topologically simpler sub-fields. Each sub-field is covered by a structured grid, but logic must be introduced into the flow solver at inter-zonal boundaries resulting artificially from the partitioning. The unconstrained unstructured grid approach can cover the flow field with a single grid, perhaps the ensemble of grid points from a multi-block grid, and needs not to worry about artificial interior boundaries. But, because the ordering of grid points is no longer simple, additional logic must be devised and computer memory reserved to determine the neighbors of each nodal point.

The key advantage of the structured grid is its ease in facilitating the use of efficient block matrix structured algorithms for solving the equations governing the flow. In addition, for solving the Navier-Stokes equations with a turbulence model, only about 30 to 40 words of memory are required per node point, a factor of as small as one fifth that used in some unstructured grid calculations. The key advantage of the unstructured grid is the removal of constraints placed on structured grids, particularly for complex body surface geometries. It is far easier to use a set of unstructured triangular elements to uniformly and completely cover a wing-fuselage-appendage-tail-body configuration, than rectangular elements. The grid definition along a body surface is a major problem today for structured grids. On the other hand, it is very difficult to check to see if a three dimensional unstructured grid is good or not because, unlike the structured grid, there are no natural grid surfaces, i.e. "i", "j" or "k" planes, within the flow volume to view.

Perhaps the winning candidate for the future will be a hybrid structured-unstructured combination mesh, where the goal is to maximize the advantages and minimize the disadvantages of each individual approach. Key advances in hybrid grids are now being made by K. Nakahashi of the University of Osaka Prefecture, Japan, who uses an unstructured surface grid to completely cover a complex three dimensional body configuration that is then extended out in a structured manner to discretize the flow volume about the body [6] and by K. Powell of the University of Michigan who uses an overall structured "parent" grid that can be locally refined by lower level "child" grids that can still be further refined if required by still lower level grids to as fine a level as desired for resolving both internal features of the flow and boundary geometries [7].

2.2. Upwind vs. central differencing

Central differencing for hyperbolic equations had two strikes against it, (1) it was unstable if used explicitly and (2) it is blind to saw tooth oscillations in the solution and needs added dissipation to control them. But it is simple to use and is naturally second order accurate. It took A. Jameson to remove the first strike by incorporating it into a Runge-Kutta formulation that was itself previously abandoned by many for use in solving partial differential equations. Upwind differencing is naturally stable and of only first order accuracy unless additional points (than those immediately adjacent, i.e. i-1,i,i+1) are used, and was useless for conservation law systems of equations with characteristic speeds of mixed sign, subsonic flows, until Steger and Warming devised flux splitting in the late 1970's. Since then many improvements have been made, notably by Roe with his flux difference vector splitting, and Harten and Yee with their TVD methods. The net result of this development are upwind methods of very high precision at shock waves but with much more complicated logic, including flux limiters that serve as switches triggered by local flow conditions, than central difference methods.

Upwind differencing can be viewed as a version of central differencing with an added dissipation term. However, the term required for this equivalence is not the one usually chosen in central difference methods and worse it is set rather arbitrarily according to the "looks of" and the stability the solution. It is hard to tell how much additional dissipation is too much and humans should not be trusted to control it. On the other hand, the dissipation in upwind methods is more natural in that is not added explicitly as a term or set of terms. Instead it results from the choice of numerical domains of dependence for each characteristic speed and is controlled by the flux limiters with apparently less chance of human fiddling. Dissipation is the key element in numerical methodology and respect for is paramount.

The main advantage of central differencing is its relative simplicity that facilitates the construction of efficient algorithms including the use of multigrid procedures to accelerate convergence. It has performed very well for transonic flows. The main advantages of upwind differencing is its accurate representation of shock waves, good performance in supersonic and hypersonic flows, and the implementation of boundary conditions because of its inherent mechanism for discriminating between incoming and outgoing signals. It also increases the diagonal dominance of the block matrix equations to be solved and thus enhances the stability of implicit methods used for their solution. However, unlike central difference methods, relatively expensive Jacobian matrices must be calculated even when used explicitly and flux limiters must be calculated throughout the flow field even though they may be used only at shock waves. The upwind schemes carry around much excess baggage in flow field regions not requiring it. In addition, the limiter switches can slow convergence and inhibit the use of multigrid procedures.

The development of central difference methods appears to have reached maturity while upwind difference methods are still in a flurry of activity. The foundation of most upwind methods to date has been based on one dimensional characteristic theory. This 1-D basis and then application in multi-dimensions has raised criticism that is spurring activity now in the development of truly multidimensional upwind techniques. It should be remembered that P. Goorjian pioneered this area of research several years ago [5] by considering the rotated Riemann problem in the stream-wise and normal to the flow line coordinate directions and then following the flow of entropy, vorticity, and acoustic wave phenomena within this coordinate system.

2.3. Turbulence modeling vs. large eddy simulation

The most widespread turbulence model used in numerical flow field simulation came not from the turbulence modeling community but from the CFD community, those who were actually performing the calculations. This model is the algebraic Baldwin-Lomax model. A model receiving much praise today, the Johnson-King one equation model, also has closer links to the CFD community than the turbulence community. The turbulence community has generated several two equation models that have not yet received widespread acceptance from the developers of Navier-Stokes codes. The main reason for this is that although these models can perform very well under certain conditions, their performance in general is not sufficiently good enough to warrant the added cost and numerical difficulties they introduce. Although significant progress is continuing to be made in the understanding of turbulent phenomena, the situation of a lack of impact on CFD by the turbulence community has essentially remained in a steady state for the last two decades and may remain so for the next two. The expectation of a model that could reduce the uncertainty in the treatment of general turbulent phenomena to reasonable engineering accuracy, perhaps of the order of one percent, is probably unrealistic for some time if at all.

An alternative to Navier-Stokes plus turbulence modeling that has the potential to reduce turbulence uncertainty to less than one percent is large eddy simulation (LES), in which the Navier-Stokes equations are solved alone, or with an unsophisticated turbulence model for sub grid scale effects, on a fine enough grid to resolve all significant turbulence effects. This has been relatively impractical to date except for the simplest of flows and far from practical for complex flows such as the flow past a fighter aircraft at high angles of attack and at flight Reynolds numbers. However, computer resources are becoming more powerful each year and it will be practical within the present decade to simulate through LES many lower Reynolds number viscous flows of engineering interest in three dimensions, such as those, perhaps, occurring within turbo-machinery (see [4]).

The key advantage of turbulence modeling is that the algebraic, one and two equation models can be incorporated now into Navier-Stokes high Reynolds number codes and executed on present computer resources. The uncertainty of model reliability is its chief drawback. Realistically there is no alternative to it for the simulation of flow past aerospace configurations at flight Reynolds numbers until there is a massive increase in computer resource power.

2.4. Approximate factorization and indirect relaxation vs. direct inversion

Approximate factorization by which a complex implicit multi-dimensional matrix equation is split into simpler one dimensional factors has been the work horse procedure for solving the Navier-Stokes equations for the last decade and a half. Approximately factored algorithms use efficient block tridiagonal matrix procedures for each factor. However, the factorization procedure itself introduces error that can severely limit the CFL number, or efficiency of the resulting algorithm. It performs very well for viscous high Reynolds number flows with highly refined grids near body surfaces, but because of its CFL limitations is not more efficient for solving the Euler equations than explicit procedures such as the Runge-Kutta algorithms devised by A. Jameson. Additionally, it suffers from the distinction that it is known to be theoretically unconditionally unstable in three dimensions. However, this curse does not materialize in practice, probably because it is used with added dissipation terms.

Approximately factored algorithms are direct in that they require no iterations within a single time step advance. Indirect relaxation schemes have been devised to avoid the error penalties introduced by factorization. These schemes also use block tridiagonal matrix inversion but in only one direction and use relaxation, usually some form of Gauss-Seidel relaxation, for representing terms from the other coordinate directions. Iterations are used within each time step until the solution is relaxed to acceptance. These schemes can achieve high numerical efficiency, or equivalently high CFL numbers, much higher than approximately factored schemes for some problems. However, they suffer from the uncertainties of when to stop the sub-iteration process during each time step, asymmetries introduced by the relaxation procedure, and uncertainty in stability parameters.

Both approximate factorization and indirect relaxation procedures represent a form of cheating or short cutting to avoid solving the implicit block matrix equation approximating the flow equations directly. Approximate factorization changes them into a factored form and indirect relaxation solves them by representing some of their terms with data lagging in time behind other fully updated terms. Direct solution of the matrix equation without factorization or relaxation can obtain steady state solutions in only a few iterations, but is computationally intensive even in two dimensions and with present computer resources is impractical in three dimensions. As the computer resources improve there should be a steady shift toward the direct solution procedure.

2.5. Implicit and multigrid procedures on massively parallel computers

Numerical algorithms need to continually adapt to the evolution of computer architecture. Soon a major shift is expected to occur toward massively parallel computers. Initially we can expect that the best algorithms for these new machines will be the simpler less efficient algorithms of today, such as explicit central difference schemes. The more sophisticated and efficient algorithms, those using block tridiagonal implicit procedures, will be more difficult or impossible to adapt. It will appear at first as a major step forward in hardware, a shift backwards in CFD software, and probably a net increase in computer resources. With time new algorithms that never could have been conceived of for use on serial computers will then hopefully evolve for the efficient use of massively parallel computer hardware.

The major difficulties will be partitioning the computational problem to the multitude of processors, mapping both the grid and the algorithm to the processors with the goal of minimizing interprocessor communication. Explicit algorithms will be the easiest for both structured and unstructured grids and, particularly if multigrid procedures can be adapted, should be fairly efficient for solving the Euler equations on massively parallel computers. But for Navier-Stokes equations the hardware can not be expected to overcome the inefficiency of explicit algorithms and implicit algorithms will be required. Each grid block, structured or unstructured, may have to zoned or partitioned further to form small enough sub-block elements so that implicit algorithms - approximately factored, indirect relaxation, or direct solution - can be used by a single processor or small combination of processors to update all nodal points within the sub-block element. Considerable algorithm development will be required.

3. Conclusion

Several issues concerning future direction in CFD were discussed above. Although it is desirable from a management point of view to make choices on the future directions, the view of the CFD landscape is presently unclear, its computer environment is rapidly changing, and decisions to eliminate some approaches now could be premature.

Acknowledgement

The author would like to thank the following individuals for their time and generosity in sharing their views of the future of CFD. M. D. Salas (who set me straight on a number of key issues), J. L. Thomas, and R. Biedron of NASA Langley Research Center, E. Turkel and R. Radespiel of ICASE at NASA Langley Research Center, T. Holst, L. Schiff, and T. Pulliam of NASA Ames Research Center, and D. R. Chapman of Stanford University. All the viewpoints that seem reasonable to the reader are those of the above, the author takes full credit for the rest.

References

[1] MacCormack, R. W., "Solution of the Navier-Stokes in three dimensions," AIAA Paper No. 90-1520, 1990.

[2] Chapman, D. R., Mark, H., and Pirtle, M., "Computers vs. wind tunnels for aerodynamic flow simulation," Astronautics and Aeronautics, April 1975.

[3] Chapman, D. R., "A perspective on aerospace CFD past and future developments," Aerospace America, January, 1992.

[4] Garner, P., Meredith, P., and Stoner, R., "Areas of future CFD development as illustrated by transport aircraft applications," Proceedings of the 10th Computational Fluid Dynamics Conference, AIAA Paper No. 91-1527, Honolulu, Hawaii, June 24-27, 1991.

[5] Goorjian, P. M., "Algorithm developments for the Euler equations with calculations of transonic flows," AIAA Paper No. 87-0536, 1987. See also Goorjian, P. M., A New Algorithm for the Navier-Stokes Equations Applied to Transonic Flows Over Wings, AIAA Paper No. 87-1121-CP.

[6] Nakahashi, K., "Panel discussion on CFD and complex grids," held at the 4th International Symposium on Computational Fluid Dynamics, Sept. 9-12, 1991, Davis, CA.

[7] Powell K., "Presentation on multi-resolution and subcell resolution schemes," at the Workshop on Algorithmic Trends in CFD for the 1990's, Sept. 15-17, 1991, Hampton, VA.

ADVANCES FOR THE SIMULATION OF COMPRESSIBLE VISCOUS FLOWS ON UNSTRUCTURED MESHES

O. Pironneau

(Université Paris 6 & INRIA)

ICASE meeting, Hampton USA, 15-20 September1991.

Abstract .

Compressible viscous flows are generally treated numerically as a superposition of viscous effects on inviscid flows governed by the Euler equations. However powerful this approach is it will be shown on a few examples that there is more to compressible viscous flows than this superposition. Along the way some of the recent findings relevant to Finite Volume and Finite Element Methods will be reviewed.

INTRODUCTION

The general equations describing a real reacting gas consists of the Navier-Stokes equations with modified laws of states, viscosities functions of the temperature and reaction-diffusion equations for the chemistry. When the Knudsen number is not small one may even have to go back to the Boltzmann equations instead of Navier-Stokes. Applications of such systems of equations are numerous and include high altitude airplanes, space shuttles, combustion engines and to some extend meteorology and oceanography when the shallow water equations are used.

Needless to says that mathematical results for such complex systems are scarce. Among the questions that one would like to solve are:

- Existence, uniqueness and regularity results for the solution.
- Convergence of the numerical schemes
- Validity of the approximations such as Euler's, PNS, equilibrium chemistry
- Precision of turbulence models...

These are not just of theoretical interest because it is almost impossible to give a proof of convergence of a numerical method without having some existence results for the continuous system. Too many hypotheses need to be made for the convergence proof to be convincing. While there is work for several generations of applied mathematicians to come, there is no doubt that some progress have been made in the past decade for nonlinear systems related to the Navier-Stokes equations. We begin by reviewing some of the results available. Then the discretization schemes will be discussed; then the problem of proving convergence. Finally we will discuss some recent problems that have arisen in conjunction with domain decomposition.

1. COMPRESSIBLE FLOW EQUATIONS: STATEMENT OF THE PROBLEM

Existence results for large initial data is a very necessary prerequisite to numerical simulation of a system of Partial Differential Equations. Experience shows that much of the difficulties are solved once the existence problem is solved and convergence of numerical methods usually follows.

1.1 Navier-Stokes equations as a modified system of conservation laws

The Navier-Stokes equations can be written as

$$\frac{\partial W}{\partial t} + \nabla.F(W) - \nabla.G(W, \nabla W) = 0 \qquad (1)$$

where $W = \{\rho, \rho u_1, \rho u_2, \rho u_3, \rho E\}$ and where $G(W, \nabla W), F(W)$ are 5×5 matrices. When the viscous terms are neglected ($G = 0$) it is a system of conservation laws (Euler equations) with one degenerate direction.

This symbolism is very attractive because it stresses the connection between Navier-Stokes equations and systems of conservation laws or even viscosity solutions of these (Kruskov [2]). However it presents the Navier-Stokes equations as a non linear parabolic problem for W, which it is not. This can be seen from the boundary conditions that are necessary to have a well posed problem with a locally unique solution.

By classical methods Matsumura-Nishida [32] and Valli[33] have shown that the Navier-Stokes equations are well posed with given initial data ($W|_{t=0}$ given) and given boundary data on the velocity u and energy E on the whole boundary and given density ρ on the inflow boundary only. (This result requires small initial data; an existence result in the large is not known so far). If the system was truly parabolic then W could be given on the whole boundary Γ of the domain occupied by the fluid, but this is not the case.

From this point of view Boltzmann's equation is a better problem as an existence result in the large is known (Diperna-lions [46]).

For Euler equations it is true that the most important contribution of applied mathematics is the theory of Conservation Laws (Lax [1]) and the concept of viscosity solutions (Kruskov[2]). These results address the well posedness of Euler equations and their precision as an approximation to the Navier-Stokes equations.

The theory of conservation laws is (almost) complete in one space dimension (existence, regularity and convergence of numerical methods), but still very much an open mathematical problem in two and three dimensions. However, coupled with a finite volume approach, it is at the root of most sophisticated upwinding techniques for their numerical simulations (van Leer [4], Osher et al [5]...).

The equations are integrated on a cell volume σ

$$\frac{\partial}{\partial t} \int_\sigma W + \int_{\partial \sigma} F(W)n = 0 \qquad (2)$$

and the theory of one dimensional Riemann problems is applied to the boundary integral in the direction of the flow.

Entropy inequalities must be added to the Euler equations for well posedness. Entropy functionals generalize the physical notion of entropy and has been proved to be a powerful tool to study the convergence of numerical methods (Bardos et al [3], Johnson et al [6]...). It is used also to study the propagation of singularities in the flow and there is hope that it may lead to valuable informations for turbulence modeling (Serre[7], Majda[8]).

In several space dimensions, so far the most powerful method (Diperna[9] and Tartar[36]), can be used only on a *scalar* conservation law. Let us summarize briefly the result.

Consider (3) when the solution depends on a small parameter h such as the mesh size.

$$\partial_t c^h + \nabla.f_h(c^h) = 0 \quad in \ R^2 \times (0,T) \quad c^h(x,0) = c_0(x) \quad in \ R^2. \quad (3)$$

where f_h is a nonlinear function from $R \to R$ with $\partial f_h/\partial c \neq 0$. The solution c^h is uniquely defined if it satisfies the entropy inequalities:

$$\partial_t \eta(c^h,k) + \nabla.q(c^h,k) \leq 0 \quad \forall k \in R^d \qquad (4)$$

where $\eta(\lambda,k) = |\lambda - k|$ and $q(\lambda,k) = sgn(\lambda - k)[f_h(\lambda) - f_h(k)]$. Note that (4) contains equation (3).

It can also be approached by solving (Kruskov[2])

$$\partial_t c_\nu^h + \nabla.f_h(c_\nu^h) - \nu\Delta c_\nu^h = 0 \quad in \ R^2 \times (0,T) \quad c_\nu^h(x,0) = c_0(x) \ in \ R^2.$$

It is also the mean of a cinetic equation (Lions et al [44])

$$\partial_t \phi + A(x,v)\nabla_x \phi = \partial_v m \qquad (5)$$

where $A(v) = \partial f_h(v)/\partial c$ and m is a positive bounded measure supported by the shocks.

The main new result on (3) is that if the solution is bounded (in $L^\infty(R^2 \times (0,T))$) for all h (TVD is no longer necessary), then $\lim_{h\to 0} c^h = c^0$ the solution of (3)(4) with $h = 0$.

Thus the convergence of a numerical scheme could follow once it is shown to be stable as done in [6] for the first time for a finite element method . Such an analysis extends the famous theorem known in Finite Difference as *Stability+consistency $\to$ convergence* .

But it gives precise conditions of stability which must be verified by the schemes and consistency in the sense of (2).

To stress the importance of Entropy inequalities Hughes et al advocate the use of entropy variables in numerical schemes.

1.2 Other formulations of the Navier-Stokes equations

1. One promising point of view is to consider the Navier-Stokes system as a limit of the Boltzmann equation when the mean free particle path goes to zero. This approach has lead to new algorithms, based on (5) in the case of (3)) which may turn out very effective in the future (Deshpande[57], Perthame et al [58])

2. For the specialist of incompressible flow simulation, the Navier-Stokes equations are a modification of those that governed incompressible fluids. Thus, following Bristeau et al [21] the classical formulation is more feasible:

$$\partial_t \rho + \nabla.\rho u = 0$$
$$\partial_t(\rho u) + \nabla.(\rho u \otimes u) + \nabla p - \nu \Delta u - \lambda \nabla(\nabla.u) = 0$$

coupled with an equation for the pressure itself coupled with an equation for E.

3. Another approached followed by Akai-Ecer[59], Eldabaghi et al[60] is to consider only the stationary case and use Clebsch variables: $u = \nabla\phi + \nabla \times \psi$. Then the Navier-Stokes equations are considered as a perturbation of the transonic equation coupled with a transport equation for the vorticity and an elliptic equation for ψ. While this approach leads to fast algorithms for the Euler equations, the approximation of the viscous terms is rather complex.

2. UNSTRUCTURED MESHES

There is a new debate between boundary fitted meshes and auxiliary domain approaches as suggested by Young et al [15], Kuznetsov [51] and others. The later is not a good mesh by usual standard but because it is structured over most of the domain so it can have many more points, preconditioners are easy to build and the Finite Difference Data Structure can be used almost everywhere (figure 2.1).

Among the boundary fitted method are those which use triangles and tetraedras and allow, at least theoretically, a random distribution of elements or cells.

Unstructured meshes are more adapted to industrial applications especially in 3 dimensions where it is too difficult to construct orthogonal meshes. They will win over the structured mesh-auxiliary domain approach only if it is shown that it is better say to solve a problem with 3000 well located points rather than 30,000 points whose locations are not dictated by the geometry of the problem.

Triangular or tetraedral meshes can be generated in arbitrary domains by the advancing front method (Lohner[10], Perraire et al[11]) or by the Delaunay-Voronoi criterium. The generation of Delaunay-Voronoi meshes is no longer slow and possible now on any domain (Baker [12], Georges et al [13]). It is also possible to generate these meshes with prescribed density of vertices in prescribed regions (Peraire et al [11], Hecht et al [14]),(see figures 2.2, 2.3, 2.4).

Automatic Quadrangular mesh generators have also been developed (Young et al [15] and Jacquotte [16]). In [15] they have been shown to be completely general if hanging mid-side nodes are allowed.

Unstructured meshes also allow adaptivity to the solution during its computation. The theory of adaptive local mesh refinement is somewhat complete for incompressible fluids (Babuska [18], Bank [19], Verfurth [20]) but still in its infancy for compressible flows, even though it is currently in use in industry (Periaux et al [21], Young et al [15]).

The criteria for mesh refinement should be based on a local error estimate. For example, for the incompressible Navier-Stokes equations it is known that the difference between the computed solution u_h and the exact solution u is related to the mesh size h by (Bernardi [22] for example)

$$|\nabla(u - u_h)|_0 \leq ch|u.\nabla u|_0 \qquad (10)$$

Thus one way is to divide the mesh when $|u_h.\nabla u_h|$ is large (Erikson et al [43]).

Many other criteria have been proposed. In (Babuska et al [18] and Verfurth [20]) the mesh is refined when the norm of the strong residual error $||u_h.\nabla u_h + \nabla p_h - \nu\Delta u_h||$ is large. In Verfurth [20] the local error is also estimated by solving a local Stokes problem with the residual error on the right hand side and/or with a higher order method, and the convergence is shown.

There are also reasons to believe that some singularities like boundary layers or shear layers do not require a fine mesh for all variables. For example, in a boundary layer the pressure is continuous; only the velocity has a strong gradients. Thus algorithmic speed up could be obtained by refining only some of the variables (Achdou et al[52]).

In all cases the elements are subdivided. Thus the successive meshes are imbedded in each other and multigrid can be used to solve quickly the linear systems (Bank et al [39], Mavriplis [53]). But it is also possible to remesh completely the domain and redistribute the vertices according to the local errors as in (Hecht et al [14]) ; then an interpolation procedure is needed but the main advantage lies in the possibility to adapt the mesh to non isotropic singularities such as shocks and boundary layers. Time dependent singularities can even be followed by derefining/refining the mesh at every time step (Lohner [10]). Again this methodology is used for hypersonic flows but it will not be systematic until a complete error analysis of the numerical method used is available.

Structured meshes on auxiliary domains have many advantages; it is usually not necessary to store the matrices of the linear systems, multigrid can be used, parallel computing is easier...

Composite meshes have been used at Boeing (Young et al [15]) where the matrices are stored only in regions where the mesh is not structured. Theoretical studies (Scott[23]) show that multigrids can be used even if the grids are not imbeded into one another. Only parallel computing on machines like the CM (see Fezoui [24] for example) are more difficult with unstructured meshes, so far.

3. DISCRETIZATION

3.1 Flux based methods.

The Flux based Finite volume method is most successful for the discretization of (1) on unstructured meshes. Control volumes σ are chosen, the elements (Jameson [62], Lohner [10], Mavriplis[53]) or the Voronoi polygons of the vertices (or an approximation of them (Dervieux[42])).

Then (1) is integrated over σ :

$$\frac{\partial}{\partial t} \int_\sigma W + \int_{\partial\sigma} F(W)n + + \int_{\partial\sigma} G(W, \nabla W)n = 0$$

Usually W is assumed continuous piecewise linear on the triangulation and either an artificial viscosity is added to G or F is changed into F_h via upwinding. In the later case right and left values of W are obtained on $\partial\sigma$ by projecting W on the space of piecewise constant functions over the cells σ.

Time discretization is done either with an explicit scheme (Jameson[62], Mavriplis[53], Lohner[10]) or with a semi-implicit scheme which requires the solution of a linear or non-linear system at each time step. For these it seems that the Linear and the nonlinear GMRES algorithm (Saad[65]) performs well once the proper preconditioner is found (Mavriplis et al [64]). Figures 2.3, 2.4 showed results obtained with a flux based method with upwinding and an implicit time scheme for the Navier-Stokes equation with a turbulence model

Technically this method is also a finite element method in the class of Petrof-Galerkin methods (the test space , here piecewise constant functions, is different from the approximation space, here the continuous piecewise linear functions). Other finite element approximations have been proposed with the flux formulation of (1) (see Chavent et al[67], Cockburn[66] for example)

There is some amount of arbitrariness in the imposition of the boundary conditions. What is usually done is to set the known components of W to their boundary values in $F_h(W)$ during the iterative process and similarly for G. But the computation of $\nabla W|_\Gamma$ may pause a problem.

Nicolaides [17] showed that Voronoi meshes are well adapted to flux based Finite Volume Methods because of the orthogonality between the Voronoi polygons and the edges of the mesh. In particular it solves the problem of how to apply boundary conditions when the viscous terms are present.

To illustrate the point consider the simple PDE with u, a, f vector valued in R^2 ($\nabla.a = 0$) :

$$a\nabla u - \Delta u = f \quad in \quad \Omega, \quad u|_\Gamma = 0 \quad (\Gamma = \partial\Omega) \tag{6}$$

When it is discretized by taking for degrees of freedom the fluxes of u on the edges of a triangulation the normal components $u.n$ are the unknown at the mid-edges (figure 3.1); but since the tangential component $u.\tau$ are not known, it is not obvious how to impose the boundary condition.

In [17] convergence is shown (for the incompressible Navier-Stokes equations also) if one proceeds as follows. Rewrite the equation as

$$a\nabla u + \nabla \times \omega - \nabla p = f, \quad \omega = \nabla \times u \quad p = \nabla.u, \tag{7}$$

and approximates ω at the vertices and p at the center of the triangles by using:

$$\omega = \frac{1}{|V|} \int_V \nabla \times u = \int_{\partial V} u \times n \quad p = \frac{1}{|T|} \int_T \nabla.u = \int_{\partial T} u.n. \tag{8}$$

When V is the Voronoi polygon of a vertex intersected with Ω and T a triangle, these formulae extend up to the boundary and fixing $u.\tau = u \times n$ is done through a given boundary integral in the formulation. Indeed (7) integrated gives ($\nabla.a = 0$)

$$\int_{\partial T} [(u.n)a + \omega \times n - p.n] = \int_T f, \tag{9}$$

$$\int_V \omega = \int_{\partial V} u.\tau, \quad \int_T p = \int_{\partial T} u.n$$

and on $\partial V \cap \Gamma$ $u.\tau$ is know; all other integrals are approximated by Gauss quadratures.

Adaptation of this method to the compressible Navier-Stokes equations has not been done yet. It requires to take p and ω as temporary variables even though they may be eliminated when the non linear system is built.

3.2 Extension of methods developed for incompressible fluids.

For incompressible flows the Navier-Stokes equations are

$$\nabla.u = 0 \quad \partial_t u + \nabla.(u \otimes u) + \nabla p - \nu \Delta u = 0.$$

Both equations may not be discretized with the same Finite Element spaces but rather as

$$\int_\Omega q_h \nabla.u_h = 0$$

$$\partial_t \int_\Omega u_h v_h + \int_\Omega v_h \nabla.(u_h \otimes u_h) + \int_\Omega v_h \nabla p_h$$

$$+ \nu \int_\Omega \nabla u_h : \nabla v_h = 0$$

for all $q_h \in Q_h$ and all $v_h \in V_h$.

After a semi-implicit dicretization in time one finds that the degrees of freedom $\{U, P\}$ of $\{u_h, p_h\}$ are solution of a linear system of the type

$$\begin{pmatrix} A & B \\ B^T & 0 \end{pmatrix} \begin{pmatrix} U \\ P \end{pmatrix} = \begin{pmatrix} F \\ 0 \end{pmatrix}$$

The system may not have a solution but it does if Q_h and V_h are compatible in the sense of the inf-sup LBB condition (see (20)).

So for compressible fluids one retains the idea of dual approximations for the density and velocity. Experience shows that the energy can be approximated in the same space as the density.

Bristeau et al[21], Rogé et al [68] have computed flows with a discretization of the equations with $\sigma_h = ln\rho_h$, E_h in Q_h the finite element space of piecewise linear continuous functions on a triangular mesh T_h and u_h in V_h the finite element space of piecewise linear continuous functions on the triangular mesh obtained by dividing each triangle into 4 sub triangles by joining the mid-edges. After a fully implicit time discretization the set of nonlinear discrete equations are solved by the nonlinear GMRES algorithm. Upwinding is not necessary to obtain results such as in Figures 3.2, the viscosity of the fluid is sufficient.

4. CONVERGENCE

It must be admitted that not much is known mathematically on the convergence of numerical methods for compressible flows.

Some partial results are available, for scalar hyperbolic laws as was already mentioned (Johnson et al [6]) . Figure 4.1 shows a Euler flow computed with the method which is shown to be convergent in some cases in [6].

To illustrate the difficulties characteristic to the Navier-Stokes equations which do not exists with inviscid flows a partial result is recalled (Bernardi et al [34], Pironneau et al [35]).

Schemes should work at all regimes including the low Reynolds number regime. When the inertial terms are neglected in the Navier-Stokes equations the systems reduces to (for clarity we assume $\lambda = 0$) :

$$\partial_t \rho + \nabla.\rho u = 0 \quad \nabla p - \mu \Delta u = 0 \quad -\kappa \Delta \theta = \frac{\mu}{2}|\nabla u + \nabla u^T|^2 \quad p = \rho \theta$$
$$(15)$$

Even for this simpler problem existence and convergence of approximations are not known except in the hypothetical situation where κ is small thus giving θ constant; the system would then reduce to the following nonlinear problem:

$$-\nu \Delta u + c \nabla \rho = f, \quad \nabla.(\rho u) = 0, \quad in \ \Omega \qquad (16)$$

for which a valid set of boundary conditions is

$$u|_\Gamma = u_\Gamma \quad \rho|_{\Gamma^-} = \rho_\Gamma \quad \Gamma^- = \{x \in \Gamma : u_\Gamma(x).n(x) < 0\} \qquad (17)$$

The variational formulation must take the boundary condition on ρ in a weak sense (a valuable information which could be remembered for the Navier Stokes equations):

Find $u \in H_0^1(\Omega)^d$ and $\rho \in L^2(\Omega)$ such that

$$\nu \int_\Omega \nabla u \nabla v + c \int_\Omega v \nabla \rho = \int_\Omega f v, \quad \forall v \in H_0^1(\Omega)^2 \qquad (18)$$

$$-\int_\Omega \rho u \nabla \phi + \int_{\Gamma^-} \phi \rho_\Gamma u_\Gamma . n = 0, \quad \forall \phi \in H^1(\Omega) \quad with \quad \phi|_{\Gamma - \Gamma^-} = 0 \qquad (19)$$

Then one can show existence for smooth data satisfying: $\int_\Gamma u_\Gamma . n \geq 0$, $\Gamma^- \neq \varnothing$.

To approximate the system by FEM one simply replaces $H_0^1(\Omega)^d$ by V_{0h} and $H^1(\Omega)$ and $L^2(\Omega)$ by Q_h, two spaces of piecewise polynomial continuous functions. Convergence can be shown if V_{0h} is an internal approximation of $H_0^1(\Omega)^2$ and Q_h is an internal approximation of $L^2(\Omega)$ satisfying the same inf-sup condition as for incompressible fluids :

$$\inf_{q_h \in Q_h} \sup_{v_h \in V_h} \frac{\int_\Omega v_h \nabla q_h}{(\int_\Omega |\nabla v_h|^2 \int_\Omega q_h^2)^{1/2}} \geq \beta > 0. \qquad (20)$$

It has been shown numerically that oscillations develop when this condition is not satisfied (for example if u and ρ are approximated with the same degree on the same mesh); but upwinding also smoothes these oscillations in most cases (figure3.3).

Computational results for (18)(19) can be found in [41]

5. NEW AND OLD PROBLEMS IN DOMAIN DECOMPOSITION

There are two reasons to use domain decompositions: either because it is efficient on multi-processors and each domain is allocated into a processor, and/or because the flow is structured and approximations to the full Navier-Stokes equations can be made in subdomains.

Domain decomposition for multiprocessors is an algorithmic problem where one must prove that the matching conditions at the artificial boundaries yield convergence (Glowinski et al [25]). Several iterative procedure have been studied (Neumann or Dirichlet condition, least square instead of Schwartz's, switch to mixed finite element methods, overlapping versus non overlapping domains...)

The method has been shown to be efficient even on sequential computers in the case of elongated domains; it is ideal for medium or coarse grain parallelism but may work also for fine grain parallelism. For example, Cowsar[54] has solved a Darcy flow with FEM and multigrid or Conjugate gradient on the iPSC860. Speed up deteriorates only when there are less than 5 nodes in each domain/processor. In such cases a domain allocation algorithm should be used(Farhat et al [55])

The decomposition of a compressible flow into a PNS or boundary layer region, a potential or Euler region and a viscous recirculating region is used routinely. However very little is know about the error attached to this procedure; Prandtl's boundary layer approximation is not even justified for incompressible flows in the general case.

Brezzi et al [26], Achdou et al[27] have proved that PNS is indeed a feasible approximation to the full system of equations around slender bodies; the outcome of the theory is also an adaptive approach to define the region of validity of PNS. Exactly as in the case of water with ice the basic idea is to make the free boundary part of the unknown. For this one uses a cut-off function on the viscous terms ($\chi(\phi) = sgn(\phi) \max\{\delta, |\phi|\}$ and solve:

$$\frac{\partial W}{\partial t} + \nabla.F(W) - \nabla.G(W, \chi(\nabla W)) = 0. \tag{11}$$

This new system is Euler when the cut-off is active on all derivatives $\partial W_j/\partial x_i$, PNS when the cut-off is active on $\partial W_j/\partial x_1$ (x_1 is the direction of the flow) and Navier-Stokes when it is not. The theory is complete for incompressible fluid (existence, uniqueness and convergence when $\delta \to 0$). Numerically this implies that whenever an approximation to the full system is used one should compute the terms neglected, check that they are small , and put them back in if they are not.

For hypersonic flows, Coron et al[28], Sone et al [45], among others, have studied how to match a Boltzmann zone with a full Navier-Stokes region.

Chapman et al [29] showed the connection between Boltzmann equation and the Navier-Stokes equations; the viscosity obtained however tends to zero with the mean free path; another expansion proposed by Bardos et al [30] relates the viscosity to the Mach number as well.

These asymptotic expansions make it possible to know when and where the Navier-Stokes equations should be used in place of Boltzmann's equation; $\nabla_x.(u\rho S)$ or $\nabla\theta/\theta$ are two such criteria. Then the problem is to transform outflow Navier-Stokes data, ρ, u, θ, into a density of particles ϕ for Boltzmann's equation and vice-versa. Coron showed that it is not always possible unless a kinetic layer is introduced which requires the solution of a one dimensional Boltzmann-like problem for χ :

$$(v_1 + u_1)\partial_\eta \chi(\eta, v) + L\chi = 0 \qquad (12)$$

where u_1 is the Navier-Stokes velocity normal to the interface, L is the linearized collision operator about the maxwelian at zero mean velocity and Navier-Stokes temperature.

The same technique can be used to model a Knudsen layer near a solid wall. Coron[28] and Golse[46] obtain:

$$u.n = 0 \quad \rho u.\tau = c_1 \frac{\partial u.\tau}{\partial n} + c_2 \frac{\partial \theta}{\partial \tau} \qquad (13)$$

$$\rho(\theta - \theta_{wall}) = c_3 \frac{\partial u.n}{\partial n} + c_4 \frac{\partial \theta}{\partial n} + c_5 \nabla.(u - u.nn) \qquad (14)$$

Sone et al[45] have obtained similar nonlinear conditions by studying numerically the bifurcations of the nonlinear system. These conditions confirm the conjectures of Gupta et al[69] based on physical grounds.

REFERENCES

[1] P.D. Lax : Shock waves and Entropy. In Contribution to Nonlinear functional Analysis. E. Zarantanello ed. Academic Press. p 603-634, 1971.

[2] S. Kruskov: First Order Quasi-linear equations with Several Space Variables. Math Sb. **123** , p228-255, 1970.

[3] C. Bardos, A. Leroux, J.C. Nedelec: First order quasi-linear equations with boundary conditions. Comm PDE **4** (9), p1017-1034, 1979.

[4] B. van Leer: Toward the Ultimate Conservative Scheme III. Upstream Centered Finite Difference Schemes for Ideal Compressible flows. J. Comp. Phys. **23** , p263-275, 1977.

[5] S. Osher, E. Tadmor: On the Convergence of Difference Approximations to Scalar Conservation Laws. Math Comp. **50** , 181, p19-51, 1981.

[6] C. Johnson, A. Szepessy: On the Convergence of a Finite Element Method for a Nonlinear Hyperbolic Conservation Law. Math. Comp. **49** p427-444, 1987.

[7] D. Serre: Oscillations non-linéaires hyperboliques de grandes amplitiudes. Internal report 33, ENS-Lyon, 1990.

[8] A. Majda: Compressible Fluid Flow and Systems of Conservation Law in Several Space Variables. Springer series in Applied Mathematical Sciences **53** , 1984.

[9] R. DiPerna: Measure-valued solutions to Conservation Laws. Arch. Rat. Mech. Anal. **88** , p223-270, 1985.

[10] R. Lohner: 3D grid generation by the advancing front method. In Laminar and Turbulent flows. C. Taylor, W.G. Habashi, H. Hafez eds. Pinneridge Press, 1987.

[11] J. Peraire, J. Peiro, L. Formaggia, K. Morgan, O. Zienkiewicz: Finite Element Euler Computations in three dimensions. Int. J. Numer. Meth. in Eng. 26: 2135-2159, 1988.

[12] T.J. Baker: Automatic mesh generation for complex three dimensional regions using a constrained Delaunay triangulation, Eng. Comp. 5: 161-175, 1989.

[13] Georges, P.L., Hecht, F. , Saltel, E. 1990. Automatic 3D Mesh Generation with Prescribed Mesh Boundaries. IEEE Trans.

Magnetics, 26, 2: 771-774.

[14] Hecht , F., Vallet, M. G. 1991. Non Isotropic Automatic Mesh Generation. (To Appear).

[15] D.P. Young, R.G. Melvin, M.B. Bieterman, F. T. Lohnson, S.S. Samant: A locally refined grid FEM: Application to CFD and computational physics. J. Comp. Physics. **92** ,1,p1-66, 1991.

[16] O. Jacquotte: A mechanical model for a new generation method in CFD. Comp. Meths. Appl. Mech. Eng. **66** , p323-338, 1988.

[17] R.A. Nicolaides: Analysis and Convergence of the MAC scheme. SIAM. J. Numer. Ana. (to appear, 1992).

[18] I. Babuska, W.C. Rheinbolt: A posteriori estimates for the finite element method. Int. J. Numer. Methods in Engrg. 12 : 1597-1615, 1978.

[19] R. Bank, N. Welfeirt: A posteriori estimates and automatic mesh refinement for the Stokes problem (to appear in SIAM numer Anal).

[20] R. Verfurth: A posteriori error estimators and adaptive mesh-refinement techniques for the Navier-Stokes equations. In Gunzburger, M. D. Nicolaides, R. A. (Eds) Incompressible CFD-Trends and Advances. Cambridge University Press. 1991.

[21] M.O. Bristeau, R. Glowinski, B. Mantel, J. Periaux, G. Rogé: Adaptive Finite Element Methods for three dimensional viscous flow simulation in aerospace engineering. Proc 11th Int. Conf. on Numerical Methods in Mechanics. Williamsburg, Virginia 1988.

[22] C. Bernardi, G. Raugel: A conforming finite element method for the time-dependent Navier-Stokes equations. SIAM J. Numer. Anal. 22,455-473, 1985.

[23] Scott, R. 1990. Zhang, S. Higher Dimensional Nonnested Multigrid methods. Research Report UH/MD-84. University of Houston.

[24] L. Fezoui: Numerical simulation of Euler equations on the Connection Machine. INRIA report 1990.

[25] R. Glowinski , M.F. Wheeler: Domain decomposition and Mixed Finite Element Methods for Elliptic Problems. In *Domain Decomposition Methods for PDE* . R. Glowinski G.H. Golub, G. Meurant, J. Periaux eds. SIAM Philadelphi, p350-369 1988.

[26] F. Brezzi, C. Canuto, A. Russo: A self-adaptive formulation for the Euler/Navier-Stokes coupling. Comp. Methods in Appl. Mech. and Eng. 73 317-330, 1989.

[27] Y. Achdou, O. Pironneau: The χ method for the Navier-Stokes equations. Note CRAS to appear, 1991.

[28] F. Coron : Hypersonic flow in rarefied gas using Boltzmann and Navier-Stokes equations. (to appear) 1991.

[29] S. Chapman, T. Cowling: The mathematical theory of nonuniform gases. Cambridge University Press, 1951.

[30] C. Bardos, F. Golse, D. Levermore: Fluid dynamic limits of Kinetic equations II: Convergence proofs for the Boltzmann equations. Note CRAS **309-I** , 727-732, 1989.

[31] R.J. Diperna, P.L. Lions: On the Cauchy problem for the Boltzmann equation: Global existence and weak stability results. Annals of MAth. **130** , 321-366, 1989.

[32] A. Matsumura, A. Nishida, T. 1980. The initial value problem for the equations of motion of viscous and heat conductive gases. J. Math. Kyoto Univ. 20, 67-104.

[33] A. Valli: An existence theorem for compressible viscous flow. Mat. It. Anal. Finz. Appl. 18-C, 317-325, 1981.

[34] C. Bernardi, O. Pironneau: Existence And Approximation Of Compressible Viscous Isothermal Stationary Flows. Note CRAS, 1990.

[35] O. Pironneau, J. Rappaz: Numerical Analysis for Compressible Isentropic Viscous Flows. IMPACT. 1:109-137.1989.

[36] L. Tartar: The compensated compactness method applied to systems of conservation laws.In *Systems of Nonlinear PDEs* , J. Ball ed. NATO series ASI. C. Reidel Publishing Co. 1983.

[37] J. A. Desideri M.V. Salvetti, M.C. Ciccoli: Developments in hypersonic reactive flow computations. Hermes meeting, Amsterdam, nov 1990 (to appear).

[38] P.L. Lions: Résolution de problèmes elliptiques quasi-linéaire. Arch. Rat. Mech. Anal. **74** p335-353, 1980.

[39] Bank, R., Dupont, T., Yserentant, H. 1987. The Hierarchical Basis Multigrid Method. Konrad-Zuse-Zentrum preprint SCV-87-2.

[40] Bernardi, C., Raugel, G. 1985. A conforming finite element

method for the time-dependent Navier-Stokes equations. SIAM J. Numer. Anal. 22,455-473.

[41] Bristeau, M.O. Bernardi, C., Pironneau, O. Vallet, M.G. 1990. Numerical Analysis for Compressible Isothermal Flows. Proc. Symp. on Applied Math. Venice.

[42] Dervieux, A. 1985. Steady Euler simulations using unstructured meshes. Von Karman Lecture notes series 1884-04.

[43] Erikson, K., Johnson, C. 1988. An Adaptive Finite Element Method for Linear Elliptic Problems. Math Comp. 50: 361-383.

[44] P.L. Lions, B. Perthame, E. Tadmor: Kinetic formulation of scalar conservation laws. Note CRAS to appear. 1991.

[45] Y. Sone, K. Aoki, I. Yamashita: A study of unsteady strong condensation on a plane condensed phase. Proc 15th Int Symp.on Rarefied Gas Dynamics. V. Boffi ed. Vol II p 323-333.1987.

[46] F. Glose: Application of the Boltzmann equation within the context of upper atmosphere vehicle aerodynamics. Proc. 8th Int. Conf. on Comp. Meth. in Applied Sciences and Eng. R. Glowinski ed. North Holland 1988.

[47] A. Majda: Compressible Fluid Flow and systems of Conservation Laws in Several Space Variables. Applied Math. Sciences series, 53, Sringer, 1984.

[48] H. Babovsky, R. Illner:Convergence Proof for Nanbu's Simulation Method for the full Boltzmann equation. SIAM Journal of Numerical Analysis,26,(1989),1,pp.45-65.

[49] B. Perthame, R. Sanders: The Neumann Problem for Nonlinear Second Order Singular Perturbation Problems. SIAM J. Math. Anal. 19 ,2, 295-311, 1988.

[50] R. Sanders, A. Weiser: A high resolution Staggered Mesh Approach for Nonlinear Hyperbolic Systems of Conservation Laws. To appear in J. of Comp. Physics, 1991.

[51] Y. Kuznetsov: Fictitious domain methods (to appear)

[52] Y. Achdou, O. Pironneau, R. Glowinski: Tuning the mesh for a stream function vorticity formulation of the Navier-Stokes equations INRIA research report. Oct. 1991.

[53] D. J. Mavriplis: Three dimensional unstructured multigrid for the Euler equations. Icase Report 91-41.

[54] L. Cowsar Parallel Domain Decomposition Method for Mixed FEM for Elliptic PDE. Rice University Internal report TR90-37. Nov 1990

[55] C. Farhat L. Fezoui, S. Lanteri CFD with irregular grids on the connection machine Parallel CFD conf. Stuttgart, June 1991.

[56] T.J.R. Hughes, L.P. Franca, M. Mallet: A new Finite Element Formulation for CFD: Symmetric form of the Compressible Euler and Navier Stokes equations. Comp. Methods. in Applied Mech. and Eng. **54** , p223-234. 1986.

[57] S.M. Deshpande: A second order accurate kinetic theory based method for inviscid compressible flows. NASA Technical Paper 2613 (1986).

[58] B. Perthame: Boltzmann type scheme and the entropy property. SIAM J. Numer Anal.

[59] H. Akay, A. Ecer: Compressible viscous flows with Clebsch variables. in *Numerical methods in laminar and turbulent flows* . C. Taylor, W. Habashi, M. Hafez ed. Pineridge press 1989-2001, 1987.

[60] F. Eldabaghi, O. Pironneau, J. Periaux, G. Poirier: 2D/3D Finite Element Solution of the Steady Euler equation by vector correction. Int. J. for Num Methods in Fluids. Vol 7, p1191-1209. 1987.

[61] M. Hafez, W. Habashi, P. Kotiuga: conservative calculation of non isentropic transonic flows. *AIAA* . **84** 1929. 1983.

[62] A. Jameson, T. Baker, N.P. Weatherhill: Calculation of inviscid transonic flow over a complete aircraft. AIAA paper 86-0103. Reno, Jan. 1986.

[63] R. Lohner: 3D grid generation by the advancing front method. In *Laminar and turbulent flows* . C. Taylor, W.G. Habashi, H. Hafez eds. Pinneridge press, 1987.

[64] V. Venkatakrishnan, D. Mavriplis: Implicit Solvers for Unstructure Meshes. ICASE report 91-40. 1991.

[65] Y. Saad: Krylov subspace methods for solving unsymmetric linear systems. *Math of Comp* . **37** 105-126, 1981.

[66] B. Cockburn, S. Hou, C-W Shu: The R-K local projection discontinuous Galerkin FEM for conservation laws. Math of Comp.Vol 54, No 190, p545-581. 1990.

[67] G. Chavent, B. Cockburn, G. Cohen, J. Jaffré: A discontinuous $> FEM$ for nonlinear hyperbolic equations. IN *Innovative Methods in Engineering* . Springer 1984.

[68] M.O. Bristeau, R. Glowinski, L. Dutto, J. Periaux, G. Rogé: Compressible flow calculation using compatible approximations. Int. J. Numer. Methods in Fluids. vol 11, No 6, p719-750, 1990.

[69] R. Gupta, C. Scott, J. Moss: NASA Technical paper 2452. 1985.

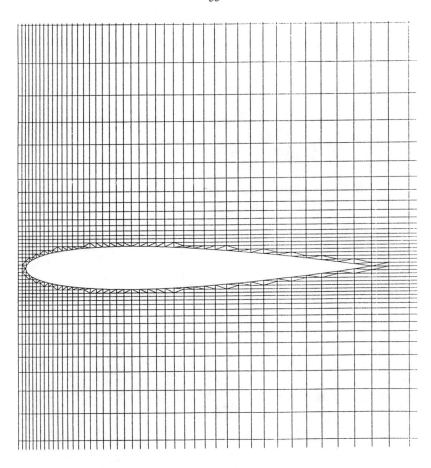

Figure 2.1 A NACA 012 profile discretized with a mesh which is boundary fitted only near the profile and structured elsewhere. (Courtesy of M. Baspalov, INRIA).

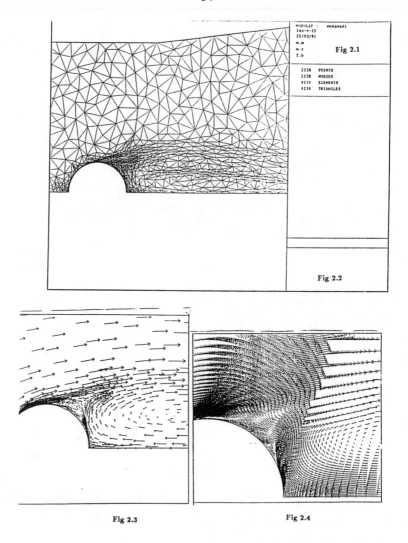

Fig 2.1

Fig 2.2

Fig 2.3

Fig 2.4

Figure 2.2, 2.3, 2.4 The figure shows two computation of the same flow. Right: with a standard mesh; left with an optimized mesh. The mesh on the right (fig 2.2) has 4 times more points and yet the same quality of details is present in the solution computed on the left.

The Navier-Stokes equations with a turbulence model $(k - \epsilon)$ is simulated at Mach 0.5 and Re=10.000. (Courtesy of B. Mohamadi, INRIA.)

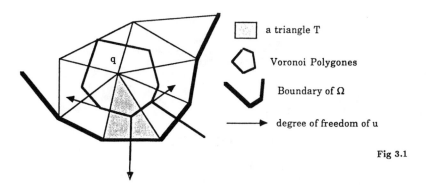

a triangle T

Voronoi Polygones

Boundary of Ω

degree of freedom of u

Fig 3.1

Figure 3.1 Triangular mesh and Voronoi polygons. For finite volume methods, it is possible to apply no slip conditions rigorously even though only the normal components only of the velocity is available at the mide-edges. For this one defines the vorticity at the vertices by a contour integral of the velocity u on the Voronoi polygons and $\nabla.u$ at the center of the triangle by a contour integral of u.n on the triangle. At the boundary the tangential component of u is known and so the contour integral can be computed even though the boundary is not part of the Voronoi polygon.

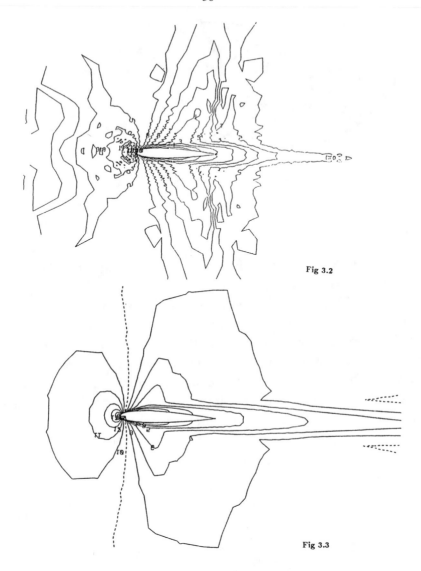

Fig 3.2

Fig 3.3

Figure 3.2, 3.3 The flow around a profile at $M_\infty = 0.85$ and Re=2000 is computed by a method which does not use artificial viscosity nor upwinding; top: by a finite element method which does not satisfy the inf-sup condition (linear approximation for all the variables); bottom: by a method that satisfies the inf-sup condition (linear element for ρ, θ and linear on the mesh where each triangle has been divided into 4 elements for the velocity u). Thus the oscillations where not due to the lack of upwinding.(Courtesy of M.O. Bristeau, INRIA.)

Fig 4.1   Fig 4.2

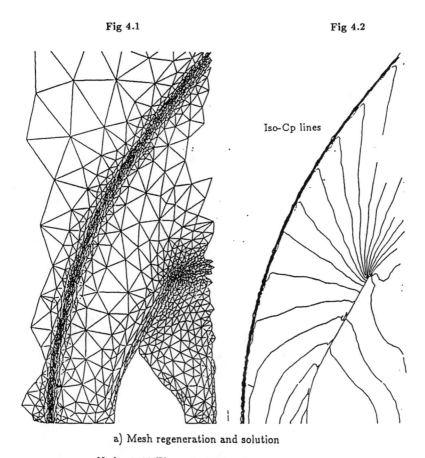

Iso-Cp lines

a) Mesh regeneration and solution

Nodes 2199 Elements 4256

Figure 4.1,4.2

Solution of the Euler equations at Mach 5. The mesh has been adapted to the solution with the non isotropic Voronoi mesh generated discussed in the paper. The numerical method is the finite element method with least square regularization and shock capturing for which Johnson-Szepessy [6] could prove convergence for a scalar nonlinear conservation law (Courtesy of M.G. Mallet, Dassault Aviation).

SOME CHALLENGES IN MASSIVELY PARALLEL COMPUTATION

Professor Martin H. Schultz[1]

Department of Computer Science
Yale University
New Haven, Connecticut 06510

1. Introduction

The search for faster, more accurate, less costly computational solutions to hard problems, such as those in the study of fluid dynamics, leads us to seriously consider the use of modern ultrafast supercomputers. It turns out that the world of supercomputers is currently undergoing a major revolution which bids to accelerate the previously rapid rate of performance improvements while at the same time significantly improving price/performance. Moreover, with the adoption of the High Performance Computing and Communication Initiative by the federal government, this revolution is likely to gain momentum.

This revolution is being fueled by the movement toward massively parallel computer architectures based on commodity, inexpensive RISC (Reduced Instruction Set Computer) CPUs, very dense memories, and parallel secondary storage devices. The technology of these components is advancing so rapidly that contemporary uniprocessor RISC workstations in the $5000-$15000 price range have more compute power than the supercomputer of the recent past, eg. the Cray 1. Of course, when we go (massively) parallel, we get extraordinary power from the parallel CPUs, parallel memories, and parallel secondary storage devices. Virtually all vendors of high performance computers have committed to producing such architectures within the next few years. Moreover, several vendors have announced plans to have parallel systems which will have a peak capacity of 100s of gigaflops in the 1992-1993 time frame and teraflops in the 1995 time frame.

Interestingly enough, the very existence of such powerful, scalable, parallel supercomputers has itself stimulated our appetite for even more computational power. Every new generation of parallel

---
[1]This work was supported in part by Grant N0014-91-J-1576.

supercomputers has caused us to broaden our perspectives of what may be computationally tractable and interesting and created a significant new demand for future generations of new systems. Thus we have the push of microelectronics technology from below and the pull from applications scientists and engineers from above. As these systems become sufficiently fast, we can start to seriously think about using them for engineering design work to optimize designs as distinct from understanding the science of various physical phenomena. A good example is the Boeing 777 which I am told is being totally designed on computers (as distinct from understanding the science of fluid dynamics over say a wing of the Boeing 777).

Massively parallel supercomputers are sufficiently unique in the world of computers and the problems to which people aspire to apply them are sufficiently difficult, that the advent of such systems has promoted the growth of a new scientific/engineering discipline called Computational Science. This discipline which is devoted to the intelligent application of massively parallel supercomputers to real world problems involves:

aspect of problem	discipline
Modelling	Application Discipline
Mathematical Analysis	Applied Mathematics
Algorithms	Computer Science

Each discipline brings its own set of paradigms to bear and makes an important contribution. Moreover, all these contribution are critical for the entire enterprise to be successful. The synergy among the various subdisciplines is such as to guarantee that the whole is much more than the sum of its parts.

While Computational Science has played an important role in the study and application of science and engineering, it will play an even bigger role when our computational simulations are faster, more accurate, and cost less. Of course, computational simulation will never replace physical experiments; but it can become a full and equal partner in the development and application of science.

Now that I have given you the "good news", I have to give you the "bad news." In order to make effective use of all these latent power, we must invent new massively parallel algorithms for our mathematical models and we must do parallel programming to implement these new parallel algorithms. These are both difficult, but not impossible tasks. However, they do involve Computer Science in a very major way. Indeed one might say that it is the Computer Science aspects of the problem that are the chief stumbling block to realizing the potential of this wonderful technology. To be sure, on close inspection it may turn out that there aren't really good massively parallel algorithms for our current mathematical models and this may by itself force a reconsideration of the modeling and mathematical analyses of the problem at hand in order to find other mathematical formulations which will have more effective massively parallel algorithms. Thus these new radically different computer architectures will probably provoke a reexamination of the very foundations of scientific disciplines. In other words, there is not the slightest chance that Computer Science alone will obsolete the other sciences and engineering.

In this talk we present two specific examples of direct interest to computational fluid dynamics. The first concerns results on the parallel solution of very large sparse linear systems on multiprocessors. This is joint work with Mark Chang, Carl Scarbnick, and Andrew Sherman, all of SCIENTIFIC Computing Associates Inc.

We have developed a software package, MP-PCGPAK2, that solves large, sparse linear systems of equations in parallel. Such problems are the compute intensive kernel for many large scale scientific and engineering computations. Good examples, are the numerical solution of partial differential equations of fluid dynamics by implicit finite difference or finite element methods.

The parallel linear solvers we have implemented are based on the preconditioned conjugate gradient for positive definite, symmetric systems and preconditioned GMRES for systems that are indefinite and/or nonsymmetric. GMRES is a generalization of the conjugate gradient method that was invented by Youcef Saad and myself in the 1980s [1]. The current implementation of MP-PCGPAK2 was developed by SCIENTIFIC Computing Associates Inc. [2] and is based on message passing. Versions of MP-PCGPAK2 exist for the Intel iPSC/860 and nCUBE/2 hypercubes, but are not portable.

MP-PCGPAK2 can be used to solve any nonsingular sparse matrix problem. We do NOT make any assumption about the distribution of the nonzeroes in the matrix. There are three main types of computations in MP-PCGPAK2: (1) a sparse matrix vector product; (2) the solution of two systems with sparse triangular matrices; and (3) "overhead" computations such as inner products needed to carry out the iteration.

We have used this package to solve some very large linear systems arising from finite difference discretizations of model elliptic partial differential equations. In a matter of a few minutes we can solve linear systems with more than 5 million variables on the 128 node iPSC/860 at Oak Ridge National Laboratory and systems with more than 10 million variables on the 1024 node nCUBE/2 at Sandia National Laboratory. With nCUBE's new operating system (version 3.0) and compilers, we should be able to solve these problems in about a minute. This solver is the only general purpose sparse matrix solver that I know of which can effectively handle systems of this magnitude. Such a capability is a prerequisite to a successful attack on many of the Grand Challenge problems which depend upon being able to solve coupled systems of partial differential equations on grids at least as large as 100x100x100.

The second specific example involves the solution of time dependent partial differential equations, namely the shallow water equations. This coupled system of three PDEs is a well known primitive model of atmospheric flow. Ashish Deshpande, a graduate student at Yale, and I have parallelized the standard explicit time differencing for evolving the solution of the shallow water equations in time. Our emphasis has been on developing a parallel program which is both efficient and portable over a range of parallel architectures.

In order to achieve our goal, we have used Linda which is a portable parallel programming language invented by David Gelernter of Yale University [3].

Linda is a coordination language which when added to any language like C provides a portable, parallel programming language. The Linda model consists of a multiprocessor with a virtual shared associative memory, which is called "tuple space", and six operations for manipulating data in the "tuple space." It is extraordinarily simple to express BOTH data parallel algorithms and control parallel algorithms in the Linda model. Porting a Linda code from one parallel platform to another involves only a recompilation.

Typically one starts with a traditional sequential program and then parallelizes it by adding appropriate Linda commands. In practice this is similar to adding compiler annotations or directives in order to vectorize a code. Usually we need to add a relatively small number of lines of code. From my point of view, the chief advantage of Linda over other schemes for writing portable parallel codes is its easy of use. In fact, it is so conceptually natural that many people find it to be useful tool for developing complex sequential software.

We have used a commercial version of Linda for C (called C-Linda) from SCIENTIFIC Computing Associates to solve a set of partial differential equations (the shallow water equations) on a rectangular, doubly periodic domain using second order finite difference approximations. The same, identical program was executed in a variety of parallel and distributed environments. We have obtained benchmark results for the Intel hypercubes (distributed memory machines), the Sequent (a shared memory machine), and networks of workstations. We have evaluated the performance of Linda on all these platforms and compared it to that of implementations using alternative methods (message passing in NX/2 on the hypercubes and PVM on the network of the distributed memory machines). Our results show that Linda parallel programs are easy to write and understand and can provide very good efficiency on all machines. The efficiency of the Linda codes in all cases is comparable to the efficiency of the alternatives on those machines. The details of these experiments will appear in a paper to be presented at the Supercomputer '92 Conference during November, 1992. We believe that the results for this problem are representative of the results for applying Linda to implement parallel explicit time marching schemes for initial value problems for PDEs.

References

[1] Y. Saad and M. H. Schultz, 1986. "GMRES: A generalized minimal residual algorithm for solving nonsymmetric linear systems," *SIAM J. Sci. Statist. Comput.* **7**, p. 856.

[2] *SCIENTIFIC, MP-PCGPAK2 User's Guide*, Version 1.0, SCIENTIFIC Computing Associates Inc., 265 Church St, New Haven, CT 06510, April 1991.

[3] N. Carriero and D. Gelernter, 1990. *How to Write Parallel Programs: A First Course*, Cambridge MIT Press.

ACCELERATION TECHNIQUES

LOCAL PRECONDITIONING OF THE EULER EQUATIONS AND ITS NUMERICAL IMPLICATIONS

Bram van Leer

Department of Aerospace Engineering
The University of Michigan
Ann Arbor, MI 48109-2140

1 Introduction

In many problems of fluid dynamics the task is to obtain the steady flow pattern that results for large times. One way of finding such time-asymptotic solutions of the flow equations is to march in time from a suitable set of initial values, while applying the appropriate boundary conditions, until all time derivatives have become vanishingly small. Numerically this can be achieved, for instance, by applying an explicit finite-volume scheme over and over, until the residual has dropped below some small threshold value.

Explicit marching schemes are attractive, because of their simplicity. They lend themselves easily to vectorization and parallelization, require little storage, and carry over to unstructured grids. Their chief drawback is the limitation on the allowable time-step. For the Euler equations this is given by the Courant-Friedrichs-Lewy (CFL) criterion: the largest stable time-step locally is of the order of the shortest time it takes any of the acoustic waves to cross a computational cell.

The local stability limit on the time-step may vary greatly throughout the computational domain, because of spatial variations in the wave speed as well as in the cell size and shape. This suggests the technique of "local time-stepping", i.e. using the largest stable time-step, or some fixed fraction of it, in each computational cell, rather than one globally stable value for all cells. The temporal accuracy of the marching scheme is lost, but this is acceptable as long it does not affect the steady state. Schemes with this property are easily found; for instance, multi-stage schemes preserve the steady state.

1.1 A fundamental problem

The method of local time-stepping is without doubt the most widely used acceleration method in marching to a steady solution. Nevertheless, it does not address the essential problem of *the stiffness of the Euler equations.*

The hyperbolic system of the Euler equations becomes stiff when the speeds of the waves described by these equations become vastly different, i.e., the ratio of the largest to the smallest characteristic speed, the so-called *characteristic condition number*, becomes large. This happens in almost incompressible (low Mach-number) flow, stagnation flow and transonic flow. In any of these situations the convergence speed of a marching method deteriorates: while the fastest wave always crosses one cell per "local" time-step, the slowest wave hardly advances.

What is needed is a different time-step for each of the waves described by the Euler equations; thus, the time-step becomes matrix-valued. I have named this technique "characteristic time-stepping" in an earlier paper [1]; it is equivalent to preconditioning the Euler equations with a locally evaluated matrix. Local time-stepping could be called preconditioning with a locally evaluated scalar.

The concept of characteristic time-stepping was introduced at the NASA Transonic Symposium in 1988 [2]. Since then, our knowledge of this subject has increased by an order of magnitude. Most of it has been compiled by Wen-Tzong Lee (University of Michigan, Ann Arbor) in his doctoral thesis [3], but it is still expanding at a rapid pace. I shall expose our most striking findings in the sections below; first, though, a further discussion of the stiffness problem, and a review of work by others, are due.

1.2 Three flow regimes

The characteristic speeds depend on the direction of wave propagation. The largest range of characteristic speeds is found in the flow direction, where they equal $q - c$, q and $q + c$; here q is the flow speed and c is the sound speed. The characteristic condition number

K, which equals the ratio of the largest to the smallest wave speed, can be expressed in terms of the Mach number M. For a significant range of Mach numbers its value exceeds 3, in which case we shall consider the equations "stiff". With regard to the range of values of the characteristic condition number, three flow regimes can be distinguished.

1. *Subsonic flow regime*: for $M < 0.5$ the flow speed q is the smallest characteristic speed, hence $K = \frac{q+c}{q} = \frac{M+1}{M} > 3$. The subsonic stiffness is caused by slow shear and entropy waves. K is large especially in the following situations:

   - almost incompressible flow ($M \to 0$ everywhere); in addition to slow convergence to a steady state, loss of accuracy has been reported [4, 1];

   - stagnation regions;

   - near a solid surface in viscous flow.

2. *Transonic flow regime*: for $0.5 \le M \le 2$ the acoustic speed $q-c$ is the smallest characteristic speed, hence $K = \frac{q+c}{|q-c|} = \frac{M+1}{|M-1|} \ge 3$. The transonic stiffness is caused by the slow backward-moving acoustic wave. K is large especially in the following situations:

   - near a sonic point, causing slow convergence to steady transonic solutions;

   - inside a transonic numerical shock profile, causing the slow motion of a transonic shock to its steady position.

3. *Supersonic flow regime*: for $M > 2$ the acoustic speed $q-c$ still is the smallest characteristic speed, hence $K = \frac{q+c}{q-c} = \frac{M+1}{M-1} < 3$. Supersonic stiffness is not much of a problem since K tends to 1 as M increases, making preconditioning unnecessary.

1.3   Previous work on local preconditioning

To overcome the stiffness of the Euler equations, several authors have studied and suggested preconditioning methods. Earlier work in this area uses the term "pseudo-unsteady" (PUS) equations, referring to the fact that the time-evolution is not described correctly by the preconditioned equations. Most of the research effort was in removing the stiffness in the low-speed regime.

An early example of a PUS method is Chorin's [5] artificial compressibility method for the computation of steady incompressible flow, in which the time-derivative of the pressure is added to the continuity equation $\nabla \cdot \mathbf{u} = 0$. The coefficient of this term can be adjusted for optimal convergence to a steady state in numerical applications.

Turkel [6] argued that similar terms should also be added to the momentum equations, which is equivalent to preconditioning Chorin's PUS equations. These preconditioned equation systems for incompressible flow are then generalized by Turkel for compressible flow, with emphasis on the low-speed regime; this leads to a two-parameter family of preconditioning matrices. During our search for preconditioning matrices we found that Turkel's family actually has one member that is effective also in the transonic flow regime; this was overlooked by Turkel. Turkel's paper is purely on theory; it contains no numerical examples.

Feng and Merkle [7] favor one particular preconditioning matrix for low-speed flow included in Turkel's family of matrices. One attractive feature of their work is that they numerically demonstrate the effect of convergence acceleration through preconditioning. Their tests cover several combinations of preconditioning matrices and time-marching methods, for inviscid incompressible and compressible flow, and even viscous compressible flow. The marching methods they test are fourth-order Runge-Kutta (explicit) and Approximate Factorization (implicit).

The only significant work on preconditioning dealing with the entire Mach-number range is that of Viviand [8], who again thinks in terms of PUS equations. Viviand starts from the iso-energetic Euler equations, resulting from the assumption of homogeneous total specific enthalpy (stagnation enthalpy); this system is known to be hyperbolic. Viviand analyses a four-parameter family of preconditionings; his analysis, however, is more concerned with issues such as finding preconditioned systems that are in full conservation form, or have simple expressions for the characteristic speeds, than with reducing the characteristic condition number. Viviand's contribution to convergence acceleration is to design for the largest permissible time-step. In the optimal PUS equations he derives, the maximum characteristic speed always equals the flow speed, but the range of the characteristic speeds is not discussed, and no numerical results are presented.

2 Preconditioning matrices for the Euler equations

2.1 Preconditioning in one space dimension

For the one-dimensional flow equations, optimal preconditioning is trivial. The equations for quasi-one-dimensional flow in a channel with variable cross-section can be written in the form

$$\frac{\partial \mathbf{U}}{\partial t} = -\frac{\partial \mathbf{F}(\mathbf{U})}{\partial x} + \mathbf{s}(\mathbf{U}, x) = \mathbf{Res}(\mathbf{U}), \tag{1}$$

or, giving up the conservation form,

$$\frac{\partial \mathbf{U}}{\partial t} = -\mathbf{A}(\mathbf{U})\frac{\partial \mathbf{U}}{\partial x} + \mathbf{s}(\mathbf{U}, x) = \mathbf{Res}(\mathbf{U}), \tag{2}$$

where $\mathbf{A}(\mathbf{U})$ is the Jacobian of $\mathbf{F}(\mathbf{U})$. The characteristic speeds are the eigenvalues of the matrix $\mathbf{A}$; they equal $q - c$, q and $q + c$. From the latter form of the equations it is clear that the preconditioning matrix

$$\mathbf{P} = |q||\mathbf{A}|^{-1}, \tag{3}$$

equalizes the characteristic speeds; that is, the preconditioned equations

$$\frac{\partial \mathbf{U}}{\partial t} = |q||\mathbf{A}|^{-1}\mathbf{Res} \tag{4}$$

have all characteristic speeds equal to q, hence $K = 1$.

This preconditioning is unique for supersonic flow: there are no other preconditioning matrices that achieve the optimal condition number $K = 1$. For subsonic flow there is a multi-parameter set of matrices that can do the job; a useful one-parameter family is discussed in detail by Lee [3]. The matrix (3) is a member of this family, but it is a *different* member that allows generalization to multi-dimensional preconditioning. This is one reason why finding an optimal multi-dimensional preconditioning matrix is not a trivial task.

2.2 Preconditioning in two space dimensions

In two space dimensions it is most convenient to write the Euler equations in terms of the symmetrizing variables $\tilde{\mathbf{U}}$, with

$$
d\tilde{\mathbf{U}} = \begin{pmatrix} \frac{dp}{\rho c} \\ du \\ dv \\ dp - c^2 d\rho \end{pmatrix}, \tag{5}
$$

and to choose a coordinate system that is initially aligned with the flow direction, so that $u = q$ and $v = 0$ in the coefficient matrices. The equations then take the form

$$
\frac{\partial \tilde{\mathbf{U}}}{\partial t} = -\tilde{\mathbf{A}}(\tilde{\mathbf{U}})\frac{\partial \tilde{\mathbf{U}}}{\partial x} - \tilde{\mathbf{B}}(\tilde{\mathbf{U}})\frac{\partial \tilde{\mathbf{U}}}{\partial y} = \mathbf{Res}(\tilde{\mathbf{U}}), \tag{6}
$$

with

$$
\tilde{\mathbf{A}} = \begin{pmatrix} u & c & 0 & 0 \\ c & u & 0 & 0 \\ 0 & 0 & u & 0 \\ 0 & 0 & 0 & u \end{pmatrix}, \quad \tilde{\mathbf{B}} = \begin{pmatrix} 0 & 0 & c & 0 \\ 0 & 0 & 0 & 0 \\ c & 0 & 0 & 0 \\ 0 & 0 & 0 & 0 \end{pmatrix}. \tag{7}
$$

The matrices $\tilde{\mathbf{A}}(\tilde{\mathbf{U}})$ and $\tilde{\mathbf{B}}(\tilde{\mathbf{U}})$ do not have the same eigenvectors, and therefore can not be diagonalized simultaneously. This means Equation (6) can not be written as a system of coupled scalar convection equations; otherwise, optimal preconditioning would be as easy as in the case of one space dimension.

The derivation of a local preconditioning matrix that narrows the range of characteristic speeds therefore is not obvious. Since there is less freedom among preconditioning matrices for supersonic flow, we limited our search first to this flow regime. It appeared that an optimal matrix could be found among symmetric matrices of the form

$$
\mathbf{P} = \begin{pmatrix} P & R & 0 & 0 \\ R & Q & 0 & 0 \\ 0 & 0 & Z & 0 \\ 0 & 0 & 0 & 1 \end{pmatrix}, \tag{8}
$$

and that this form worked for subsonic flow as well. Two free parameters can be eliminated by requiring that the propagation of shear disturbances with the flow speed q should not be affected by the preconditioning; this leads to the two-parameter form

$$
\mathbf{P} = \begin{pmatrix} \frac{X}{\beta^2} M^2 & -\frac{X}{\beta^2} M & 0 & 0 \\ -\frac{X}{\beta^2} M & \frac{X}{\beta^2} + 1 & 0 & 0 \\ 0 & 0 & X & 0 \\ 0 & 0 & 0 & 1 \end{pmatrix}.
\tag{9}
$$

The parameters β and X can be chosen so as to manipulate the shape of acoustic wave fronts generated by the preconditioned Euler equations. In the subsonic regime an acoustic wave front produced by a point disturbance is no longer a circle but becomes an ellipse; its center does not move with the flow speed. The shape and placement of the ellipse are a function only of β, while the size of the ellipse is proportional to X. In supersonic flow β can be chosen such that the entire acoustic front collapses onto the two steady Mach waves.

The choices of β and X that yield the optimal characteristic condition number are listed below.

$$
\beta = \begin{cases} \sqrt{1 - M^2}, & M < 1, \\ \sqrt{M^2 - 1}, & M \geq 1; \end{cases}
\tag{10}
$$

$$
X = \begin{cases} \sqrt{1 - M^2}, & M < 1, \\ \sqrt{1 - M^{-2}}, & M \geq 1. \end{cases}
\tag{11}
$$

The choice of β for subsonic flow is such that the acoustic front produced by a point disturbance has forward-backward symmetry with respect to a *fixed* frame instead of a frame moving with the flow velocity. There may, however, be a practical use for asymmetric wave fronts; these can be made to fit into cells with a high aspect-ratio, possibly removing a numerical source of stiffness. This matter is still under investigation.

The optimal preconditioning is perfect for $M \geq 1$, that is, $K = 1$, and reduces the condition number from $(M + 1)/\min(M, 1 - M)$ to $1/\sqrt{1 - M^2}$ for $M < 1$. All further analysis of matrices with more degrees of freedom than (8) indicates that we can not improve upon this result.

2.3 Preconditioning in three space dimensions

When moving from two to three dimensions, some further deterioration of the condition number has to be accepted. The reason is that in three dimensions a new kind of shear wave is possible, namely, one

	original	preconditioned		
	equations	1-D	2-D	3-D
supersonic	$\frac{M+1}{M-1}$	1	1	$\frac{1}{\sqrt{1-M^{-2}}}$
subsonic	$\frac{M+1}{\min(1-M,M)}$	1	$\frac{1}{\sqrt{1-M^2}}$	$\frac{1}{\sqrt{1-M^2}}$

Table 1: Characteristic condition number for the Euler equations before and after preconditioning.

that rotates the flow velocity; this wave mode can not be separated from the acoustic waves when manipulating the Euler equations. In consequence, when the acoustic waves are slowed down by a factor X in the supersonic case, the 3-D shear wave is slowed down too. The condition number therefore deteriorates to $1/\sqrt{1-M^{-2}}$ for $M \downarrow 1$.

The optimal preconditioning matrix for the 3-D Euler equations in a streamline-aligned coordinate system is similar to its two-dimensional version (9), namely,

$$
\mathbf{P} = \begin{pmatrix}
\frac{X}{\beta^2}M^2 & -\frac{X}{\beta^2}M & 0 & 0 & 0 \\
-\frac{X}{\beta^2}M & \frac{X}{\beta^2}+1 & 0 & 0 & 0 \\
0 & 0 & X & 0 & 0 \\
0 & 0 & 0 & X & 0 \\
0 & 0 & 0 & 0 & 1
\end{pmatrix}, \tag{12}
$$

where β and X are still given by Equation (11).

2.4   Summary of results

The effect of preconditioning on the characteristic condition number is summarized in Table 1 and Figure 1.

3   Numerical implementation

Before the above matrices can be applied in practice, a number of issues must be dealt with. These are

- *Avoiding the divergence of the matrix or the time-step.*
  The matrices all diverge for $M = 1$, while the allowable time-step diverges for $M = 0$. For these values of M a cap has to be put on the matrix elements and on the time-step to be used.

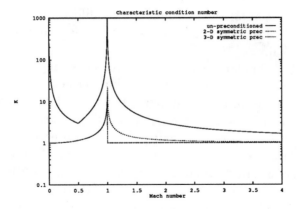

Figure 1: 2-D/3-D condition number as a function of Mach number.

- *Modifying the numerical flux function.*
  Since the preconditioning is derived on the basis of the differential equations, it may clash with the numerical truncation error, when applied to a discrete approximation of the differential equations. For upwind discretizations based on Roe's approximate Riemann solver we found a severe stability restriction at low Mach numbers, forcing us to modify the numerical flux function. The modified flux function, based on an approximate Riemann solver for the *preconditioned* Euler equations, turned out to improve both the stability and the accuracy of the numerical discretization.

- *Transforming the matrices.*
  The matrices shown above are appropriate for flow aligned with the x-axis and for the symmetrizing flow variables; two similarity transformations are needed to put these matrices into the form appropriate for any flow angle and for the conserved flow variables.

A short discussion of the flux-function modification can be found in [1]; for a full discussion of all three topics the reader is referred to Lee's thesis [3].

Numerical simulations including preconditioning so far have been based on first-order upwind differencing, either by flux-difference splitting (Roe [9]) or by flux splitting (Van Leer [10]). The test cases include channel flows and external flows. Significant convergence acceleration through preconditioning has been observed, in agreement with theoretical predictions. In addition, for low Mach numbers, deterioration of accuracy as observed by Volpe [4] appears to be avoided with the use of the modified upwind flux function.

A numerical example demonstrating convergence acceleration is presented in Figures 2 and 3. Shown here are the Mach-number contours of transonic flow ($M = 0.85$, angle of attack 1°) over a NACA 0012 airfoil, computed with the preconditioned scheme, followed by the residual histories for both this computation and a similar computation without preconditioning. The preservation of accuracy for low-speed flow is illustrated by Figures 4, 5 and 6, showing Mach-number contours for Mach 0.0001 flow over a circular bump, computed with the original Euler scheme, the preconditioned Euler scheme, and a potential-flow solver, respectively.

More numerical examples are given in [1]; a more complete set of experiments is described in [3].

4   Further analysis

4.1   Turkel's preconditioning

Turkel [11] presented a preconditioning matrix with two free parameters we may call α_T and β_T; it is intended for use at low Mach numbers. For the equations in primitive variables $\mathbf{V} = (\, p,\, u,\, v,\, \rho\,)^T$ it is presented in the form of its inverse

$$\left(\mathbf{P_V^T}\right)^{-1} = \begin{pmatrix} \frac{1}{\rho\beta_T{}^2} & 0 & 0 & 0 \\ \frac{\alpha_T u}{\rho\beta_T{}^2} & 1 & 0 & 0 \\ \frac{\alpha_T v}{\rho\beta_T{}^2} & 0 & 1 & 0 \\ 0 & 0 & 0 & 1 \end{pmatrix}. \tag{13}$$

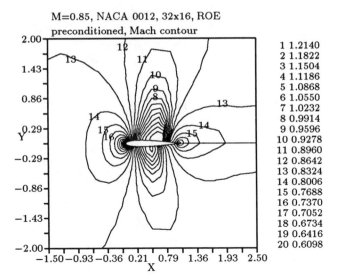

Figure 2: Steady solution of the Euler equations for flow over a NACA 0012 airfoil at $M_\infty = 0.85$, $\alpha = 1.0°$; shown are Mach-number contours. Matrix preconditioning was applied in combination with local time-stepping. Numerical flux function used: modified Roe

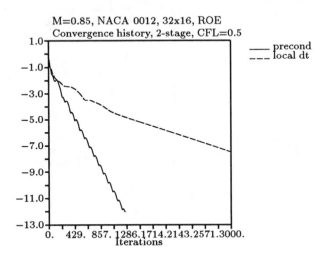

Figure 3: Residual-convergence histories for the computation of the solution of Figure 2 and a similar computation without preconditioning.

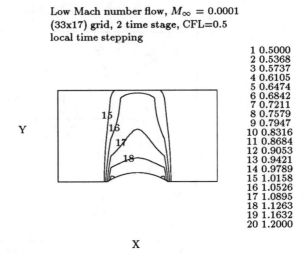

Figure 4: Mach-number contours for steady solution of the Euler equations for flow over a circular bump in a semi-infinite plane, with $M_\infty = 0.0001$. Numerical flux function used: original Roe; no preconditioning.

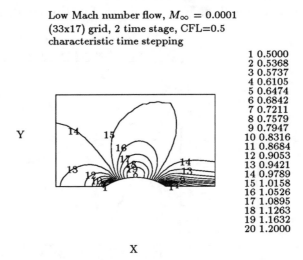

Figure 5: Mach-number contours for the steady solution of the Euler equations for flow over a circular bump in a semi-infinite plane, with $M_\infty = 0.0001$. Numerical flux function used: modified Roe; preconditioning was applied.

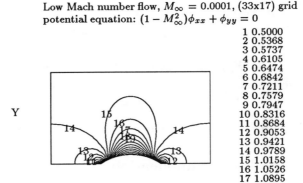

Figure 6: Mach-number contours of the steady potential flow over a circular bump in a semi-infinite plane, with $M_\infty = 0.0001$.

For stream-aligned coordinates and symmetrizing state variables $\tilde{\mathbf{U}}$ the matrix transforms into

$$\mathbf{P}_{\tilde{\mathbf{U}}}^T = \begin{pmatrix} \beta_T{}^2 & 0 & 0 & 0 \\ -\alpha_T M & 1 & 0 & 0 \\ 0 & 0 & 1 & 0 \\ \rho c(\beta_T{}^2 - 1) & 0 & 0 & 1 \end{pmatrix}. \tag{14}$$

Note that this preconditioning matrix is not symmetric.

An analysis of the characteristic speeds after preconditioning shows that Turkel's matrix produces exactly the same wave fronts as our symmetric matrix (9), if the parameters are chosen as follows:

$$\beta_T = M\frac{X}{\beta}, \tag{15}$$

$$\alpha_T = 1 + M^2\frac{X^2}{\beta^2} - \left(1 + \frac{M^2 - 1}{\beta^2}\right)X. \tag{16}$$

To generate the optimal characteristic condition number, Turkel's parameters must be chosen as

$$\beta_T = \begin{cases} M, & M < 1, \\ 1, & M \ge 1; \end{cases} \tag{17}$$

$$\alpha_T = \begin{cases} 1 + M^2, & M < 1, \\ 2\left(1 - \sqrt{1 - M^{-2}}\right), & M \geq 1. \end{cases} \qquad (18)$$

This result was not known to Turkel, although he did derive the relation

$$\alpha_T = 1 + \beta_T^2,$$

yielding forward-backward symmetry of the acoustic front for $M < 1$.

4.2 Multi-grid application

Marching schemes can be used in a multi-grid strategy, provided they can efficiently damp high-frequency error modes. Van Leer, Tai and Powell [12] presented and tested multi-stage schemes with optimized high-frequency damping, for the time integration of convection or convection-diffusion equations. Their procedure for optimizing the high-frequency damping is a geometry exercise in the complex plane: putting the zeros of the multi-stage amplification factor on top of the locus of eigenvalues of the discrete spatial operator. The optimum is achieved for one specific value of the time-step; finding this value is part of the design process. The derivation is straightforward for one-dimensional convection and readily extends to two and three dimensions, but the extension of the theory to a nonlinear hyperbolic system such as the multi-dimensional Euler equations fails, because of the large spread of the characteristic speeds.

The local preconditioning technique described above removes this spread, except for Mach numbers close to 1. The effect of the preconditioning on discretizations of the spatial Euler operator is a strong concentration of the pattern of eigenvalues in the complex plane. Again, by strategically placing zeros in the complex plane, multi-stage schemes can be designed that systematically damp most high-frequency waves admitted by the particular discrete operator.

The effect of preconditioning on the locus of eigenvalues of the first-order upwind Euler discretization is demonstrated by Figures 7 (original operator) and 8 (preconditioned operator), for $M = 0.5$ and flow aligned with the x-direction.

Thus, preconditioning can speed up the multi-grid process in two ways:

- by accelerating the single-grid scheme used in the process;

- by making the single-grid scheme a more reliable smoother.

The existence of these two separate effects was demonstrated by Tai [13] for multi-grid acceleration of a one-dimensional flow computation; for multi-dimensional computations it remains to be demonstrated. For more details see [14] or [3].

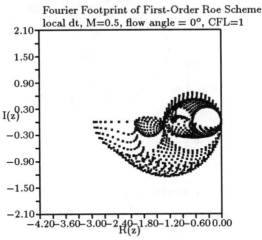

Figure 7: Fourier footprint of the first-order upwind approximation of the spatial Euler operator, for $M = 0.5$, and flow speed aligned with the grid.

5   Conclusions and outlook

Local preconditioning of the multi-dimensional Euler equations can reduce the spread of the characteristic speeds from a factor $(M + 1)/\min(M, |M - 1|)$ to a factor $1/\sqrt{1 - \min(M^2, M^{-2})}$. There is mounting evidence that the latter value is the lowest attainable. There appears to be a multi-parameter family of preconditioning matrices that can achieve the optimal characteristic condition number.

Numerical experiments in which optimal preconditioning is applied to an explicit upwind discretization of the two-dimensional Euler equations, show that it significantly increases the rate of convergence to a steady solution. Another effect of the preconditioning is that it strongly clusters the eigenvalues of the discrete spatial operator. Most convergence-acceleration methods will benefit from such a clustering.

Our present research efforts are divided over further analyzing and testing preconditioning matrices. Three important theoretical

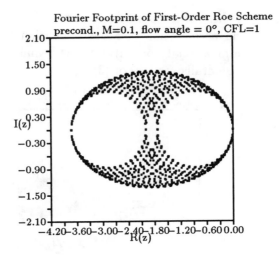

Figure 8: Fourier footprint (symbols) of the preconditioned first-order upwind approximation of the spatial Euler operator, for $M = 0.5$, and flow speed aligned with the grid. The frequencies included in the footprint are $\beta_x \in [0, \pi]$, $\beta_y \in [0, \pi]$.

issues are

- charting the entire family of optimal preconditioning matrices and their properties;

- matching the acoustic wave front to the shape of a computational cell by preconditioning;

- analyzing the interaction of Euler preconditioners with the viscous terms of the Navier-Stokes equations.

The results of these investigations may have far-reaching consequences for the development of CFD algorithms in the 1990s.

References

[1] B. van Leer, W. T. Lee, and P. L. Roe, "Characteristic time-stepping or local preconditioning of the Euler equations," in *AIAA 10th Computational Fluid Dynamics Conference*, 1991.

[2] B. van Leer, "Euler solvers for transonic applications," in *NASA Transonic Symposium: Theory, Application, and Experiment*, 1989.

[3] W.-T. Lee, *Local Preconditioning of the Euler Equations.* PhD thesis, University of Michigan, 1991.

[4] G. Volpe, "On the use and accuracy of compressible flow codes at low Mach numbers," AIAA Paper 91-1662, 1991.

[5] A. J. Chorin, "A numerical method for solving incompressible viscous flow problems," *Journal of Computional Physics*, vol. 2, 1967.

[6] E. Turkel, "Preconditioned methods for solving the incompressible and low speed compressible equations," *Journal of Computational Physics*, vol. 72, 1987.

[7] J. Feng and C. L. Merkle, "Evaluation of preconditioning methods for time-marching systems," AIAA Paper 90-0016, 1990.

[8] H. Viviand, "Pseudo-unsteady systems for steady inviscid flow calculations," in *Numerical Methods for the Euler Equations of Fluid Dynamics*, 1985.

[9] P. L. Roe, "The use of the Riemann problem in finite-difference schemes," *Lecture Notes in Physics*, vol. 141, 1980.

[10] B. van Leer, "Flux-vector splitting for the Eulerequations," *Lecture Notes in Physics*, vol. 170, 1982.

[11] E. Turkel, "Preconditioned methods for solving the incompressible and low speed compressible equations." ICASE Report 86-14, 1986.

[12] B. van Leer, C. H. Tai, and K. G. Powell, "Design of optimally-smoothing multi-stage schemes for the Euler equations," in *AIAA 9th Computational Fluid Dynamics Conference*, 1989.

[13] C.-H. Tai, *Acceleration Techniques for Explicit Euler Codes.* PhD thesis, University of Michigan, 1990.

[14] B. van Leer, W. T. Lee, P. L. Roe, K. G. Powell, and C. H. Tai, "Design of optimally-smoothing schemes for the Euler equations," *Journal of Applied and Numerical Mathematics*, 1991.

INCREMENTAL UNKNOWNS IN FINITE DIFFERENCES

Roger Temam

Laboratoire d'Analyse Numérique,
Université Paris-Sud, Orsay
and
Institute for Scientific Computing and Applied Mathematics
Indiana University.

ABSTRACT

Incremental Unknowns have been introduced in (Temam, 1990) as a means to approximate inertial manifolds when finite differences are used. They seem to have a much broader interest and in fact they produce a new efficient concept in finite differences. We describe here their utilization for the solution of linear elliptic problems where they appear as a preconditioner; and their utilization for time dependent problems where they allow a differentiated treatment of the small scale and large scale components of a flow and they lead to improved CFL stability conditions.

Introduction

Our aim in this article is to present the Uncremental Unknowns (IU) method. This is a new concept in finite differences. It consists in replacing the nodal values of the unknown functions on the fine grids by proper increments to the values on the coarse grids. The IU method has emerged from the study of time dependent nonlinear problems, but despite its initial motivation, the IU method seems to have also some interest in linear algebra for the resolution of linear elliptic problems, and it appears then as a preconditioner.

In Sec. 1 we describe some of the motivations ; in Sec. 2 we present the IU method and show how it relates to other existing methods. Numerical results are given in Sec. 3. In Sec. 4 we give some theoretical results concerning the utilization of IU for linear elliptic problems. Finally in Sec. 5 we show briefly how the method applies to (linear or nonlinear) evolution problems.

Content

1. The Motivations
2. The Incremental Unknowns
3. Numerical Implementation
4. Theoretical Results.
5. Evolution Equations.

1. The Motivations

There are several motivations for the introduction and utilization of incremental unknowns ; they correspond also to directions in which we intend to pursue the developement of the method :

1.1. Turbulence and nonlinear dynamics

Turbulent flows are considered as superposition of small and large eddies and the understanding of the interaction of small and large eddies is an important part of the understanding of turbulence.

In an attempt to address this difficulty problem from the point of view of dynamical systems theory the concepts of Inertial Manifolds and Approximate Inertial Manifolds have been recently introduced (see e.g. Foias-Sell-Temam (1985, 1988), Mallet-Paret-Sell (1988), Foias-Manley-Temam (1987, 1988), M. Kwak (1991)). Essential in this approach is the decomposition of u into two sets of unknowns

$$u = y + z, \qquad (1.1)$$

corresponding to the small scale and the large scale components of the flow. An inertial manifold is then the graph of a function

$$z = \Phi(y).$$

On such a manifold the law

$$z(t) = \Phi(y(t)) \qquad (1.2)$$

is enforced, which defines a dependence of high frequencies on the low frequencies ; the high frequencies are said to be slaved (exactly or approximately) by the low frequencies.

When spectral (or pseudo-spectral) methods are used, u is represented by a series

$$u(x,t) = \sum_{j=1}^{\infty} \hat{u}_j(t) w_j(x), \qquad (1.3)$$

and, for a given cut-off value m, we obtain a decomposition like (1.1) by setting

$$y(x,t) = \sum_{j=1}^{m} \hat{u}_j(t) w_j(x), \quad z(x,t) = \sum_{j=m+1}^{\infty} \hat{u}_j(t) w_j(x). \quad (1.4)$$

The law (1.2) is expressed by relations

$$\hat{u}_j(t) = \Phi_j(\hat{u}_1(t), ..., \hat{u}_m(t)), \quad j \geq m + 1.$$

When finite differences are used, all nodal values of the function play the same role and there is no obvious way to decompose the function into its small scale and large scale components. As we will see Incremental Unknowns provide a means to achieve a decomposition like (1.1).

1.2. Large scale scientific computing and multilevel methods

Large scale scientific computing allows the utilization of a large number of spatial variables ; discretizations involving millions of variables are not anymore exceptional in CFD. In the case of spectral discretizations, when many modes are used, the magnitude of the high frequencies is much smaller than that of the low frequencies, by several orders of magnitude :

$$|\hat{u}_{10^6}(t)| << |\hat{u}_1(t)|, \quad |z(t)| << |y(t)|.$$

Hence the discretized system is *stiff* and this stiffness is a new problem in numerical analysis which arises just by the fact that we can afford using a large number of spatial discretization variables. This problem did not exist ten years ago and will probably become more acute when we will have access to the giga or teraops computers. We feel that for the development of performant algorithms this stiffness problem should be addressed.

This problem was one of the motivations for the development of the Nonlinear Galerkin Method (see e.g. Marion-Temam (1989, 1990), Temam (1989), Jolly-Kevrekidis-Titi (1990)) which is based on :

(i) The utilization of proper approximations of the attractor, associated to approximate inertial manifolds.

(ii) The decomposition of the unknown function u into two arrays of unknowns as in (1.1) or more generally into a larger number of arrays of unknowns :

$$u = y + z_1 + ... + z_r. \tag{1.5}$$

(iii) A systematically differentiated treatment of the small scale and large scale components of the flow, in a way which can be also appropriate for parallel computing.

In a similar manner, when trying to implement such algorithms with finite differences we were lead to the introduction of the incremental unknowns. In finite differences all nodal values play the same role while we need here to distinguish the small scale and large scale components of the flow.

2. The Incremental Unknowns

In this section we present the IU in the context of linear elliptic problems.

2.1. Space Dimension One

We start with the very simple two-point boundary value problem in space dimension one :

$$-u'' = f, \quad 0 < x < 1, \tag{2.1}$$

$$u(0) = u(1) = 0. \tag{2.2}$$

We set $h = 1/2N$, where N is an integer, and consider the discretization of (1.1), (1.2) with mesh h : the approximate values of u, $u_j \simeq u(jh)$, $j = 0, ..., 2N$, satisfy

$$2u_j - u_{j-1} - u_{j+1} = h^2 f_j, \quad j = 1, ..., 2N - 1, \tag{2.3}$$

$$u_0 = u_{2N} = 0, \tag{2.4}$$

where $f_j = f(jh)$.

The incremental unknowns for this problem consist of the numbers $y_{2i} = u_{2i}$, $i = 0, ..., N$ and, at points $2i+1$, the numbers

$$z_{2i+1} = u_{2i+1} - \frac{1}{2}(u_{2i} + u_{2i+2}), \quad i = 0, ..., N - 1. \tag{2.5}$$

Thus z_{2i+1} is the increment of u to the average of the values at the neighboring points, $2i$ and $2i + 2$; hence by Taylor's formula, z_{2i+1} is small, of order h^2.

The space dimension one case is particular. The incremental unknowns lead to a decoupling of the system ; indeed at $j = 2i + 1$, (2.3) reduces to

$$z_{2i+1} = \frac{1}{2}h^2 f_{2i+1}, \quad i = 0, ..., N - 1, \tag{2.6}$$

so that these incremental values are now explicit. At $j = 2i$, (2.3) using (2.5) and (2.6) becomes

$$2u_{2i} - u_{2i-1} - u_{2i+1} = h^2 f_{2i}$$
$$2y_{2i} - y_{2i-2} - y_{2i+2}$$
$$= 2h^2 f_{2i} + 2(z_{2i+1} + z_{2i-1})$$
$$2y_{2i} - y_{2i-2} - y_{2i+2}$$
$$= h^2(2f_{2i} + f_{2i+1} + f_{2i-1}) \tag{2.7}$$

The system consisting of (2.7) and (2.4) is similar to the system consisting of (2.3), (2.4) but involves half as many unknowns. We can of course repeat the procedure. If we start with $h = 1/2^\ell N$, $\ell, N \in \mathbb{N}$, then after ℓ steps we reduce the initial system involving $2^\ell N$ unknowns to a similar one involving N unknowns. This reduction is similar to the cyclic reduction method of G. Golub and R.S. Varga (see. e.g. Birkhoff-Varga-Young (1962), Golub-Van Loan (1989)). In higher dimension the algebra is not as remarkable as in dimension one but other important aspects of the IU are still present.

2.2. Space Dimension Two

In space dimension two or more, incremental unknowns are defined similarly : two nested grids at least are necessary ; on the coarse grid, the incremental unknowns are the nodal values of the function ; and at the points of the fine grid not belonging to the coarse grid the IU are increments to the average value of u at the closest coarse grid points.

Consider for instance the Laplace operator with Dirichlet boundary condition in the square $\Omega = (0, 1)^2$:

$$-\Delta u = f \quad \text{in } \Omega, \tag{2.8}$$

$$u = 0 \quad \text{on } \partial\Omega. \tag{2.9}$$

We set again $h = 1/2N$, where N is an integer and consider the usual five-points discretization of the Dirichlet problem on the grid of mesh h :

$$\left(2u_{\alpha,\beta} - u_{\alpha-1,\beta} - u_{\alpha+1,\beta} + \left(2u_{\alpha,\beta} - u_{\alpha,\beta-1} - u_{\alpha,\beta+1}\right)\right.$$
$$= h^2 f_{\alpha,\beta}, \tag{2.10}$$

$$u_{\alpha,\beta} = 0 \quad \text{if } \alpha \text{ or } \beta = 0 \text{ or } 2N. \tag{2.11}$$

Here $f_{\alpha,\beta} = f(\alpha h, \beta h)$ and $u_{\alpha,\beta}$ is the approximate value of u at $(\alpha h, \beta h)$.

We then consider the coarse grid with mesh $2h = 1/N$ and introduce the incremental unknowns. Those consist of the nodal values $y_{2i,2j} = u_{2i,2j}$ at the points of the coarse grid $(2ih, 2jh)$,

$i, j = 0, ..., N$. The fine grid points that do not belong to the coarse grid are of three sorts (see Fig. 2.1) :

$$
\begin{array}{ccc}
\times & \circ & \times \\
2i, 2j+2 & 2i+1, 2j+2 & 2i+2, 2j+2 \\
\\
\circ & \circ & \circ \\
2i, 2j+1 & 2i+1, 2j+1 & 2i+2, 2j+1 \\
\\
\times & \circ & \times \\
2i, 2j & 2i+1, 2j & 2i+2, 2j
\end{array}
$$

Fig. 2.1. *Coarse grid* ($\times$) *and fine grid* ($\circ$) *points.*

(a) Points of type $(2ih, (2j+1)h)$, at the middle of two vertical coarse grid points ; the IU is then

$$
z_{2i,2j+1} = u_{2i,2j+1} - \frac{1}{2}(u_{2i,2j} + u_{2i,2j+2}),
$$
$$
i = 1, ...N-1, \quad j = 0, ..., N-1.
$$

(b) Points of type $((2i+1)h, 2jh)$ at the middle of two horizontal grid points ; the IU is then

$$
z_{2i+1,2j} = u_{2i+1,2j} - \frac{1}{2}(u_{2i,2j} + u_{2i+2,2j}),
$$
$$
i = 0, ...N-1, \quad j = 1, ..., N-1.
$$

(c) Points of type $((2i+1)h, (2j+1)h)$, at the center of a square of edge $2h$, the vertices of which are coarse grid points ; the IU is then

$$
z_{2i+1,2j+1} = u_{2i+1,2j+1} - \frac{1}{4}(u_{2i,2j} + u_{2i+2,2j} +
$$
$$
+ u_{2i,2j+2} + u_{2i+2,2j+2}), \quad i, j = 0, ...N-1.
$$

It follows from Taylor's formula that the increments z are small, of order h^2. When several level of grids are used corresponding to meshes $h_d = h_0/2^d$, then the z at the level d are of order $h_0^2/2^{2d}$ and thus decay exponentially fast as d increases. We describe in more details in Sec. 3 the implementation of the IU method when several levels of grids are used.

2.3. <u>Comparison with existing methods. Other aspects of IU</u>.

The introduction of the incremental unknowns amounts to a linear change of variables ; thus, in the language of linear algebra, it yields a preconditioner.

Like the multigrid method the IU method is based on the utilization of several nested grids in finite differences. However, as we will see, the resolution of the linear system is different : the unknowns remain the nodal values when the multigrid method is used while we make use here of the conjugate gradient method which is not usually the case with multigrid methods.

The IU method presented here can be also compared to the multilevel methods asociated with the utilization of hierarchical basis multigrid conditioners in finite elements (see e.g. H. Yserentant (1986, 1990), M. Dryja and O.B. Widlund (1991), J. Xu (1989, 1990, 1991), J. Atanga and D. Silvester (1991)). In fact the incremental unknowns are the same as the components of the function in the Q_1–hierarchical basis. However the IU method is developped strictly in the context of finite differences. It is simple to implement and it does not produce the drawbacks (and advantages) of finite elements methods. Also, in a forthcoming article, we will present several alternate definition of incremental unknowns which are not associated with finite elements (see Chen-Temam (1991a)).

Although we emphasize here the utilization of IU for linear problems we recall that their primary motivation is nonlinear problems.

Some other aspects of incremental unknowns are noteworthy. Firstly IU provides a new insight in mesh refinement. Indeed when local mesh refinement is considered in conjunction with domain decomposition (see Figure 2.2), the utilization of incremental unknowns attaches <u>small</u> <u>quantities</u> to the new added nodes, instead of nodal values of the same order of magnitude as the other ones.

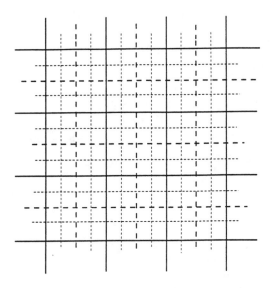

Fig. 2.2. Local mesh refinement
and incremental unknowns.

Also incremental unknowns are a promosing tool for parallel computing. Assume that the domain Ω under consideration is divided into subdomains Ω_i (see Figure 2.3). Then the (high frequencies) incremental unknowns attached to the subdomains Ω_i are expected to have little interaction ; they are suitable for parallel treatment.

Ω	Ω_2	Ω_3	

Fig. 2.3. Domain decomposition for incremental
unknowns and parallel computing.

3. Numerical Implementation

When two levels of grids are used let us rewrite equations (2.10), (2.11) in the form of a linear system

$$AU = F, \tag{3.1}$$

where $U, F \in \mathbb{R}^{(2N-1)^2}$ denote the $u_{\alpha,\beta}$, $f_{\alpha,\beta}$, $\alpha, \beta = 1, ..., 2N - 1$, properly ordered, and A is the well known symmetric positive definite matrix associated with the Dirichlet problem.

We then introduce the incremental unknowns described above. When two level of grids (two nested grids) are used, we denote by

$$\bar{U} = \begin{pmatrix} Y \\ Z \end{pmatrix} \tag{3.2}$$

the vector of incremental unknowns, and we denote by S the transfer matrix :

$$U = S\bar{U}. \tag{3.3}$$

Thus (3.1) is rewritten

$$AS\bar{U} = F$$

or after multiplying by ^tS on the left,

$$\bar{A}\bar{U} = \bar{F}, \tag{3.4}$$

$$\bar{A} = {}^tSAS, \quad \bar{F} = {}^tSF.$$

Obviously $\bar{A}$ is still symmetric positive definite. However the matrix $\bar{A}$ is complicated and we do not make it explicit. The resolution of (3.4) by Gauss elimination method or by Jacobi or SSOR method is not appropriate. The matrix S is simple and computing the product of $\bar{A}$ with $\bar{U}$ amounts to computing the product $S\bar{U}$ and then the product of this vector by A; it necessitates about the same number of operations as computing the product of A with a vector. For this reason the resolution of (3.4) can easily be performed by conjugate gradient methods.

Let us assume that the unknowns $u_{\alpha,\beta}$ in (3.1) have been ordered as

coarse grid points/other fine grid points.
Then this induces a splitting of U, F, A of the form

$$U = \begin{pmatrix} U_c \\ U_f \end{pmatrix}, \quad F = \begin{pmatrix} F_c \\ F_f \end{pmatrix},$$

$$A = \begin{pmatrix} A_{cc} & A_{cf} \\ A_{fc} & A_{ff} \end{pmatrix};$$

and it is easily seen that the general form of S and S^{-1} is :

$$S = \begin{pmatrix} I & 0 \\ S_{fc} & I \end{pmatrix}, \quad S^{-1} = \begin{pmatrix} I & 0 \\ -S_{fc} & I \end{pmatrix}.$$

When several level of grids (several nested grids) are used, we order the $u_{\alpha,\beta}, f_{\alpha,\beta}$ in the form :

$$U = \begin{pmatrix} U_c^1 \\ U_f^1 \\ \cdot \\ \cdot \\ \cdot \\ U_f^j \end{pmatrix}, \quad F = \begin{pmatrix} F_c^1 \\ F_f^1 \\ \cdot \\ \cdot \\ \cdot \\ F_f^j \end{pmatrix}$$

Here U_c^1, F_c^1 correspond to the coarse grid points (mesh h_0); U_f^1, F_f^1 correspond to the points of the second coarsest grid (mesh $h_0/2$) not belonging to the coarsest grid. Finally U_f^j, F_f^j correspond to the points of the finest grid (mesh $h_0/2^j$) not belonging to the previous ones.

We then introduce the incremental unknowns described in Sec. 2 at each level. We set

$$\bar{U} = \begin{pmatrix} Y_1 \\ Z_1 \\ \cdot \\ \cdot \\ \cdot \\ Z_j \end{pmatrix}$$

and denote by S the transfer matrix

$$U = S\bar{U}.$$

Therefore as in (3.4)

$$ASar{U} = F,$$
$$ar{A}ar{U} = ar{F}, \tag{3.5}$$
$$ar{A} = {}^{t}SAS, \quad ar{F} = {}^{t}SF.$$

The structure of S remains relatively simple and the computation of the product by $ar{A}$ by a vector $ar{U}$ necessitates about the same number of operations as the product of A by a vector.

As we will see below an important difference between (3.5) and (3.1) is that the condition number of $ar{A}$ is much smaller than that of A, respectively $O\left(\left(\log \frac{1}{h}\right)^2\right)$ and $O(h^2)$ where $h = \frac{h_0}{2^j}$ is the finest grid. We know from classical results (see e.g. O. Axelson and V.A. Baker (1984)) that at most $O(j)$ and $O(2^j)$ iterations are needed to reduce the energy norm of the error by a given factor if the Conjugate Gradient Method is used.

The figures hereafter (Fig. 3.1 and 3.2) borrowed from Chen-Temam (1991a) show the results of numerical tests using the multigrid method, the conjugate gradient method on (3.1), the conjugate gradient method on (3.5).

Our computational results show that the Incremental Unknown Method is comparable with the multigrid method and it is easy to program.

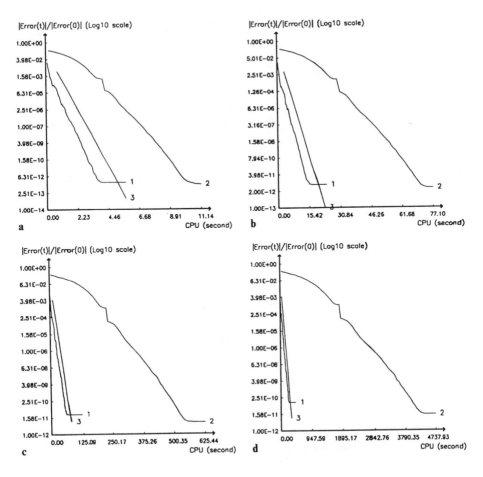

Fig. 3.1(a − d).

Fig. 3.1(a-d). Comparison between the Incremental Unknown method (line 1), Conjugate Gradient method (line 2) and Multigrid method (line 3). The relative L^2 norm of the error is plotted against the CPU time used. N is the number of grids in 1-direction and the initial guess Error (0) is the value of $\sin[16xy(1-x)(1-y)]$ at grid point. a $N = 256$ $(j = 8)$. b $N = 512$ $(j = 9)$. c $N = 1024$ $(j = 10)$. d $N = 2048$ $(j = 11)$.

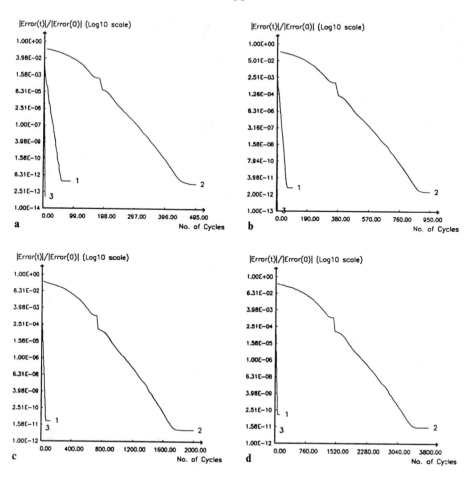

Fig. 3.2(a − d).

Fig. 3.2(a-d). Comparison between the Incremental Unknown method (line 1), Conjugate Gradient method (line 2) and Multigrid method (line 3). The relative L^2 norm of the error is plotted against the iteration cycle (or number). N is the number of grids in 1-direction and the initial guess Error (0) is the value of $\sin[16xy(1 - x)(1 - y)]$ at grid point. a $N = 256$ ($j = 8$). b $N = 512$ ($j = 9$). c $N = 1024$ ($j = 10$). d $N = 2048$ ($j = 11$).

4. Theoretical Results (Linear elliptic problems)

Two types of theoretical results have been derived for the IU method applied to linear elliptic problems. They concern energy estimates and estimates on condition number of the matrix.

4.1. Energy Estimates

We assume for simplicity that two levels of grids are used. We have indicated that, by Taylor's formula, the z components of the incremental unknonws are expected to be small, of order h^2. By using energy methods (which are essential for nonlinear problems), we can recover a priori estimates of this type, at least for their L^2 norm (root mean square norm).

In the *one-dimensional case* when the meshes are $h = 1/2N$ and $2h$, we find with straightforward calculations :

$$\sum_{i=0}^{N-1} (z_{2i+1})^2 \leq \text{const} \times h, \qquad (4.1)$$

$$\frac{1}{4h} \sum_{i=0}^{N-1} (y_{2i+2} - y_{2i})^2 \leq \text{const}, \qquad (4.2)$$

where the constants are independent of h; (4.1) shows that the L^2 norm of the increment z is bounded by ch and (4.2) is an L^2 estimate on the discrete derivative of the coarse grid function y.

In the two-dimensional case, the computations are more involved ; they yield (see M. Chen and R. Temam (1991a)).

$$\sum_{i,j=0}^{N-1} \left\{ z_{2i+1,2j+1}^2 + z_{2i+1,2j}^2 + z_{2i,2j+1}^2 \right\} \leq \text{const}, \qquad (4.3)$$

$$\sum_{i=0}^{N-1} \sum_{j=1}^{N-1} \left\{ y_{2i+2,2j} - y_{2i,2j} \right\}^2 \leq \text{const}, \qquad (4.4)$$

$$\sum_{i=1}^{N-1} \sum_{j=0}^{N-1} \left\{ y_{2i,2j+2} - y_{2i,2j} \right\}^2 \leq \text{const}, \qquad (4.5)$$

where the constants are independent of h; (4.3) shows that the root mean square norm of z is bounded by ch and (4.4), (4.5) are L^2 estimates on the discrete derivatives of the coarse grid function y.

The behaviour of the incremental quantities $z_{i,j}$ is shown for a test case in Fig. 4.1.

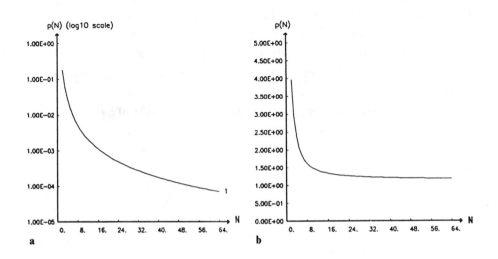

Fig. 4.1(a − b).

Fig. 4.1(a-b). The incremental unknowns $z_{i,j}$ are small. From a and b, one can see that
$$\sum_{i,j=0}^{N-1} \left\{ z_{2i+1,2j+1}^2 + z_{2i+1,2j}^2 + z_{2i,2j+1}^2 \right\} / \sum_{i,j=0}^{2N+1} u_{i,j}^2 \le ch^2,$$
where c is a constant which is independent of h. Here the solution u is the value of $x^{1.6}y(1-x)\sin\left[(1-y)^{1/2}\right]$ at grid points.

a) $P(N) = \sum_{i,j=0}^{N-1} \left\{ z_{2i+1,2j+1}^2 + z_{2i+1,2j}^2 + z_{2i,2j+1}^2 \right\} / \sum_{i,j=0}^{2N+1} u_{i,j}^2.$

b) $P(N) = \sum_{i,j=0}^{N-1} \left\{ z_{2i+1,2j+1}^2 + z_{2i+1,2j}^2 + z_{2i,2j+1}^2 \right\} / h^2 \sum_{i,j=0}^{2N+1} u_{i,j}^2.$

4.2. Condition number

For a theoretical justification of the good results obtained with the utilization of incremental unknowns, we have studied the condition number of the matrix $\bar{A}$ in (3.4) and compared it to the condition number of the matrix A in (3.1).

We assume that j levels of grids are used with meshes h_0, $h_0/2, ..., h_0/2^j$, and we denote by A_d and $\bar{A}_d$ the matrices A and $\bar{A}$ when d level of grids are used (coarsest mesh h_0, finest mesh $h_0/2^j$). Hence $\bar{A}_0$ is the same as A_0.

One can show (see M. Chen and R. Temam (1991b)) that the condition number of $\bar{A}_j$ is bounded as follows

$$\text{cond } (\bar{A}_j) \leq \text{const } (j+1)^2 \text{cond } (A_0), \qquad (4.6)$$

where cond (A_0) is the condition number of A_0, and the constant is independent of j. Thus for increasing j, cond $(\bar{A}_j)$ increases as $\left(\log \frac{1}{h_j}\right)^2$ $(h_j = h_0/2^j)$ while cond (A_j) is known to increase as $c \left(\frac{1}{h_j}\right)^2 \simeq c2^{2j}$.

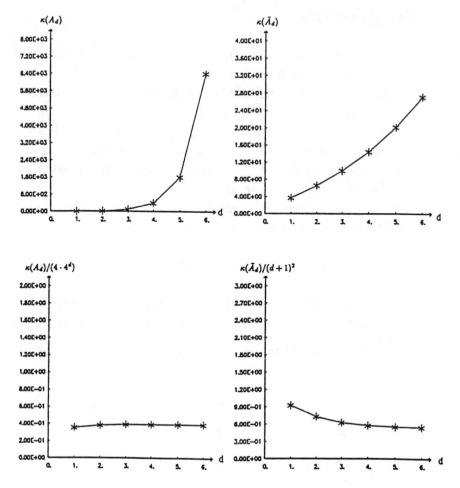

Fig. 4.2(a − d). Comparison between the condition numbers $\kappa(A_d)$ and $\kappa(\bar{A}_d)$. A_d is the matrix in the 5 − point finite difference equation associated with the Laplace operator.

Figure 4.2 gives a comparison of the condition numbers of A_d and $\bar{A}_d$ for the matrices associated with the Dirichlet problem in the square.

5. Evolution Equations

In this final section we sketch the utilization of IU in evolution problems, initiated in (Temam, 1989).

We start with a linear evolution equation such as the heat equation, which is written as an abstract evolution equation in a Hilbert space H :

$$\frac{du}{dt} + \mathcal{A}u = f, \tag{5.1}$$

$$u(0) = u_0. \tag{5.2}$$

After spatial discretization by finite differences with mesh h we obtain a (finite dimensional) differential system

$$U_h' + A_h U_h = F_h \tag{5.3}$$

$$U_h(0) = U_{h_0} \tag{5.4}$$

Now if $h = h_0/2^j$ and incremental unknowns are used as indicated before, then we write

$$U_{h_j} = S_{h_j} \bar{U}_{h_j}$$

$$\bar{U}_{h_j} = \begin{bmatrix} Y_1^j \\ Z_1^j \\ \cdot \\ \cdot \\ \cdot \\ Z_j^j \end{bmatrix} = \begin{bmatrix} Y^j \\ Z^j \end{bmatrix},$$

and dropping the indices j, h_j we can derive for Y and Z a coupled system

$$Y' + \bar{A}_{YY}Y + \bar{A}_{YZ}Z = \bar{F}_Y, \tag{5.5}$$

$$Z' + \bar{A}_{ZY}Y + \bar{A}_{ZZ}Z = \bar{F}_Z. \tag{5.6}$$

This system must then be discretized in time. It is natural to treat differently the Y corresponding to the coarse structures and the Z corresponding to the fine structures (which are small in magnitude). In particular if we use a discretization scheme explicit (or semi-explicit) in Y and implicit in Z, the CFL stability condition is that required by the Ys ; thus the time mesh can be larger and this is one of the reasons for improved efficiency.

For nonlinear problems, with say, a quadratic nonlinearity as in Navier-Stokes equations, we obtain a coupled $Y - Z$ system

$$Y' + \bar{A}_{YY}Y + \bar{A}_{YZ}Z + \bar{B}_Y(Y + Z, Y + Z) = \bar{F}_Y, \quad (5.7)$$
$$Z' + \bar{A}_{ZY}Y + \bar{A}_{ZZ}Z + \bar{B}_Z(Y + Z, Y + Z) = \bar{F}_Z. \quad (5.8)$$

One can neglect the evolution of Z (Z' being smaller than Y' and than the other terms in equation (5.8)). Also Z can be neglected in the B term in (5.8) (see Temam, 1989), and we obtain a simplified form of (5.8)

$$\bar{A}_{ZY}Y + \bar{A}_{ZZ}Z + \bar{B}_Z(Y, Y) = \bar{F}_Z, \quad (5.9)$$

in which Z can be solved in terms of Y. By reporting this value of Z in (5.7) we obtain a large eddy simulation model for Y (see Temam, 1989, 1991). This large eddy simulation model of the Smagorinsky type is, by contrast with Smagorinsky's, valid for **nonhomogeneous flows**. Details will be given elsewhere.

Concluding Remarks

Incremental unknowns are at the crossroad of several concepts : linear algebra, preconditioners, multigrid methods, finite elements hierarchical bases on the linear side ; dynamical systems, inertial manifolds, large eddies simulations on the turbulence and nonlinear side. They are appropriate for parallel computing and are related to wavelets which appear in other classes of IU which have not been described here ("oscillating" or "wavelet-like IU", see (Chen-Temam 1991c)).

There are good hopes that incremental unknowns will develop as a useful and powerful tool in computing in the 90's.

References

- J. Atanga and D. Silvester, 1991, Preconditionning techniques for the numerical solution of the Stokes problem, Proc. of the Second International Conference on the Application of Super-computers in Engineering (ASE91), Boston, to appear.

- O. Axelsson and V.A. Baker, 1984, *Finite Element Solution of Boundary Value Problems: Theory and Computation*, New York Academic Press.

- G. Birkhoff, R.S. Varga and D. Young, 1962, Alternating implicit methods, *Advances in Computers*, vol. 3, F. Alt and M. Rubinoff, Eds., Academic Press.

- M. Chen and R. Temam, 1991a, Incremental unknowns for solving partial differential equations, *Num. Math.*, vol. 59, pp. 255-271.

- M. Chen and R. Temam, 1991b, Incremental unknowns in finite differences : condition number of the matrix, *SIAM J. of Matrix Analysis and Applications (SIMAX)*, to appear.

- M. Chen and R. Temam, 1991c, Nonlinear Galerkin method in finite difference case and wavelet-like incremental unknowns, to appear.

- M. Dryja and O.B. Widlund, 1991, Multilevel additive methods for elliptic finite element problems, preprint.

- C. Foias, O. Manley and R. Temam, 1987, Sur l'interaction des petits et grands tourbillons dans les écoulements turbulents, *C.R. Acad. Sci. Paris*, Série I, 305, pp. 497-500 ; 1988, On the interaction of small and large eddies in two-dimensional turbulent flows, *Math. Model. and Numer. Anal. (M2AN)*, vol. 22, pp. 93-114.

- C. Foias, G. Sell and R. Temam, 1985, Variétés inertielles des équations différentielles dissipatives, *C.R. Acad. Sci. Paris*, Série I, 301, pp. 139-142 ; 1988, Inertial manifolds for nonlinear evolutionary equations, *J. Diff. Equ.*, vol. 73, pp. 309-353.

- G.H. Golub and C.F. Van Loan, 1989, *Matrix Computations*, The John Hopkins University Press, Baltimore.

- M.S. Jolly, I.G. Kevrekidis and E.S. Titi, 1990, Approximate inertial manifolds for the Kuramoto-Sivashinsky equation : analysis and computations, *Physica D*, vol. 44, pp. 38-60.

- M. Kwak, 1991, Finite dimensional inertial forms for the $2D$ Navier-Stokes equations, IMA Preprint, 1991, University of Minnesota.

- S.F. McCormick (Ed.), 1987, *Multigrid Methods*, SIAM, Philadelphia.

- J. Mallet-Paret and G.R. Sell, 1988, Inertial manifolds for reaction-diffusion equations in higher space dimension, *J. Amer. Math. Soc.*, vol. 1, pp. 805-866.

- M. Marion and R. Temam, 1989, Nonlinear Galerkin methods, *SIAM J. Numer. Anal.*, vol. 26, pp. 1139-1157.

- M. Marion and R. Temam, 1990, Nonlinear Galerkin methods ; The finite elements case, *Numerische Mathematik*, vol. 57, pp. 205-226.

- R. Temam, 1988, *Infinite Dymensional Dynamical Systems in Mechanics and Physics*, Springer-Verlag, New-York, Applied Mathematical Sciences Series, vol. 68.

- R. Temam, 1989, Attractors for the Navier-Stokes equations, Localization and approximations, *J. Fac. Sci. Tokyo*, Sec. IA, vol. 36, pp. 629-647.

- R. Temam, 1990, Inertial manifolds and multigrid methods, *SIAM J. Math. Anal.*, vol. 21, pp. 154-178.

- R. Temam, 1991, Approximation of attractors, large eddies simulation and multi-scale methods, *Proc. of Royal Soc. A*, Special issue commemorating the work of A.N. Kolmogorov in probability and turbulence.

- H. Yserentant, 1986, On the multilevel splitting of finite element spaces, *Numer. Math.*, vol. 49, pp. 379-412.

- H. Yserentant, 1990, Two preconditionners based on the multilevel splitting of finite elements, *Numer. Math.*, vol. 58, pp. 163-184.

- J. Xu, 1989, Theory of multilevel methods, Ph.D. Thesis, Cornell, Rep. AM-48, Penn. State U.

- J. Xu, 1990, Iterative methods by space decomposition and subspace correction : a unifying approach, Rep. AM-67, Penn. State U.

- J. Xu, 1991, A new class of iterative methods for nonselfadjoint or indefinite problems, Penn. State U.

SUPERCOMPUTER IMPLEMENTATIONS OF PRECONDITIONED KRYLOV SUBSPACE METHODS[1]

Youcef Saad

University of Minnesota, Computer Science Department
4-192 EE/CSci Building, 200 Union Street S.E.
Minneapolis, Minnesota 55455

ABSTRACT

Preconditioned Krylov subspace methods are among the preferred iterative techniques for solving large sparse linear systems of equations. As computer architectures are evolving and problems are becoming more complex, iterative techniques are undergoing several mutations. In particular, there has been much recent work on new preconditioners that yield higher parallelism or on new implementations of the standard ones. In the past, a conservative and well understood approach has consisted of porting standard preconditioners to the new computers. However, in addition to their limited parallelism, such preconditioners have other drawbacks. The simple ILU(0) preconditioner may fail to converge for realistic problems arising from applications such as Computational Fluid Dynamics. The more robust analogues ILU(k) [30] or ILUT(k) [40] that allow more fill-in are not always a good alternative since they tend to be sequential in nature. In this paper we discuss these issues and give an overview of the standard approaches. Then we will propose a number of alternatives. It will be argued that some approaches based on multi-coloring can offer good compromises between generality and efficiency.

1   Introduction

Iterative methods for solving large sparse linear systems of equations arising from the discretization of partial differential equations, have been steadily gaining popularity in many scientific and engineering areas. In the past, direct methods have often been preferred for 2-dimensional problems, particularly on computers with large memories. However, there is currently a general consensus that as three-dimensional models will gradually become commonplace, iterative

---

[1]This research was supported by The Minnesota Supercomputer Institute.

methods will become almost mandatory. The memory and the computational requirements for 3-dimensional problems or 2-dimensional PDE's with several degrees of freedom per grid point, may cause serious challenges to the most efficient direct solvers available today. This and other factors are driving researchers to consider alternatives among iterative methods.

A secondary reason why iterative methods are currently gaining ground is that they are far easier than direct methods to implement efficiently on high-performance computers. The key to the success in obtaining good performance from iterative solvers is to attempt to satisfy the following three requirements.

1. Use good implementations of the low level primitives, such as matrix-vector multiplications, inner products, and vector combinations.

2. Select a preconditioner that allows a good level of parallelism and implement it efficiently.

3. Select a preconditioner that also has good 'intrinsic' qualities, i.e., that requires few iterations to converge.

Typically, (1) is easy to achieve, for example, via the use of available libraries supplied for the computer under consideration. In contrast there is often a conflict between (2) and (3). Preconditioners that allow a large degree of parallelism such as diagonal preconditioners, typically cause a large increase in the number of iterations when compared with their sequential counterparts, such as the Incomplete LU factorization (ILU). For example, a rather popular technique is to reorder the equations by coloring the unknowns in such a way that no two unknowns of the same color are related by an equation. In the simplest case of the 5-point matrix arising from the centered difference discretization of the Laplacean in 2 or 3 dimensional spaces, only two colors are needed, commonly referred to as red and black. If the unknowns of the same color are numbered consecutively, then a large degree of parallelism is available in the preconditioning phase. The drawback of this approach is that if an ILU type preconditioned Krylov subspace method is used then the number of iterations may increase substantially, often defeating the benefits gained from the higher degree of parallelism. Although it was observed [13] that there are orderings that have precisely the opposite effect these are not too well understood, and it is unlikely that any specific rules can be derived for the general case where the matrix arises from the discretization of coupled Partial Differential Equations as is the case in CFD.

We will show that much can be done in recovering the good convergence properties by mixing some preconditionings that satisfy (1) (e.g., multi-color SOR / SSOR) while enforcing (2) by increasing the accuracy of the preconditioner (e.g., several steps in SOR/SSOR or higher level of fill-in in ILU).

We will start by giving an overview of the standard techniques used in implementing preconditioned Krylov subspace methods. Then we will discuss some of the newer issues related to multi-coloring principles.

2   Preconditioned Krylov Subspace Methods

Consider a linear system of the form

$$Ax = b, \tag{1}$$

where A is a large sparse nonsymmetric real matrix of size N. A number of conjugate gradient-type iterative techniques based on projection processes on so-called Krylov subspaces have been proposed in recent years to solve such systems. A small subset of references in this area is [6, 8, 14, 19, 18, 25, 22, 28, 29, 32, 55, 37, 42, 44]. GMRES is a technique introduced in [44] for solving general large sparse nonsymmetric linear systems of equations by minimizing the 2-norm of the residual vector $b - Ax$ over x in the Krylov subspace

$$K_m = \text{Span}\{r_0, Ar_0, \ldots, A^{m-1}r_0\},$$

where r_0 is the initial residual vector $b - Ax_0$.

The idea of preconditioning is simply to transform the above system, e.g., by multiplying it through with a certain matrix M^{-1}, into one that will be easier to solve by a Krylov subspace method. For example, when the preconditioner M is applied to the right, we will be solving instead of (1), the preconditioned linear system

$$(AM^{-1})(Mx) = b. \tag{2}$$

A brief description of the right-preconditioned GMRES method follows, for details see [44].

ALGORITHM 2.1 GMRES – with right Preconditioning

1. **Start:** *Choose x_0 and a dimension m of the Krylov subspaces. Define an $(m+1) \times m$ matrix $\bar{H}_m$ and initialize all its entries $h_{i,j}$ to zero.*

2. **Arnoldi process:**

   (a) *Compute $r_0 = b - Ax_0$, $\beta = \|r_0\|_2$ and $v_1 = r_0/\beta$.*

   (b) *For $j = 1, ..., m$ do*

      - *Compute $z_j := M^{-1}v_j$*
      - *Compute $w := Az_j$*
      - *For $i = 1, \ldots, j$, do* $\quad \begin{cases} h_{i,j} := (w, v_i) \\ w := w - h_{i,j}v_i \end{cases}$
      - *Compute $h_{j+1,j} = \|w\|_2$ and $v_{j+1} = w/h_{j+1,j}$.*

   (c) *Define $V_m := [v_1, ..., v_m]$.*

3. **Form the approximate solution:** *Compute $x_m = x_0 + M^{-1}V_m y_m$ where*
   $$y_m = \text{argmin}_y \|\beta e_1 - \bar{H}_m y\|_2 \text{ and } e_1 = [1, 0, \ldots, 0]^T.$$

4. **Restart:** *If satisfied stop, else set $x_0 \leftarrow x_m$ and goto 2.*

The Arnoldi loop simply constructs an orthogonal basis of the preconditioned Krylov subspace $\text{Span}\{r_0, AM^{-1}r_0, \ldots, (AM^{-1})^{m-1}r_0\}$ by a modified Gram-Schmidt process, in which the new vector to be orthogonalized is defined from the previous vector in the process.

An interesting variant is one that allows the preconditioner to fluctuate from step to step in the inner GMRES iteration. This can be important in some applications. To derive such a variant, we first observe that the last step in the above algorithm forms the solution as a linear combination of the preconditioned vectors $z_i = M^{-1}v_i, i = 1, \ldots, m$. Because these vectors are all obtained by applying the same preconditioning matrix M^{-1} to the v's, we need not save them. We only need to apply M^{-1} to the linear combination of the $v's$, i.e., to $V_m y_m$. If we allowed the preconditioner to vary at every step, i.e., if z_j is now defined by

$$z_j = M_j^{-1}v_j$$

we may think of saving these vectors to use them in up-dating x_m in step 3. This suggests the following 'flexible' variant of the previous algorithm.

ALGORITHM **2.2 Flexible variant of preconditioned GMRES (FGMRES)**

1. **Start:** *Choose x_0 and a dimension m of the Krylov subspaces. Define an $(m+1) \times m$ matrix $\bar{H}_m$ and initialize all its entries $h_{i,j}$ to zero.*

2. **Arnoldi process:**

   (a) *Compute $r_0 = b - Ax_0$, $\beta = \|r_0\|_2$ and $v_1 = r_0/\beta$.*

   (b) *For $j = 1, ..., m$ do*
   - *Compute $z_j := M_j^{-1} v_j$*
   - *Compute $w := Az_j$*
   - *For $i = 1, \ldots, j$, do* $\begin{cases} h_{i,j} := (w, v_i) \\ w := w - h_{i,j} v_i \end{cases}$
   - *Compute $h_{j+1,j} = \|w\|_2$ and $v_{j+1} = w/h_{j+1,j}$.*

   (c) *Define $Z_m := [z_1, ..., z_m]$.*

3. **Form the approximate solution:** *Compute $x_m = x_0 + Z_m y_m$ where*
   $y_m = \text{argmin}_y \|\beta e_1 - \bar{H}_m y\|_2$ *and* $e_1 = [1, 0, \ldots, 0]^T$.

4. **Restart:** *If satisfied stop, else set $x_0 \leftarrow x_m$ and goto 2.*

The approximate solution obtained from this modified algorithm satisfies a few simple properties [39]. In particular, the approximate solution x_m minimizes the residual norm $\|b - Ax_m\|_2$ over $x_0 + \text{Span}\{Z_m\}$. In addition, if at a given step j, we have $Az_j = v_j$ (i.e., if the preconditioning is 'exact' at step j) then the approximation x_j is exact. Furthermore, if $M_j = M$ for $j = 1, \ldots, m$ then the method is clearly equivalent to the standard GMRES algorithm, right-preconditioned with M.

The possible applications of this added flexibility are numerous. We list below just a few possibilities.

1. Use of *any* iterative techniques as a preconditioner: Block-SOR, SSOR, ADI, Multi-grid, etc... but also GMRES, CGNR, CGS etc... Thus, an iterative method from another class (e.g., CGNR, CGNE) can be used as a preconditioner to GMRES. This may lead to some interesting hybrid methods [39].

2. Use of chaotic relaxation type preconditioners (e.g., in a parallel computing environment)

3. Mixing preconditioners to solve a given problem. In the standard version, one can use a new preconditioner at each outer iteration. Here we can even change it at every step. One obvious application is when the preconditioner is not exactly defined or is subject to nonnegligible perturbations every time it is computed.

Recently, Tezduyar and co-authors [47] used the approach in (3.) by applying two types of preconditioners alternatively at each FGM-RES step to mix the effects of "local" and "global" interactions. Tezduyar et al. reported good success with this procedure, much better than using any of the two preconditioners by itself.

Preconditioners of particular interest within this framework are relaxation type techniques. As an example, the SSOR preconditioning matrix defined by

$$M_{SSOR}(A) = (D - \omega E)D^{-1}(D - \omega F)$$

in which $-E$ is the strict lower part of A, $-F$ its strict upper part, and D its diagonal, has some important advantages some of which will become clear later. In the context of preconditioning, it is customary to just take $\omega = 1$ as the gains from selecting an optimal ω are typically small. However, it is clear that one can use different values of ω at each step of FGMRES and this can open up the possibility of using heuristics to determine the best ω dynamically, by simply monitoring convergence. Alternatively, a selection of a small number of different ω's can be used cyclically, instead of fixing arbitrarily ω to one as is usually done. Similar ideas can be employed for the Alternating Direction Iterative (ADI) method.

3 Implementations on High Performance Computers

Looking at a typical conjugate gradient type algorithm we observe that the main operations are the following.

1. Setting up of the preconditioner;

2. Matrix vector multiplications;

3. Vector updates;

4. Dot products;

5. Preconditioning operations.

In the above list potential difficulties may arise in setting-up the preconditioner (1) and in the solution of linear systems with M, i.e., operation (5). The rest causes no major difficulties.

3.1 Sparse matrix-vector products

Matrix-vector product operations are relatively easy to implement efficiently on most computers. The first observation that has been made in this context is that this operation can be performed by diagonals when the matrix is regularly structured, i.e., when it consists of a few diagonals [27]. The matrix can be stored in a rectangular array $DIAG(1 : n, 1 : ndiag)$ and the offsets of these diagonals from the main diagonal may be stored in a small integer array $IOFF(1 : ndiag)$.

A number of generalizations of this formats for general sparse matrices have been proposed, the first of which is the ELLPACK-ITPACK format [33, 59]. Assuming that the maximum number of nonzero elements per row $jmax$ is small we can store the entries of the matrix in a real array $C(1 : n, 1 : jmax)$, the i-th row of which contains the nonzero elements of the i-th row of A. We also need to an integer array $JC(1 : n, 1 : jmax)$ to store the column numbers of each entry of C. Then the matrix-vector multiplication can be implemented by a code of the form

```
    DO 10 J=1, JMAX
        DO 20 I=1, N
            Y(I) = Y(I) + C(I,J)*X(JC(I,J))
20      CONTINUE
10  CONTINUE
```

A disadvantage of the above loop, as compared with the one used with the diagonal storage is the presence of indirect addressing. Also, if the number of nonzeros per row varies substantially, then many zero elements must be stored unnecessarily, in order to fill the rows that have less than $jmax$ nonzero elements. Several variations have been developed to alleviate the latter difficulty. For example, observing that the nonzero elements of many sparse matrices tend to concentrate in a few diagonals, one can extract a small number of diagonals, store them as was described above for structured matrices and store the remaining elements using a general sparse matrix storage scheme. The complexity of this conversion is on the order of the nonzero elements and is therefore affordable. The payoff could be substantial, especially on vector machines that perform well on long vectors. Here again many zero elements may have to be added to fill the diagonals and so we must be careful when assessing performances.

There are several 'generalized banded formats' similar to the one described above. The following scheme related to a data structure introduced by Melhem [31], can be viewed as a more general alternative to the ELLPACK format. We start by reordering the rows of the matrix in decreasing order number of nonzero elements. Then, a new data structure is built by constructing what we call "jagged diagonals" (j-diagonals). We store as a dense vector the leftmost element from each row, together with an integer vector containing the column positions of each element. This is followed by the second jagged diagonal consisting of the second nonzero element in each row, and so on. As we build more and more of these diagonals, their length decreases, since the last rows are shorter than the first ones because of the way they have been sorted. The number of j-diagonals is equal to the number of nonzero elements of the first row, i.e., to the largest number of nonzero elements per row. The nonzero elements are stored by successive j-diagonals in a one-dimensional array A. If we denote by $JD(j)$ the pointer to the beginning of the j-th jagged diagonal, and by $JA(k)$ the column position of the nonzero element stored in $A(k)$, then the following code will multiply a matrix A by a vector x.

```
      DO 10 J=1, NDIAG
         K1 = JD(J)
         K2 = JD(J+1)-1
         LEN = K2-K1+1
         Y(1:LEN) = Y(1:LEN) + A(K1:K2)*X(JA(K1:K2))
10    CONTINUE
```

Relative to the original, unpermuted, matrix A this will compute a permutation of the vector Ax. We must permute the result back to the original ordering after completion of the above program. This extra overhead is often worth the effort.

Techniques based on variants of the jagged-diagonal or ELLPACK-ITPACK format are the most commonly used in the general context of high performance implementation of Krylov subspace methods for both linear system solutions or eigenvalue problems. Thus, commercial packages and software for iterative methods tend also to rely on these formats [23, 35].

3.2 Implementations of preconditioning operations

Each step of a preconditioned iterative method requires the solution of a linear system of equations

$$Mz = y.$$

Typically M is the product of a lower and an upper triangular matrix, often having the same sparsity pattern as the lower and the upper triangular parts of the original matrix. We consider in this section a number of different ways of performing this operation which is critical to the performance of the preconditioned conjugate gradient method. We only consider lower triangular systems of the form

$$Lx = b. \qquad (3)$$

Without loss of generality we will assume that L is unit lower triangular. If not the matrix can be scaled before the CG iteration is started so as to save N multiplications per CG step. The forward sweep for solving a lower triangular system with coefficients $al(i,j)$ and right-hand-side b is as follows,

ALGORITHM **3.1 Forward elimination for a sparse triangular system**

```
Do i=1, n
    for (all j such that al(i,j) is nonzero) do
        x(i) = x(i)-al(i,j) * x(j)
    enddo
enddo
```

Let us assume that the matrix is stored row-wise in a general sparse format, using the standard sparse storage scheme [12, 38],

- AL : real array containing the nonzero elemnts of L, stored by rows;

- JAL: integer array containing the column indices of each of the elements in array AL;

- IAL: integer pointer array with $IAL(i)$ pointing to the start of row i in AL, JAL.

and that the right-hand-side occupies the same array x as the solution. Then the above algorithm translates into the following code segment:

```
do i=2, n
   do j=ial(i),ial(i+1)-1
      x(i)=x(i)-al(j) * x(jal(j))
   enddo
enddo
```

The outer loop corresponding to the variable i is sequential. The j loop is essentially a sparse dot-product of the ith row of L and the (dense) vector x. We may split this dot product among the processors and add the partial results at the end. However, typically the length of the vector involved in the dot product is very short and so this approach is quite inefficient in general on vector or parallel computers. We briefly describe two approaches for breaking the sequential nature of the above implementation, one for regularly structured matrices and the other for general sparse matrices.

3.3   Level scheduling: the regular case

Let us assume that the linear system (1) arises from the discretization of a partial differential equation of the form

$$\mathcal{L}u = f$$

on the unit square $[0, 1]$ x $[0, 1]$ with, for example, Dirichlet type boundary conditions, where $\mathcal{L}$ is a non-self adjoint elliptic partial differential operator. A standard centered finite difference discretization of the above equation using n interior points on each side of the square yieds a linear system of size $N = n^2$. If the unknowns are labeled using the natural ordering, the resulting 5-point discretization of the operator $\mathcal{L}$, is a block tridiagonal matrix A, with each diagonal block being an $n \times n$ tridiagonal matrix, and the co-diagonal blocks being diagonal matrices. The corresponding incomplete factors L and U are lower and upper triangular matrices respectively with L having the same structure as the lower part of the matrix A. In the usual organization of the forward sweep for solving such triangular systems the solution is obtained one unknown at a time from $i = 1$ to $i = n$.

The stencil represented in Figure 3.1 establishes the data dependency between the unknowns in the lower triangular system solution when considered from the point of view of a grid of unknowns. In this case, it tells us that in order to compute the unknown in position (i, j) we only need the two unknowns in positions $(i - 1, j)$ and $(i, j - 1)$. As a result we can start computing x_{11} which does not depend on any other variable, and then use this to get $x_{1,2}$ and $x_{2,1}$ simultaneously. Then these two values will in turn enable us to obtain $x_{3,1}, x_{2,2}$ and $x_{1,3}$ simultaneously and so on. Thus, the computation can proceed in wavefronts. The first few steps of this wavefront algorithm are illustrated in Figure (3.2).

An important observation to make here is the maximum degree of parallelism reached, or vector length in the case of vector processing, is the minimum of n_x, n_y for 2-D problems. For 3-D problems the parallelism is of the order of the maximum size of the sets of domain points $x_{i,j,k}$, where $i + j + k = lev$, a constant level lev. See [50] for details on vector implementations. However, as can be easily seen there is little parallelism or vectorization at the beginning and at the end of the sweep. Initially the degree of parallelism is very small, starting at one then increasing to its maximum to decrease back to 1 again at the end of the sweep. The first and last few steps may take a heavy toll on achievable speed-ups.

For example, for a 4×3 grid, we can see that the levels (sets of equations that can be solved in parallel) are $\{1\}$, $\{2, 5\}$, $\{3, 6, 9\}$, $\{4, 7, 10\}$; $\{8, 11\}$; and finally $\{12\}$. The forward solution can easily be implemented for a specific problem by simply relying on the grid representation.

The idea of proceeding by *levels* or *wavefronts* is a natural one for finite difference matrices on rectangles and several authors suggested it independently [48, 49, 21, 43, 7]. However, the more general case of irregular matrices which will be discussed next is a textbook example of scheduling, or *topological* sorting and is well-known in different forms to computer scientists.

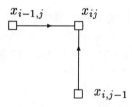

Figure 3.1 Stencil of the lower triangular matrix.

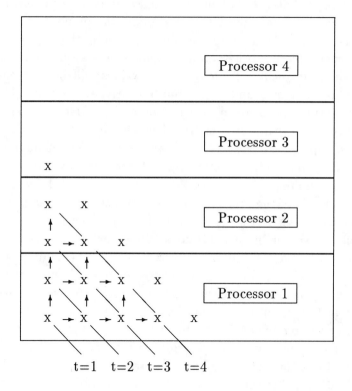

Figure 3.2 The wavefront approach for a regular grid.

3.4 Level-Scheduling for irregular grids

The simple scheme described above can be generalized to irregular grids. The technique, referred to as *level scheduling* [5, 45, 56], has as its objective to group the unknowns in subsets of unknowns that can be determined simultaneously. The idea is as follows. Consider again algorithm 3.1 for solving a unit lower triangular system The i-th unknown can be determined once all the other ones that participate in equation i become available. In the i-th step we will need the unknowns j such that $al(i,j) \neq 0$, i.e., to use graph terminology, all unknowns that are adjacent to unknown number i. Thus, we can look at the adjacency graph of the matrix and determine groups of equations that can be solved at the same time [5, 45, 56]. Recall that a vertex of the graph represents an unknown and there is a directed edge from node j to node i if $al(i,j) \neq 0$, i.e., if the unknown j is present in equation i. The edge $j \rightarrow i$ simply indicates that x_j must be known before we can attempt to determine x_i. Since L is lower triangular, the adjacency graph is a directed acyclic graph. For such graphs, it is possible and quite easy to find a labeling of the nodes that satisfy the property that when if $label(j) < label(i)$ then task j must be executed before task i. In addition, it is also easy to find which ones can be executed in parallel.

The first step of the solution algorithm consists of determining x_1 and any other unknowns for which there are no predecessors in the graph, i.e., all those unknowns x_i for which the off-diagonal elements of row i are zero. These unknowns will constitute the elements of the first level. The next step will compute in parallel all those unknowns that will have the nodes of the first level as their (only) predecessors in the graph. The following steps can be defined similarly: the unknowns that can be determined at step l, are all those that have as predecessors equations that have been determined in steps $1, 2, \ldots, l-1$. This leads naturally to the definition of a *depth* for each unknown. We define the *depth* of a vertex by performing the following loop for $= 1, 2, \ldots, n$, after initializing $depth(j)$ to zero for all j.

$$depth(i) = 1 + \max_j \{ depth(j), \text{ for all } j \text{ s.t. } al(i,j) \neq 0 \}.$$

By definition, a *level* of the graph is the set of nodes with the same depth. We can now define a data structure for the levels: a permutation $q(1:n)$ defines the new ordering and $level(i), i = 1, \cdots, nlev + 1$

points to the beginning of the i-th level in that array.

Once these level sets are found, we can proceed in two different ways. We can use the permutation vector q to permute the matrix according to the new order. In the 4×3 example mentioned in the previous subsection, this means renumbering the variables $\{1\}$, $\{2,5\}$, $\{3,6,9\}$,.. consecutively, i.e., as $\{1,2,3,....\}$. The resulting matrix after the permutation is shown in Figure 3.3.

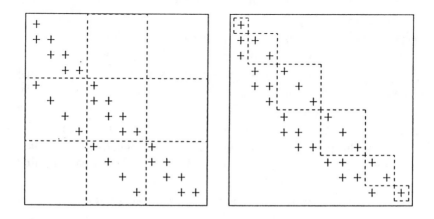

Figure 3.3 L-matrix before and after reordering by level.

An alternative is to simply keep the permutation array and use it to recognize which unknown corresponds to a given level in the solution. Then the algorithm for solving the triangular systems can be written as follows, assuming that the matrix is stored in the usual row sparse matrix format.

ALGORITHM **3.2 Forward elimination with level scheduling**

```
do lev=1, nlev
j1 = level(lev)
j2 = level(lev+1)-1
      do k = j1, j2
        i = q(k)
        do j= ial(i), ial(i+1)-1
            x(i) = x(i) - al(j) * x(jal(j))
        enddo
      enddo
  enddo
```

In [10, 9] and [4, 5] a number of experiments are presented to study the performance of level scheduling within the context of pre-conditioned conjugate gradient methods. Experiments on an Alliant FX-8 indicate that a speed-up of around 4 to 5 can be easily achieved. These techniques have also been tested for problems in Computational Fluid Dynamics [52, 53]

4   Multi-Coloring

One of the most promising general purpose approaches for solving large linear systems on massively parallel computers is to resort to multi-coloring. The general technique of multi-coloring has been known for a long time and was used in particular for understanding the theory of relaxation techniques [58, 51] as well as for deriving efficient alternative formulations of some relaxation algorithms [51, 20]. More recently, it was employed to introduce parallelism in iterative methods, see for example [3, 2, 36, 15, 34]. It is also commonly used in a slightly different form – coloring elements as opposed to nodes – in finite elements techniques [11, 54]. Coloring is especially useful in element-by-element techniques when forming the residual, i.e., in multiplying an unassembled matrix by a vector. The contributions of the elements of the same color can all be evaluated and applied simultaneously to the resulting vector [24, 17, 46]. As a background we start by describing how the standard Red-Black coloring has been exploited in the context of parallel preconditioned Krylov these two issues in turn.

subspace methods. Then we will present generalizations based on multi-coloring.

4.1 Red-Black orderings

When it is applicable red-black ordering constitutes one of the simplest and most popular ways of achieving high computational speeds on high performance computers. In its simplest form the problem addressed is to color the nodes of a simple 2-dimensional finite difference grid (5-point operator) in such a way that neighboring points have different colors. For such simple grids only two colors (referred to as Red and Black) are needed. For example, the 2-color (red-black) ordering is illustrated in Figure 4.1 for a 4×3 grid.

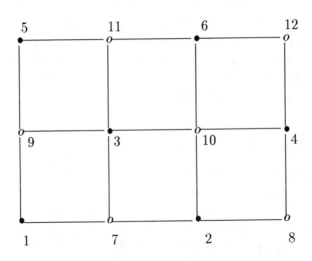

Figure 4.1 Red black labeling of a 4×3 grid. $\bullet$ = black node, o = red node.

If we reorder the unknowns in such a way as to list the red unknowns first together and then the black ones, we will obtain a system of the form

$$\begin{pmatrix} D_1 & E \\ F & D_2 \end{pmatrix} \begin{pmatrix} x_1 \\ x_2 \end{pmatrix} = \begin{pmatrix} b_1 \\ b_2 \end{pmatrix} \tag{4}$$

D_1 and D_2 are diagonal matrices. We must now address two questions. The first is to find ways of efficiently solving linear systems that are put in the above form. The second is to generalize this approach to systems that are not necessarily 2-colorable. We consider

4.2 Solution of Red-Black systems

The simplest method that can be used to solve (4) is the standard ILU(0) on this block system. The degree of parallelism here is of order N. A drawback is that often the number of iterations is higher than with the natural ordering, but the approach may still be competitive for *easy* problems.

Based on this, one can raise the interesting question as to whether or not the number of iterations can be reduced back to a competitive level by using a more accurate ILU factorization on the red-black system, e.g., ILUT [40]. Some recent experiments reveal that the answer is yes. In fact it turns out that the situation becomes reversed in that for the same level of fill-in k, the red-black ordering will outperform the natural ordering preconditioner for k large enough, in terms of number of iterations. A look at the plots of the Frobenius norms for the residual matrices $A - LU$, for the test matrix used in the experiments to be described later, reveals that indeed the incomplete factorization may become more accurate for the red-black matrices, as k increases, see Figure 4.2. The dashed line shows the Frobenius norms $\|A - LU\|_F$ for the natural ordering and the solid line shows the same norms for the Red-Black ordering using the same fill-in level. Notice that when the level of fill-in reaches 6, the quality of the RB preconditioner as measured by the F-norm becomes better than that of for the natural ordering using the same level of fill k. This reverses the situations obtained for smaller values of the fill-in level.

On the other hand a serious difficulty with this approach, from the standpoint of parallel processing, is that fill-ins will be introduced in the (2,2) blocks in L and U which loose their diagonal structure. As a result the degree of parallelism is severely reduced.

A remedy against this shortcoming of the high-accuracy ILU preconditioners is to use similar *high-level SOR or SSOR preconditioners* instead of ILUT. We must first decide what we might consider the equivalent of higher level of fill-in for the SOR and SSOR preconditioners. One way would be to perform more SOR or SSOR steps at each iteration instead of just one as is traditionally done. This can be combined with the flexible version of GMRES described earlier if needed. In fact we found that this approach worked remarkably well

on a CRAY computer. We were able to achieve much better performance than with the usual preconditioned techniques, see illustration
in a more general context in next subsection.

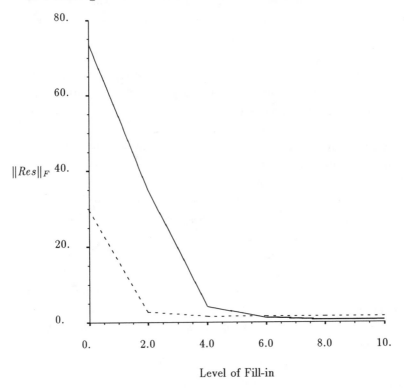

Figure 4.2 Frobenius norm of error matrix for the ILUT preconditioner as the accuracy increases. Dashed line: natural ordering,
Solid line: Red-Black ordering.

A second method that has been used in connection with the red-
black ordering is to solve the reduced system which involves only the
black unknowns, i.e., eliminate the red unknowns from the system:

$$(D_2 - FD_1^{-1}E)x_2 = b_2 - FD_1^{-1}b_1.$$

Note that this new system is again a sparse linear system with about
half as many unknowns. In addition, it has been observed that for
'easy problems' the reduced system can often be efficiently solved
with only diagonal preconditioning. The preprocessing to compute
the reduced system is highly parallel and inexpensive. In addition the

reduced system is usually well-conditioned and has some interesting properties when the original system is highly nonsymmetric [16]. We should point out that it is not necessary to form the reduced system. This strategy is more often employed when D_1 is not diagonal, such in Domain Decomposition methods, but it can also have some uses in other situations. For example, applying the matrix to a given vector x can be done using nearest-neighbor communication and is quite efficiently done without forming the Schur complement matrix. In addition this can save storage, which may be more critical in some applications.

4.3 Generalizations of the Red-Black ordering

The usual way of generalizing the red-black ordering is to resort to multi-coloring. The ultimate goal is to obtain a reduced system which is 'small' although optimality in the coloring itself is a secondary issue. A basic method for obtaining a multi-coloring of an arbitrary grid, relying on a greedy approach is quite simple to describe.

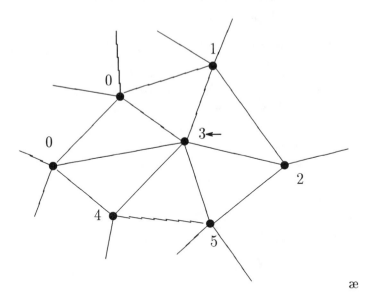

Figure 4.3 Illustration of the greedy approach for multi-coloring. The node being assigned a color is indicated by an arrow. It will be assigned color number 3, the smallest of the allowable color numbers, which are all positive integers different from 1, 2, 4, 5.

1. Initially assign color number zero (uncolored) to every node.

2. Choose an order in which to traverse the nodes.

3. Scan all nodes in the chosen order and at every node i do

$$Color(i) = \min\{k \neq 0 \mid k \neq Color(j), \forall\, j \in \text{Adj}(i))\},$$

As before, $\text{Adj}(i)$ represent the set of nodes that are adjacent to i. The color assigned to node if in step 3 is simply the smallest allowable color number which can be assigned to node i, where *allowable* means different from the colors of the neighbors and positive.

A few simple facts concerning the greedy algorithm, are the following.

- The initial order of traversal may be important in reducing the number of colors.

- If a graph is two-colorable, the algorithm will find the common 2-color (Red-Black) ordering for *Breadth-First* traversals or *Depth-First* traversals.

- The number of colors needed does not exceed the maximum degree of each node +1.

The parallelization of the algorithm is quite important and can be achieved in one of two ways. First, we can use the same algorithm and resort to some form of synchronization. The node being colored checks whether all neighbors are unlocked. As soon as they are it locks itself and gets a color assigned, and then unlocks itself. A second strategy, is to use divide-and-conquer, (similar to a 'domain-decomposition' type approach): color the subdomains – then the nodes in the interfaces. This can also be done recursively.

There has been some recent work by Jones and co-authors [26] on developing efficient parallel multi-coloring algorithms [26]. We should point out that it is important that the cost of the coloring phase be small, ideally of the order of nonzero elements. We may sacrifice optimality since we know that the degree of parallelism is often excellent even with the simplest greedy technique described above. Benantar and Flahety [11] have developed efficient methods in the context of finite element techniques based on quad-tree structures. They show that with this structure a maximum of 6 colors are required.

The idea of the greedy multi-coloring algorithm is known in Finite Element techniques (to color elements), see e.g., Berger-Brouays-Syre (1982), and Benantar and Flaherty. Wu [57] presents the greedy algorithm for coloring and uses it for SOR type iterations. Finally, the effect of coloring has been extensively studied by Adams [3, 1] and Poole and Ortega [36].

4.4 Solving multi-colored systems

We would now like to address again question on how to solve multi-colored systems. Just as for the red black ordering, we can use ILU0 or SOR, SSOR preconditioning on the reordered system. The parallelism of SOR/ SSOR is still of order N in general, but again one can anticipate a loss in the efficiency since the number of iterations is likely to increase. As a result it is much preferable to use SOR(k) / SSOR(k) as a preconditioner, where k is the number of steps in SOR/SSOR. The second class of methods consists of forming the reduced system – involving nodes of color number > 1. This system is sparse again and can be solved in a number of different ways. One idea that has been tested with good success is to multi-color the system and reduce it recursively a few number of times. However, the system becomes denser as the number of reductions increases. We should point out that if we follow this approach and seek only to get a reduced system, then the actual problem is not really to color the grid nodes but rather to separate them in two sets of nodes, one of which consists of vertices that are not connected to each other. In terms of the coloring problem, one set consists of the nodes with a selected color – e.g., the first one – and the second set contains all the other nodes.

In recent experiments [41] we have been able to reduce the execution time on a CRAY considerably by doing two levels of coloring and reduction and then using multiple-step SOR as a preconditioner for the final system. This is illustrated in the following experiments from [41].

We consider the Partial Differential Equation:

$$\Delta u + \gamma \left(x\frac{\partial u}{\partial x} + y\frac{\partial u}{\partial y} \right) + \alpha u = f \; ,$$

with Dirichlet Boundary Conditions and $\gamma = 40, \alpha = -50$; discretized with centered differences on a $27 \times 27 \times 27$ grid. This leads to

fill	ILU time	iter time	total time	iter	$\|Err.\|_2$
1	0.341E+01	0.343E+01	0.684E+01	49	0.353E-05
3	0.377E+01	0.230E+01	0.608E+01	30	0.362E-05
5	0.878E+01	0.192E+01	0.107E+02	22	0.206E-03

Table 1: Performance of GMRES(10)-ILUT(k,ϵ), for various values of k and $\epsilon = 0.001$. Natural ordering.

fill	ILU time	iter time	total time	iter	$\|Err.\|_2$
1	0.306E+01	0.742E+01	0.105E+02	*	0.114E+01
3	0.335E+01	0.565E+01	0.900E+01	80	0.625E-04
5	0.526E+01	0.383E+01	0.909E+01	51	0.226E-04

Table 2: Performance of GMRES(10)-ILUT(k,ϵ), for various values of k and $\epsilon = 0.001$. Red-Black ordering.

a linear system with $N = 25^3 = 15,625$ unknowns. The experiments have been performed on a CRAY 2.

Table 1 shows the performance of GMRES(10)-ILUT(k,ϵ), for various values of k and for $\epsilon = 0.001$. Level scheduling is used to optimize the forward and backward solves. In addition, the matrix vector multiplications that arise in the algorithm as well as in the level-scheduled forward and backward solutions are optimized by using what is referred to as the "jagged diagonal" format [38, 5]. However, the preprocessing needed to compute the incomplete factorization itself is not optimized. Table 2 shows the same run executed on the matrix obtained after the red-black ordering.

Figure 5.1 shows the iteration times for several runs on the prob-

Method	total time	iter time	iter	$\|Err.\|_2$
GMRES(20)/SOR(5)	0.777E+00	0.586E+00	17	0.373E-04
GMRES(10)/SOR(13)	0.886E+00	0.701E+00	10	0.758E-05

Table 3: Best times for Red-Black SOR/ GMRES

lem obtained from solving reduced system by a multi-colored k-step SOR preconditioner. The parameter k, the number of SOR steps used, is varied from 1 to 25. The optimal number of SOR steps is 13 for GMRES(10) and 5 for GMRES(20). Additional data for these two cases is shown separately on Table 3. Again the preprocessing is not optimized but it costs very little in this case. As a result if we compare the iteration times only, we note that this approach is about 4 times faster that the best possible performance that we could obtain from the standard ILU preconditioners. If we compared the total times then the new approach would be even more attractive. An attractive feature of this new approach is that the preprocessing – coloring and forming the reduced system and coloring again – is highly parallel and quite easy to optimize whereas high level-of-fill ILU-based preconditioners are not.

5   Future Directions

As parallel processing hardware and software is starting to mature, new methods are bound to emerge and replace older techniques that do not yield sufficient parallelism. In particular, Krylov subspace methods have an excellent potential for playing a major role in the solution of linear and nonlinear equations of computational fluid dynamics. At the same time, by providing researchers with the capability to deal with more complete models than in the past, high performance computers are indirectly causing the problems to become harder to solve by iterative methods. This trend suggests that research on *robust iterative methods* is likely to be still very active in the coming years. Indeed, the iterative methods that are available today are hardly satisfactory from the point of view of robustness. In the end, the best use of techniques based on Krylov subspace methods may very well be in conjunction with more traditional methods such as direct solvers on smaller domains (in a domain decomposition approach) or multi-level techniques, or ADI, etc.. We have seen that coupling multi-coloring with a technique as simple as SOR may yield quite good performance. This suggests that we may be able to make good use of a number of tools that are common in computer science. Thus, graph theory is essential in unraveling parallelism, especially in the realistic models that involve unstructured grids. We have seen the cases of level-scheduling and muti-coloring which are two text-book examples. These techniques are already being used by researchers in CFD see, e.g., [52, 54]. It is likely that more of these tools may find a good use in computational fluid dynamics codes in the future. This is a good example of the necessity of collaboration across several disciplines if we wish to handle the increasing complexity of the new scientific problems as well as that of the new environments in which they are to be solved.

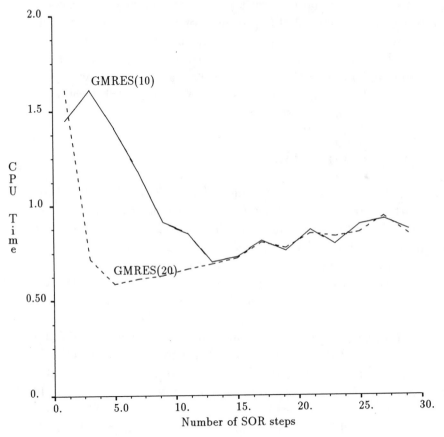

Figure 5.1 Iteration times for GMRES(10) and GMRES(20) preconditioned by SOR(k) as k varies.

References

[1] L. Adams and H. Jordan. Is SOR color-blind? *SIAM J. Sci. Statist. Comp*, 6:490–506, 1985.

[2] L. Adams and J. Ortega. A multi-color SOR Method for Parallel Computers. In *Proceedings 1982 Int. Conf. Par. Proc.*, pages 53–56, 1982.

[3] L. M. Adams. *Iterative algorithms for large sparse linear systems on parallel computers*. PhD thesis, Applied Mathematics, University of Virginia, Charlottsville, VA, 22904, 1982. Also NASA Contractor Report 166027.

[4] E. C. Anderson. Parallel implementation of preconditioned conjugate gradient methods for solving sparse systems of linear equations. Technical Report 805, CSRD, University of Illinois, Urbana, IL, 1988. MS Thesis.

[5] E. C. Anderson and Y. Saad. Solving sparse triangular systems on parallel computers. Technical Report 794, University of Illinois, CSRD, Urbana, IL, 1988.

[6] S. F. Ashby, T. A. Manteuffel, and P. E. Saylor. Adaptive polynomial preconditioning for Hermitian indefinite linear systems. *BIT*, 29:583–609, 1989.

[7] C. C. Ashcraft and R. G. Grimes. On vectorizing incomplete factorization and SSOR preconditioners. *SIAM J. Sci. Statist. Comput.*, 9:122–151, 1988.

[8] O. Axelsson. A generalized conjugate gradient, least squares method. *Num. Math.*, 51:209–227, 1987.

[9] D. Baxter, J. Saltz, M. H. Schultz, and S. C. Eisenstat. Preconditioned Krylov solvers and methods for runtime loop parallelization. Technical Report 655, Computer Science, Yale University, New Haven, CT, 1988.

[10] D. Baxter, J. Saltz, M. H. Schultz, S. C. Eisenstat, and K. Crowley. An experimental study of methods for parallel preconditioned Krylov methods. Technical Report 629, Computer Science, Yale University, New Haven, CT, 1988.

[11] M. Benantar and J. E. Flaherty. A Six color procedure for the parallel solution of Elliptic systems using the finite Quadtree structure. In J. Dongarra, P. Messina, D. C. Sorenson, and R. G. Voigt, editors, *Proceedings of the fourth SIAM conference on parallel processing for scientific computing*, pages 230–236, 1990.

[12] I. S. Duff, A. M. Erisman, and J. K. Reid. *Direct Methods for Sparse Matrices*. Clarendon Press, Oxford, 1986.

[13] I. S. Duff and G. A. Meurant. The effect of reordering on preconditioned conjugate gradients. *BIT*, 29:635–657, 1989.

[14] H. C. Elman. *Iterative Methods for Large Sparse Nonsymmetric Systems of Linear Equations*. PhD thesis, Yale University, Computer Science Dept., New Haven, CT., 1982.

[15] H. C. Elman and E. Agron. Ordering techniques for the precondiotioning conjugate gradient method on parallel computers. Technical Report UMIACS-TR-88-53, UMIACS, University of Maryland, College Park, MD, 1988.

[16] H. C. Elman and G. H. Golub. Iterative methods for cyclically reduced non-self-adjoint linear systems. Technical Report CS-TR-2145, Dept. of Computer Science, University of Maryland, College Park, MD, 1988.

[17] R. M. Ferencz. *Element-by-element preconditioning techniques for large scale vectorized finite element analysis in nonlinear solid and structural mechanics*. PhD thesis, Applied Mathematics, Stanford, CA, 1989.

[18] R. Freund, M. H. Gutknecht, and N. M. Nachtigal. An implementation of the Look-Ahead Lanczos algorithm for non-Hermitian matrices, Part I. Technical Report 90-11, Massachusetts Institute of Technology, Cambridge, Massachusetts, 1990.

[19] R. Freund and N. M. Nachtigal. An implementation of the look-ahead lanczos algorithm for non-Hermitian matrices, Part II. Technical Report 90-11, Massachusetts Institute of Technology, Cambridge, Massachusetts, 1990.

[20] R. S. Varga G. H. Golub. Chebyshev semi iterative methods successive overrelaxation iterative methods and second order Richardson iterative methods. *Numer. Math.*, 3:147–168, 1961.

[21] A. Greenbaum. Solving triangular linear systems using fortran with parallel extensions on the nyu ultracomputer prototype. Technical Report 99, Courant Institute, New York University, New York, NY, 1986.

[22] A. L. Hageman and D. M. Young. *Applied Iterative Methods*. Academic Press, New York, 1981.

[23] M. A. Heroux, P. Vu, and C. Yang. A parallel preconditioned conjugate gradient package for solving sparse linear systems on a cray Y-MP. Technical report, Cray Research, Eagan, MN, 1990. Proc. of the 1990 Cray Tech. Symp.

[24] T. J. R. Hughes, R. M. Ferencz, and J. O. Hallquist. Large-scale vectorized implicit calculations in solid mechanics on a cray x-mp/48 utilizing ebe preconditioning conjugate gradients. *Computer Methods in Applied Mechanics and Engineering*, 61:215–248, 1987.

[25] K. C. Jea and D. M. Young. Generalized conjugate gradient acceleration of nonsymmetrizable iterative methods. *Linear Algebra Appl.*, 34:159–194, 1980.

[26] M. T. Jones and P. E. Plassmann. Parallel iterative solution of sparse linear systems using ordering from graph coloring heuristics. Technical Report MCS-P198-1290, Argonne National Lab., Argonne, IL, 1990.

[27] T. I. Karush, N. K. Madsen, and G. H. Rodrigue. Matrix multiplication by diagonals on vector/parallel processors. Technical Report UCUD, Lawrence Livermore National Lab., Livermore, CA, 1975.

[28] T. A. Manteuffel. The Tchebychev iteration for nonsymmetric linear systems. *Numer. Math.*, 28:307–327, 1977.

[29] T. A. Manteuffel. Adaptive procedure for estimation of parameter for the nonsymmetric Tchebychev iteration. *Numer. Math.*, 28:187–208, 1978.

[30] J. A. Meijerink and H. A. van der Vorst. An iterative solution method for linear systems of which the coefficient matrix is a symmetric M-matrix. *Math. Comp.*, 31(137):148–162, 1977.

[31] R. Melhem. Solution of linear systems with striped sparse matrices. *Parallel Comput.*, 6:165–184, 1988.

[32] T. C. Oppe, W. Joubert, and D. R. Kincaid. Nspcg user's guide. a package for solving large linear systems by various iterative methods. Technical report, The University of Texas at Austin, 1988.

[33] T. C. Oppe and D. R. Kincaid. The performance of ITPACK on vector computers for solving large sparse linear systems arising in sample oil reservoir simulation problems. *Communications in applied numerical methods*, 2:1–7, 1986.

[34] J. M. Ortega. *Introduction to Parallel and Vector Solution of Linear Systems*. Plenum Press, New York, 1988.

[35] G. A. Paolini and G. Radicati di Brozolo. Data structures to vectorize CG algorithms for general sparsity patterns. *BIT*, 29:4, 1989.

[36] E. L Poole and J. M. Ortega. Mullticolor ICCG methods for vector computers. *SIAM J. Numer. Anal.*, 24:1394–1418, 1987.

[37] Y. Saad. Least squares polynomials in the complex plane and their use for solving sparse nonsymmetric linear systems. *SIAM J. Num. Anal.*, 24:155–169, 1987.

[38] Y. Saad. SPARSKIT: A basic tool kit for sparse matrix computations. Technical Report 90-20, Research Institute for Advanced Computer Science, NASA Ames Research Center, Moffet Field, CA, 1990.

[39] Y. Saad. A flexible inner-outer preconditioned GMRES algorithm. Technical Report 91-279, Minnesota Supercomputer Institute, University of Minnesota, Minneapolis, Minnesota, 1991.

[40] Y. Saad. ILUT: a dual strategy accurate incomplete ilu factorization. Technical Report –, Minnesota Supercomputer Institute, University of Minnesota, Minneapolis, 1991. in preparation.

[41] Y. Saad. Massively parallel preconditioned Krylov subspace methods. Technical Report –, Minnesota Supercomputer Institute, Minneapolis, Minnesota, 1991. In preparation.

[42] Y. Saad and M. H. Schultz. Conjugate gradient-like algorithms for solving nonsymmetric linear systems. *Mathematics of Computation*, 44(170):417–424, 1985.

[43] Y. Saad and M. H. Schultz. Parallel implementations of preconditioned conjugate gradient methods. Research report 425, Dept Computer Science, Yale University, 1985.

[44] Y. Saad and M. H. Schultz. GMRES: a generalized minimal residual algorithm for solving nonsymmetric linear systems. *SIAM J. Sci. Statist. Comput.*, 7:856–869, 1986.

[45] J. H. Saltz. Automated problem scheduling and reduction of synchronization delay effects. Technical Report 87-22, ICASE, Hampton, VA, 1987.

[46] F. Shakib. *Finite element analysis of the compressible Euler and Navier Stokes Equations*. PhD thesis, Aeronautics Dept., Stanford, CA, 1989.

[47] T. E. Tezduyar, M. Behr, S. K. A. Abadi, and S. E. Ray. A mixed CEBE/CC preconditionning for finite element computations. Technical Report UMSI 91/160, University of Minnesota, Minnesota Supercomputing Institute, Mineapolis, Minnesota, June 1991.

[48] H. A. van der Vorst. The performance of FORTRAN implementations for preconditioned conjugate gradient methods on vector computers. *Parallel Comput.*, 3:49–58, 1986.

[49] H. A. van der Vorst. Large tridiagonal and block tridiagonal linear systems on vector and parallel computers. *Par. Comp.*, 5:303–311, 1987.

[50] H. A. van der Vorst. High performance preconditioning. *SIAM j. Scient. Stat. Comput.*, 10:1174–1185, 1989.

[51] R. S. Varga. *Matrix Iterative Analysis*. Prentice Hall, Englewood Cliffs, NJ, 1962.

136

[52] V. Venkatakrishnan. Preconditioned Conjugate Gradient methods for the compressible Navier Stokes equations. *AIAA Journal*, 29:1092–1100, 1991.

[53] V. Venkatakrishnan and D. J. Mavriplis. Implicit solvers for unstructured grids. In *Proceedings of the AIAA 10th CFD Conference, June, 1991, HI.*, 1991.

[54] V. Venkatakrishnan, H. D. Simon, and T. J. Barth. A MIMD Implementation of a Parallel Euler Solver for Unstructured Grids. Technical Report RNR-91-024, NASA Ames research center, Moffett Field, CA, 1991.

[55] P. K. W. Vinsome. Orthomin, an iterative method for solving sparse sets of simultaneous linear equations. In *Proceedings of the Fourth Symposium on Resevoir Simulation*, pages 149–159. Society of Petroleum Engineers of AIME, 1976.

[56] O. Wing and J. W. Huang. A computation model of parallel solution of linear equations. *IEEE Transactions on Computers*, C-29:632–638, 1980.

[57] C. H. Wu. A multicolour SOR method for the finite-element method. *J. of Comput. and App. Math.*, 30:283–294, 1990.

[58] D. M. Young. *Iterative solution of large linear systems*. Academic Press, New-York, 1971.

[59] D. M. Young, T. C. Oppe, D. R. Kincaid, and L. J. Hayes. On the use of vector computers for solving large sparse linear systems. Technical Report CNA-199, Center for Numerical Analysis, University of Texas at Austin, Austin, Texas, 1985.

RECENT ADVANCES IN LANCZOS-BASED ITERATIVE METHODS FOR NONSYMMETRIC LINEAR SYSTEMS

Roland W. Freund[*]

RIACS, Mail Stop T041–5
NASA Ames Research Center
Moffett Field, CA 94035

Gene H. Golub[†]

Computer Science Department
Stanford University
Stanford, CA 94305

Noël M. Nachtigal[*]

RIACS, Mail Stop T041–5
NASA Ames Research Center
Moffett Field, CA 94035

ABSTRACT

In recent years, there has been a true revival of the nonsymmetric Lanczos method. On the one hand, the possible breakdowns in the classical algorithm are now better understood, and so-called look-ahead variants of the Lanczos process have been developed, which remedy this problem. On the other hand, various new Lanczos-based iterative schemes for solving nonsymmetric linear systems have been proposed. This paper gives a survey of some of these recent developments.

1 Introduction

Many numerical computations involve the solution of large nonsingular systems of linear equations

$$Ax = b. \tag{1.1}$$

[*]The work of these authors was supported by Cooperative Agreement NCC 2-387 between NASA and the Universities Space Research Association (USRA).

[†]The work of this author was supported in part by the National Science Foundation under Grant NSF CCR-8821078.

For example, such systems arise from finite difference or finite element approximations to partial differential equations (PDEs), as intermediate steps in computing the solution of nonlinear problems, or as subproblems in large-scale linear and nonlinear programming. Typically, the coefficient matrix A of (1.1) is sparse and highly structured. A natural way to exploit the sparsity of A in the solution process is to use iterative techniques, which involve A only in the form of matrix-vector products. Most iterative schemes of this type fall into the category of *Krylov subspace methods*: they produce approximations x_n to $A^{-1}b$ of the form

$$x_n \in x_0 + K_n(r_0, A), \quad n = 1, 2, \dots . \tag{1.2}$$

Here x_0 is any initial guess for $A^{-1}b$, $r_0 := b - Ax_0$ is the corresponding residual vector, and

$$K_n(r_0, A) := \text{span}\{r_0, Ar_0, \dots, A^{n-1}r_0\} \tag{1.3}$$

is the nth *Krylov subspace* generated by r_0 and A.

The most powerful iterative method of this type is the *conjugate gradient algorithm* (CG) due to Hestenes and Stiefel [33], which is a scheme for linear systems (1.1) with Hermitian positive definite A. Although CG was introduced as early as 1952, its true potential was not appreciated until the 1970s. In 1971, Reid [45] revived interest in the method when he demonstrated its usefulness for solving linear systems arising from self-adjoint elliptic PDEs. Moreover, it was realized (see, e.g., [7]) that the performance of CG can be enhanced by combining it with preconditioning, and efficient preconditioners, such as the incomplete Cholesky factorization [40], were developed.

Thereafter, the success of CG triggered an extensive search for CG-type Krylov subspace methods for non-Hermitian linear systems, and a number of such algorithms have been proposed; we refer the reader to [1, 51, 48, 47, 17] and the references given there. Among the many properties of CG, the following two are the most important ones: its nth iterate is defined by a minimization property over $K_n(r_0, A)$, and the algorithm is based on three-term vector recurrences. Ideally, a CG-like method for non-Hermitian matrices would have features similar to these two. It would produce iterates x_n in (1.2) that:

(i) are characterized by a minimization property over $K_n(r_0, A)$, such as the minimal residual property

$$\|b - Ax_n\| = \min_{x \in x_0 + K_n(r_0, A)} \|b - Ax\|, \quad x_n \in x_0 + K_n(r_0, A);$$

(ii) can be computed with little work per iteration and low overall storage requirements.

Unfortunately, it turns out that, for general non-Hermitian matrices, one cannot fulfill (i) and (ii) simultaneously. This result is due to Faber and Manteuffel [10, 11] who have shown that, except for a few anomalies, CG-type algorithms with (i) and (ii) exist only for matrices of the special form

$$A = e^{i\theta}(T + \sigma I), \quad \text{where} \quad T = T^H, \quad \theta \in \mathcal{R}, \quad \sigma \in \mathcal{C}, \quad (1.4)$$

(see also Voevodin [55] and Joubert and Young [35]). Note that the class (1.4) consists of just the shifted and rotated Hermitian matrices. We remark that the important subclass of real nonsymmetric matrices

$$A = I - S, \quad \text{where} \quad S = -S^T \quad \text{is real}, \quad (1.5)$$

is contained in (1.4), with $e^{i\theta} = i$, $\sigma = -i$, and $T = iS$. Concus and Golub [6] and Widlund [56] were the first to devise a CG-type algorithm for the family (1.5).

Most of the non-Hermitian Krylov subspace methods that have been proposed satisfy either (i) or (ii). Until recently, the emphasis was on requirement (i), and numerous algorithms with iterates characterized by (i) or a similar condition have been developed, starting with Vinsome's Orthomin [54]. The most widely used method in this class is the *generalized minimal residual algorithm* (GMRES) due to Saad and Schultz [49]. Of course, none of these methods fulfills (ii), and indeed, for all these algorithms work per iteration and overall storage requirements grow linearly with the iteration number n. Consequently, in practice one cannot afford to run the full version of these algorithms, and it is necessary to use restarts. For difficult problems, this often results in very slow convergence.

The second category of CG-like non-Hermitian Krylov subspace methods consists of schemes that satisfy (ii), but not (i). The archetype in this class is the classical *biconjugate gradient algorithm* (BCG), which was proposed by Lanczos [38] already in 1952 and later revived by Fletcher [12] in 1976. Since no minimization condition of type (i) holds for BCG, the algorithm can exhibit—and typically does—a rather irregular convergence behavior with wild oscillations in the residual norm. Even worse, breakdowns in the form of division by 0 may be encountered during the iteration process. In finite precision arithmetic, such exact breakdowns are very unlikely; however, near-breakdowns may occur, leading to numerical instabilities in subsequent iterations.

The BCG method is intimately connected with the *nonsymmetric Lanczos process* [37] for tridiagonalizing square matrices. In particular, the Lanczos algorithm in its original form is also susceptible to breakdowns and potential numerical instabilities. In recent years, there has been a true revival of the nonsymmetric Lanczos process. On the one hand, the possible breakdowns in the classical algorithm are now better understood, and so-called *look-ahead* variants of the Lanczos process have been developed, which remedy this problem. On the other hand, various new Lanczos-based Krylov subspace methods for solving general non-Hermitian linear systems have been proposed. Here we review some of these recent developments.

The remainder of the paper is organized as follows. In Section 2, we focus on the nonsymmetric Lanczos process; in particular, we sketch a look-ahead variant of the method and briefly discuss related work. We then turn to Lanczos-based Krylov subspace algorithms for non-Hermitian linear systems. First, in Section 3, we consider the recently proposed *quasi-minimal residual method* (QMR) and outline two implementations. In addition to matrix-vector products with the coefficient matrix A of (1.1), BCG and QMR also require multiplications with its transpose A^T. This is a disadvantage for certain applications where A^T is not readily available. It is possible to devise Lanczos-based methods that do not involve A^T, and in Section 4, we survey some of these so-called *transpose-free* schemes. In Section 5, we make some concluding remarks.

Throughout the paper, all vectors and matrices are allowed to have real or complex entries. As usual, M^T and M^H denote the transpose and conjugate transpose of a matrix M, respectively. The vector norm $\|x\| = \sqrt{x^H x}$ is always the Euclidean norm. The notation

$$\mathcal{P}_n = \{\phi(\lambda) \equiv \sigma_0 + \sigma_1\lambda + \cdots + \sigma_n\lambda^n \mid \sigma_0, \ldots, \sigma_n \in \mathcal{C}\}$$

is used for the set of all complex polynomials of degree at most n. Finally, A is always assumed to be a square matrix of order N.

2 The Nonsymmetric Lanczos Process

In this section, we consider the nonsymmetric Lanczos process. Here the matrix A is not required to be nonsingular.

2.1 A Look-Ahead Lanczos Algorithm

The Lanczos method in its original form as proposed by Lanczos [37] can break down prematurely. Taylor [52] and Parlett, Taylor, and Liu [44]—with their look-ahead Lanczos algorithm—were the first to devise a variant of the classical process that skips over possible breakdowns. We use the term look-ahead Lanczos method in a broader sense to denote any extension of the standard algorithm that circumvents breakdowns. In this section, we sketch an implementation of a look-ahead Lanczos algorithm that was recently developed by Freund, Gutknecht, and Nachtigal [18].

Given two nonzero starting vectors $v_1 \in \mathcal{C}^N$ and $w_1 \in \mathcal{C}^N$, the look-ahead Lanczos process generates two sequences of vectors $\{v_j\}_{j=1}^n$ and $\{w_j\}_{j=1}^n$ such that, for $n = 1, 2, \ldots$,

$$\text{span}\{v_1, v_2, \ldots, v_n\} = K_n(v_1, A), \qquad (2.1)$$
$$\text{span}\{w_1, w_2, \ldots, w_n\} = K_n(w_1, A^T).$$

Here, $K_n(v_1, A)$ and $K_n(w_1, A^T)$ denote the nth Krylov subspace of $\mathcal{C}^N$ generated by v_1 and A, and w_1 and A^T, respectively (cf. (1.3)). Moreover, the Lanczos vectors are constructed so that the block biorthogonality relation

$$(W^{(j)})^T V^{(k)} = \begin{cases} D^{(k)} & \text{if } j = k, \\ 0 & \text{if } j \neq k, \end{cases} \qquad j, k = 1, \ldots, l, \qquad (2.2)$$

holds. Here, the matrices $V^{(k)}$ and $W^{(k)}$ contain the Lanczos vectors built during the kth look-ahead step. More precisely,

$$\begin{aligned} V^{(k)} &= [\, v_{n_k} \quad v_{n_k+1} \quad \cdots \quad v_{n_{k+1}-1} \,], \\ W^{(k)} &= [\, w_{n_k} \quad w_{n_k+1} \quad \cdots \quad w_{n_{k+1}-1} \,], \end{aligned} \qquad k = 1, \ldots, l-1,$$

and

$$\begin{aligned} V^{(l)} &= [\, v_{n_l} \quad v_{n_l+1} \quad \cdots \quad v_n \,], \\ W^{(l)} &= [\, w_{n_l} \quad w_{n_l+1} \quad \cdots \quad w_n \,], \end{aligned}$$

where

$$1 = n_1 < n_2 < \cdots < n_k < \cdots < n_l \leq n < n_{l+1}.$$

The first vectors v_{n_k} and w_{n_k} in each block are called *regular*, and any remaining vectors are called *inner*. Note that $l = l(n)$ denotes the index of the last constructed regular vector. Furthermore, in (2.2), the blocks $D^{(k)}$ are nonsingular for $k = 1, \ldots, l-1$, and $D^{(l)}$ is nonsingular if $n = n_{l+1} - 1$.

With these preliminaries, the look-ahead Lanczos algorithm can be sketched as follows.

Algorithm 2.1 (Sketch of the look-ahead Lanczos process)

0) *Choose nonzero vectors v_1, $w_1 \in \mathcal{C}^N$.*

   Set $V^{(1)} = v_1$, $W^{(1)} = w_1$, $D^{(1)} = (W^{(1)})^T V^{(1)}$.

   Set $n_1 = 1$, $l = 1$, $v_0 = w_0 = 0$, $V_0 = W_0 = \emptyset$.

For $n = 1, 2, \ldots$, do :

1) *Decide whether to construct v_{n+1} and w_{n+1} as regular or inner vectors*

   and go to 2) or 3), respectively.

2) *(Regular step.) Compute*

$$\begin{aligned}
\mu_n &= (D^{(l)})^{-1}(W^{(l)})^T A v_n, \\
\nu_n &= (D^{(l-1)})^{-1}(W^{(l-1)})^T A v_n, \\
v_{n+1} &= A v_n - V^{(l)}\mu_n - V^{(l-1)}\nu_n, \\
w_{n+1} &= A^T w_n - W^{(l)}\mu_n - W^{(l-1)}\nu_n.
\end{aligned} \qquad (2.3)$$

   Set $n_{l+1} = n + 1$, $l = l + 1$, $V^{(l)} = W^{(l)} = \emptyset$, and go to 4).

3) *(Inner step.) Compute*

$$\begin{aligned}
\nu_n &= (D^{(l-1)})^{-1}(W^{(l-1)})^T A v_n, \\
v_{n+1} &= A v_n - \zeta_n v_n - \eta_n v_{n-1} - V^{(l-1)}\nu_n, \\
w_{n+1} &= A^T w_n - \zeta_n w_n - \eta_n w_{n-1} - W^{(l-1)}\nu_n.
\end{aligned} \qquad (2.4)$$

4) *If $v_{n+1} = 0$ or $w_{n+1} = 0$, stop. Otherwise, set*

$$V^{(l)} = \begin{bmatrix} V^{(l)} & v_{n+1} \end{bmatrix}, \qquad W^{(l)} = \begin{bmatrix} W^{(l)} & w_{n+1} \end{bmatrix},$$

$$D^{(l)} = (W^{(l)})^T V^{(l)}.$$

In [18], it is shown how one can implement Algorithm 2.1 so that only two inner products are computed at every step, for either μ_n and ν_n in (2.3), or for ν_n in (2.4). The crucial part of Algorithm 2.1 is the look-ahead strategy used in step 1). As described in [18], the decision in 1) is based on three checks. For a regular step, it is necessary that $D^{(l)}$ be nonsingular. Therefore, one of the checks monitors the size of smallest singular value of $D^{(l)}$. The other two checks attempt to ensure the linear independence of the Lanczos vectors. The algorithm monitors the size of the components μ_n and ν_n along the two previous blocks $V^{(l)}$ and $V^{(l-1)}$, respectively $W^{(l)}$ and $W^{(l-1)}$, in (2.3), and performs a regular step only if these terms do not dominate the components Av_n and $A^T w_n$ in the new Krylov spaces. Complete details of the implementation of the look-ahead Lanczos Algorithm 2.1 are given in [18].

We note that, in (2.4), ζ_n and η_n are arbitrary inner recurrence coefficients, with $\zeta_{n_k} = 0$. One possibility is to choose the Chebyshev iteration [25, 39] parameters for ζ_n and η_n. However, since the length of look-ahead steps is usually small, the choice of the inner recurrence coefficients is not crucial; in our experience, $\zeta_n = 1$ and, if $n \neq n_k$, $\eta_n = 1$, works satisfactorily. Indeed, with the look-ahead strategy proposed in [18], the algorithm performs mostly regular steps, and typically, only a few look-ahead steps of length bigger than 1 occur. In our experiments, the longest look-ahead step we encountered was of length 4.

For later use, we remark that the recurrences in (2.3) and (2.4) can be written compactly in matrix form. For example, for the right Lanczos vectors v_n, we have

$$AV_n = V_{n+1} H_n, \tag{2.5}$$

where

$$V_n := [\, v_1 \quad v_2 \quad \cdots \quad v_n \,],$$

and

$$H_n = \begin{bmatrix} \alpha_1 & \beta_2 & 0 & \cdots & & 0 \\ \gamma_2 & \alpha_2 & \ddots & \ddots & & \vdots \\ 0 & \ddots & \ddots & \ddots & & 0 \\ \vdots & \ddots & \ddots & \ddots & & \beta_l \\ \vdots & & \ddots & \ddots & \gamma_l & \alpha_l \\ 0 & \cdots & \cdots & 0 & & \gamma_{l+1} \end{bmatrix} \in C^{(n+1) \times n}$$

is a block tridiagonal matrix. The diagonal blocks α_k are square unreduced upper Hessenberg matrices, whose size is equal to the number of vectors in the corresponding block $V^{(k)}$. The matrices γ_k have only one nonzero element, in their upper right corner, and thus H_n is an upper Hessenberg matrix, with full rank

$$\text{rank } H_n = n. \tag{2.6}$$

If only regular steps 2) are performed, then the Algorithm 2.1 reduces to the classical Lanczos process. In this case, the blocks $V^{(k)}$ and $W^{(k)}$ consist of just the single vector v_k and w_k, respectively, and the orthogonality relations (2.2) now read:

$$w_j^T v_k = \begin{cases} \delta_k \neq 0 & \text{if } j = k, \\ 0 & \text{if } j \neq k, \end{cases} \quad j, k = 1, \ldots, n. \tag{2.7}$$

Moreover, H_n is just a scalar tridiagonal matrix. The condition $\delta_k \neq 0$ in (2.7) is crucial, since each step of the classical Lanczos algorithm involves a division by δ_k. The point is that one cannot guarantee $\delta_k \neq 0$, and in fact, when $\delta_k = 0$ with $v_k \neq 0$ and $w_k \neq 0$, the algorithm breaks down. Note that $\delta_k \approx 0$ signals a near-breakdown of the procedure.

Algorithm 2.1 will handle exact and near-breakdowns in the classical Lanczos process, except for the special event of an incurable breakdown [52]. These are situations when the look-ahead procedure would build an infinite block, without ever finding a nonsingular $D^{(l)}$. Taylor [52] has shown in his Mismatch Theorem that in case of an incurable breakdown, one can still recover eigenvalue information. For linear systems, an incurable breakdown would require restarting the procedure with a different choice of starting vectors. Fortunately, in practice round-off errors will make an incurable breakdown highly unlikely.

Finally, we remark that, for the important class of p-cyclic matrices A, exact breakdowns in the Lanczos process occur in a regular pattern. In this case, as was shown by Freund, Golub, and Hochbruck [16], look-ahead steps are absolutely necessary if one wants to exploit the p-cyclic structure. For details of a look-ahead Lanczos algorithm for p-cyclic matrices, we refer the reader to [16].

2.2 Historical Remarks and Related Work

The problem of breakdowns in the classical Lanczos algorithm has been known from the beginning. Although a rare event in practice, the possibility of breakdowns has certainly brought the method into discredit and has prevented many people from actually using the algorithm. On the other hand, as was demonstrated by Cullum and Willoughby [8], the Lanczos process—even without look-ahead—is a powerful tool for sparse matrix computation.

The Lanczos method has intimate connections with many other areas of Mathematics, such as formally orthogonal polynomials (FOPs), Padé approximation, Hankel matrices, control theory, and coding theory. The problem of breakdowns has a corresponding formulation in all of these areas, and remedies for breakdowns in these different settings have been known for quite some time. For example, the breakdown in the Lanczos process is equivalent to a breakdown of the generic three-term recurrence relation for FOPs, and it is well known how to overcome such breakdowns by modifying the recursions for FOPs (see [26, 9, 31] and the references given there). In the context of the partial realization problem in control theory, remedies for breakdowns were given in [36, 27]. The Lanczos process is also closely related to fast algorithms for the factorization of Hankel matrices, and again it was known how to overcome possible breakdowns of these algorithms (see [32, 22] and the references therein). However, in all these cases, only the problem of exact breakdowns was addressed.

The look-ahead Lanczos algorithm of Taylor [52] and Parlett, Taylor, and Liu [44] was the first procedure that remedies both exact and near-breakdowns. We point out that their implementation is different from Algorithm 2.1. In particular, it always requires more work per step than Algorithm 2.1, and it does not reduce to the classical Lanczos process in the absence of look-ahead steps. Furthermore, in [52, 44], details are given only for the case of look-ahead steps of size 2, and their algorithm does not generalize easily to blocks of more than two vectors.

In recent years, there has been a revival of the nonsymmetric Lanczos algorithm, and since 1990, in addition to the papers we have already cited in this section, there are several others dealing with various aspects of the Lanczos process. We refer the reader to [2, 3, 4, 22, 29, 34, 43] and the references given therein.

3   The Quasi-Minimal Residual Approach

We now return to linear systems (1.1). From now on, it is always assumed that the matrix A is nonsingular. In this section, we describe the QMR method. The procedure was first proposed by Freund [13, 15] for the case of complex symmetric matrices $A = A^T$, and then extended by Freund and Nachtigal [19] for the case of general non-Hermitian matrices.

3.1   The Standard QMR Algorithm

Recall that the nth iterate of any Krylov subspace method is of the form (1.2). If now we choose

$$v_1 = r_0 \tag{3.1}$$

in Algorithm 2.1, then, by (2.1), the Lanczos vectors $v_1, \ldots, v_n$ span $\mathcal{K}_n(r_0, A)$; hence we can write

$$x_n = x_0 + V_n z_n,$$

for some $z_n \in \mathcal{C}^n$. Together with (3.1) and (2.5), this gives the corresponding residual vector

$$r_n = r_0 - A V_n z_n = V_{n+1}(e_1 - H_n z_n), \tag{3.2}$$

where e_1 denotes the first unit vector in $\mathcal{R}^{n+1}$. As V_{n+1} is not unitary, it is not possible to minimize the Euclidean norm of the residual without expending $\mathcal{O}(Nn^2)$ work and $\mathcal{O}(Nn)$ storage. Instead, one minimizes just some weighted Euclidean norm of the coefficient vector in (3.2). More precisely, let

$$\Omega_n = \mathrm{diag}(\omega_1, \omega_2, \ldots, \omega_{n+1}), \quad \omega_j > 0, \quad j = 1, \ldots, n+1, \tag{3.3}$$

be a weighting matrix. Then $z_n \in \mathcal{C}^n$ is chosen as the solution of the least squares problem

$$\|\omega_1 e_1 - \Omega_n H_n z_n\| = \min_{z \in \mathcal{C}^n} \|\omega_1 e_1 - \Omega_n H_n z\|. \tag{3.4}$$

Note that, in view of (2.6) and (3.3), the problem (3.4) always has a unique solution. Usually, the weights in (3.4) are chosen as $\omega_j \equiv \|v_j\|$, which means that all components in

$$r_n = \left(V_{n+1} \Omega_n^{-1}\right)(\omega_1 e_1 - \Omega_n H_n z_n)$$

are treated equally.

The least-squares problem (3.4) can be solved by standard techniques based on a QR decomposition of $\Omega_n H_n$. One computes a unitary matrix $Q_n \in C^{(n+1)\times(n+1)}$ and an upper triangular matrix $R_n \in C^{n\times n}$ such that

$$Q_n \Omega_n H_n = \begin{bmatrix} R_n \\ 0 \end{bmatrix}, \tag{3.5}$$

and then obtains z_n from

$$z_n = R_n^{-1} t_n, \quad t_n = \omega_1 [\, I_n \quad 0\,] Q_n e_1, \tag{3.6}$$

which gives

$$x_n = x_0 + V_n R_n^{-1} t_n. \tag{3.7}$$

This gives the following QMR algorithm.

Algorithm 3.1 (QMR algorithm)

  0) *Choose $x_0 \in C^N$ and set $v_1 = r_0 = b - Ax_0$.*

  Choose $w_1 \in C^N$ with $w_1 \neq 0$.

 For $n = 1, 2, \ldots$, do :

  1) *Perform the nth iteration of the look-ahead Lanczos Algorithm 2.1.*

   This yields matrices V_n, V_{n+1}, and H_n which satisfy (2.5).

  2) *Update the QR factorization (3.5) of H_n and the vector t_n in (3.6).*

  3) *Compute x_n from (3.7). If x_n has converged, stop.*

We note that Q_n in (3.5) is just a product of n Givens rotations, and thus the vector t_n is easily updated in step 2). Also, as H_n is block tridiagonal, R_n also has a block structure that is used in step 3) to update x_n using only short recurrences. For complete details, see [19].

The quasi-minimization (3.4) is strong enough to obtain convergence results for QMR. One can derive error bounds for QMR that are comparable to those for GMRES. Also, it is possible to relate the norms of the QMR and GMRES residual vectors. This is in contrast to BCG and methods derived from BCG, for which no such convergence results are known. Finally, if desired, one can recover

BCG iterates from the QMR Algorithm 3.1, at the expense of only one additional SAXPY per step. For these and other properties of QMR, we refer the reader to [19, 41].

Algorithm 3.1 is only one possible implementation of the QMR method. Instead of using three-term recurrences as in the underlying look-ahead Lanczos Algorithm 2.1, the basis vectors $\{v_n\}$ and $\{w_n\}$ can also be generated by coupled two-term recurrences. Empirical observations indicate that, in finite precision arithmetic, the latter approach is more robust than the former. Details of such an implementation of the QMR method based on coupled two-term recurrences with look-ahead are presented in [20].

FORTRAN 77 implementations of the QMR Algorithm 3.1 and of the look-ahead Lanczos Algorithm 2.1 are available electronically from netlib.[1]

3.2 BCG and an Implementation of QMR without Look-Ahead

We now look at BCG in more detail. The BCG algorithm attempts to generate iterates x_n^{BCG} that are characterized by the Galerkin condition

$$x_n^{\text{BCG}} \in x_0 + K_n(r_0, A) \quad \text{and} \quad w^T(b - Ax_n^{\text{BCG}}) = 0 \quad \text{for all}$$

$$w \in K_n(w_1, A^T). \tag{3.8}$$

Unfortunately, such iterates need not exist for every n, and this is one source of possible breakdowns in BCG.

As noted already, BCG is closely related to the classical Lanczos process. More precisely, the BCG residual vectors are just scalar multiples of the right Lanczos vectors:

$$r_n = b - Ax_n^{\text{BCG}} = \theta_n v_{n+1}, \quad \theta_n \in \mathcal{C}, \quad \theta_n \neq 0. \tag{3.9}$$

In addition to r_n, the BCG algorithm also involves a second sequence of vectors $\tilde{r}_n \in K_{n+1}(\tilde{r}_0, A^T)$. Here $\tilde{r}_0 \in \mathcal{C}^N$ is an arbitrary nonzero starting vector; usually one sets $\tilde{r}_0 = r_0$ or chooses $\tilde{r}_0$ as a vector with random coefficients. The vectors $\tilde{r}_n$ are connected with the left vectors generated by the classical Lanczos process:

$$\tilde{r}_n = \tilde{\theta}_n w_{n+1}, \quad \tilde{\theta}_n \in \mathcal{C}, \quad \tilde{\theta}_n \neq 0. \tag{3.10}$$

---

[1]To obtain the codes, one needs to send a message consisting of the single line "send lalqmr from misc" to *netlib@ornl.gov.*

From (3.9) and (3.10), we have

$$\tilde{r}_{n-1}^T r_{n-1} = \tilde{\theta}_{n-1}\theta_{n-1} w_n^T v_n. \tag{3.11}$$

Recall from (2.7) that the classical Lanczos process breaks down if $w_n^T v_n = 0$ with $v_n \neq 0$ and $w_n \neq 0$. In view of (3.11), this is equivalent to

$$\tilde{r}_{n-1}^T r_{n-1} = 0, \quad r_{n-1} \neq 0, \quad \tilde{r}_{n-1} \neq 0. \tag{3.12}$$

As Algorithm 3.2 below shows, BCG also breaks down if (3.12) occurs. In addition to (3.12), there is a second source of breakdowns in BCG, namely

$$\tilde{q}_{n-1}^T A q_{n-1} = 0, \quad q_{n-1} \neq 0, \quad \tilde{q}_{n-1} \neq 0. \tag{3.13}$$

Here q_{n-1} and $\tilde{q}_{n-1}$ are the vectors generated by Algorithm 3.2 below. It can be shown (see, e.g., [46]) that a breakdown of the kind (3.13) occurs if, and only if, no Galerkin iterate x_n^{BCG} with (3.8) exists.

Unlike the BCG iterates, the QMR iterates are always well defined by (2.6). In particular, breakdowns of the kind (3.13) can be excluded in the QMR Algorithm 3.1. We stress that this remains true even if, in the QMR Algorithm 3.1, one uses the classical Lanczos process in step 1). Of course, the use of the look-ahead Lanczos Algorithm 2.1 avoids breakdowns of the first kind (3.12), except for incurable breakdowns.

As already noted, existing BCG iterates can be easily obtained from quantities generated by the QMR Algorithm 2.1. Therefore, QMR can also be viewed as a stable implementation of BCG. It is also possible to reverse the roles of the two algorithms and to get QMR iterates directly from the BCG algorithm. Such an implementation of QMR without look-ahead was derived by Freund and Szeto in [21], and is as follows.

Algorithm 3.2 (QMR without look-ahead from BCG)

0) *Choose* $x_0 \in C^N$ *and set* $x_0^{\text{QMR}} = x_0^{\text{BCG}} = x_0$.

 Set $q_0 = r_0 = b - Ax_0$, $\hat{p}_0 = 0$, $\tau_0 = \omega_1\|r_0\|$, $\vartheta_0 = 0$.

 Choose $\tilde{r}_0 \in C^N$, $\tilde{r}_0 \neq 0$, *and set* $\tilde{q}_0 = \tilde{r}_0$, $\rho_0 = \tilde{r}_0^T r_0$.

 For $n = 1, 2, \ldots$, *do :*

1) *Set* $\sigma_{n-1} = \tilde{q}_{n-1}^T A q_{n-1}$.

If $\sigma_{n-1} = 0$, stop. Otherwise, compute

$$\alpha_{n-1} = \rho_{n-1}/\sigma_{n-1},$$
$$r_n = r_{n-1} - \alpha_{n-1} A q_{n-1},$$
$$\tilde{r}_n = \tilde{r}_{n-1} - \alpha_{n-1} A^T \tilde{q}_{n-1}.$$

If BCG *iterates are desired, set*

$$x_n^{\text{BCG}} = x_{n-1}^{\text{BCG}} + \alpha_{n-1} q_{n-1}.$$

2) *Compute*

$$\vartheta_n = \frac{\omega_n \|r_n\|}{\tau_{n-1}}, \quad c_n = \frac{1}{\sqrt{1+\vartheta_n^2}}, \quad \tau_n = \tau_{n-1} \vartheta_n c_n,$$
$$\hat{p}_n = c_n^2 \vartheta_{n-1}^2 \hat{p}_{n-1} + c_n^2 \alpha_{n-1} q_{n-1},$$
$$x_n^{\text{QMR}} = x_{n-1}^{\text{QMR}} + \hat{p}_n.$$

3) *If $\rho_{n-1} = 0$, stop. Otherwise, compute*

$$\rho_n = \tilde{r}_n^T r_n, \quad \beta_n = \rho_n/\rho_{n-1},$$
$$q_n = r_n + \beta_n q_{n-1},$$
$$\tilde{q}_n = \tilde{r}_n + \beta_n \tilde{q}_{n-1}.$$

We remark that, exact for the additional updates in step 2), this algorithm is just classical BCG. Of course, unlike the QMR Algorithm 3.1, the implementation of QMR in Algorithm 3.2 can break down due to (3.12) and (3.13).

Algorithm 3.2 is only one of several possible implementations of the BCG approach; see [34, 28] for an overview of the different BCG variants. As in the nonsymmetric Lanczos process, exact and near-breakdowns in the BCG methods can be avoided by incorporating look-ahead procedures. Such look-ahead BCG algorithms have been proposed by Joubert [34] and Gutknecht [29].

4 Transpose-Free Methods

Krylov subspace methods such as BCG and QMR, which are based directly on the Lanczos process, involve matrix-vector products with A and A^T. This is a disadvantage for certain applications, where A^T

is not readily available. It is possible to devise Lanczos-based Krylov subspace methods that do not involve the transpose of A. In this section, we give an overview of such transpose-free schemes.

First, we consider the QMR algorithm. As pointed out by Freund and Zha [23], in principle it is always possible to eliminate A^T altogether, by choosing the starting vector w_1 suitably. This observation is based on the fact that any square matrix is similar to its transpose. In particular, there always exists a nonsingular matrix P such that

$$A^T P = PA. \tag{4.1}$$

Now suppose that in the QMR Algorithm 3.1 we choose the special starting vector $w_1 = P v_1$. Then, with (4.1), one readily verifies that the vectors generated by look-ahead Lanczos Algorithm 2.1 satisfy

$$w_n = P v_n \quad \text{for all} \quad n. \tag{4.2}$$

Hence, instead of updating the left Lanczos vectors $\{w_n\}$ by means of the recursions in (2.3) or (2.4), they can be computed directly from (4.2). The resulting QMR algorithm no longer involves the transpose of A; in exchange, it requires one matrix-vector multiplication with P in each iteration step. Therefore, this approach is only viable for special classes of matrices A, for which one can find a matrix P satisfying (4.1) easily, and for which, at the same time, matrix-vector products with P can be computed cheaply. The most trivial case are real or complex symmetric matrices $A = A^T$, which fulfill (4.1) with $P = I$. Another simple case are Toeplitz matrices A, i.e., matrices whose entries are constant along each diagonal. Toeplitz matrices satisfy (4.1) with $P = J$, where

$$J = \begin{bmatrix} 0 & \cdots & 0 & 1 \\ \vdots & \cdot^{\cdot^{\cdot}} & 1 & 0 \\ 0 & \cdot^{\cdot^{\cdot}} & \cdot^{\cdot^{\cdot}} & \vdots \\ 1 & 0 & \cdots & 0 \end{bmatrix}$$

is the $N \times N$ antidiagonal identity matrix. Finally, the condition (4.1) is also fulfilled for matrices of the form

$$A = T M^{-1}, \quad P = M^{-1},$$

where T and M are real symmetric matrices and M is nonsingular. Matrices of this type arise when real symmetric linear systems

$Tx = b$ are preconditioned by M. The resulting QMR algorithm for the solution of preconditioned symmetric linear system has the same work and storage requirements as preconditioned SYMMLQ or MIN-RES [42]. However, the QMR approach is more general, in that it can be combined with any nonsingular symmetric preconditioner M, while SYMMLQ and MINRES require M to be positive definite (see, e.g., [24]). For strongly indefinite matrices T, the use of indefinite preconditioners M typically leads to considerably faster convergence; see [23] for numerical examples.

Next, we turn to transpose-free variants of the BCG method. Sonneveld [50] with his CGS algorithm was the first to devise a transpose-free BCG-type scheme. Note that, in the BCG Algorithm 3.2, the matrix A^T appears merely in the update formulas for the vectors $\tilde{r}_n$ and $\tilde{q}_n$. On the other hand, these vectors are then used only for the computation of the vector products $\rho_n = \tilde{r}_n^T r_n$ and $\sigma_n = \tilde{q}_n^T A q_n$. Sonneveld observed that, by rewriting these products, the transpose can be eliminated from the formulas, while at the same time one obtains iterates

$$x_{2n} \in x_0 + \mathcal{K}_{2n}(r_0, A), \quad n = 1, 2, \ldots, \tag{4.3}$$

that are contained in a Krylov subspace of twice the dimension, as compared to BCG. First, we consider ρ_n. From Algorithm 3.2 it is obvious that

$$r_n = \psi_n(A) r_0 \quad \text{and} \quad \tilde{r}_n = \psi_n(A^T) \tilde{r}_0, \tag{4.4}$$

where ψ_n is the nth residual polynomials of the BCG process. With (4.4), one obtains the identity

$$\rho_n = \tilde{r}_0^T \left(\psi_n(A) \right)^2 r_0, \tag{4.5}$$

which shows that ρ_n can be computed without using A^T. Similarly,

$$q_n = \varphi_n(A) r_0 \quad \text{and} \quad \tilde{q}_n = \varphi_n(A^T) \tilde{r}_0,$$

for some polynomial $\varphi_n \in \mathcal{P}_n$, and hence

$$\sigma_n = \tilde{r}_0^T A \left(\varphi_n(A) \right)^2 r_0. \tag{4.6}$$

By rewriting the vector recursions in Algorithm 3.2 in terms of ψ_n and φ_n and by squaring the resulting polynomial relations, Sonneveld showed that the vectors in (4.5) and (4.6) can be updated by means of short recursions. Furthermore, the actual iterates (4.3) generated by CGS are characterized by

$$r_{2n}^{\text{CGS}} = b - Ax_{2n} = \left(\psi_n^{\text{BCG}}(A)\right)^2 r_0. \tag{4.7}$$

Hence the CGS residual polynomials $\psi_{2n}^{\text{CGS}} = \left(\psi_n^{\text{BCG}}\right)^2$ are just the squared BCG polynomials. As pointed out earlier, BCG typically exhibits a rather erratic convergence behavior. As is clear from (4.7), these effects are magnified in CGS, and CGS typically accelerates convergence as well as divergence of BCG. Moreover, there are cases for which CGS diverges, while BCG still converges.

For this reason, more smoothly converging variants of CGS have been sought. Van der Vorst [53] was the first to propose such a method. His Bi-CGSTAB again produces iterates of the form (4.3), but instead of squaring the BCG polynomials as in (4.7), the residual vector is now of the form

$$r_{2n} = \psi_n^{\text{BCG}}(A)\chi_n(A)r_0.$$

Here $\chi_n \in \mathcal{P}_n$, with $\chi_n(0) = 1$, is a polynomial that is updated from step to step by adding a new linear factor:

$$\chi_n(\lambda) \equiv (1 - \eta_n\lambda)\chi_{n-1}(\lambda). \tag{4.8}$$

The free parameter η_n in (4.8) is determined by a local steepest descent step, i.e., η_n is the optimal solution of

$$\min_{\eta \in \mathcal{C}} \|(I - \eta A)\chi_{n-1}(A)\psi_n^{\text{BCG}}(A)r_0\|.$$

Due to the steepest descent steps, Bi-CGSTAB typically has much smoother convergence behavior than BCG or CGS. However, the norms of the Bi-CGSTAB residuals may still oscillate considerably for difficult problems. Finally, Gutknecht [30] has noted that, for real A, the polynomials χ_n will always have real roots only, even if A has complex eigenvalues. He proposed a variant of Bi-CGSTAB with polynomials (4.8) that are updated by quadratic factors in each step and thus can have complex roots in general.

In the CGS algorithm, the iterates (4.3) are updated by means of a formula of the form

$$x_{2n}^{\text{CGS}} = x_{2(n-1)}^{\text{CGS}} + \alpha_{n-1}(y_{2n-1} + y_{2n}). \tag{4.9}$$

Here the vectors $y_1, y_2, \ldots, y_{2n}$ satisfy

$$\text{span}\{y_1, y_2, \ldots, y_m\} = \mathcal{K}_m(r_0, A), \quad m = 1, 2, \ldots, 2n.$$

In other words, in each iteration of the CGS algorithm two search directions y_{2n-1} and y_{2n} are available, while the actual iterate is updated by the one-dimensional step (4.9) only. Based on this observation, Freund [14] has proposed a variant of CGS that makes use of all available search directions. More precisely, instead of one iterate x_{2n}^{CGS} per step it produces two iterates x_{2n-1} and x_{2n} of the form

$$x_m = x_0 + [\,y_1 \quad y_2 \quad \cdots \quad y_m\,]z_m, \quad z_m \in C^m. \tag{4.10}$$

Furthermore, the free parameter vector z_m in (4.10) can be chosen such that the iterates satisfy a quasi-minimal residual condition, similar to the quasi-minimization property of the QMR Algorithm 3.4. For this reason, the resulting scheme is called transpose-free quasi-minimal residual algorithm (TFQMR). For details, we refer the reader to [14], where the following implementation of TFQMR is derived.

Algorithm 4.1 (TFQMR algorithm)

0) *Choose $x_0 \in C^N$.*

   Set $w_1 = y_1 = r_0 = b - Ax_0$, $v_0 = Ay_1$, $d_0 = 0$.

   Set $\tau_0 = \|r_0\|$, $\vartheta_0 = 0$, $\eta_0 = 0$.

   Choose $\tilde{r}_0 \in C^N$, $\tilde{r}_0 \neq 0$, and set $\rho_0 = \tilde{r}_0^T r_0$.

For $n = 1, 2, \ldots,$ do :

1) *Compute*

$$\sigma_{n-1} = \tilde{r}_0^T v_{n-1},$$
$$\alpha_{n-1} = \rho_{n-1}/\sigma_{n-1},$$
$$y_{2n} = y_{2n-1} - \alpha_{n-1}v_{n-1}.$$

2) *For $m = 2n - 1, 2n$ do :*

   Compute

$$w_{m+1} = w_m - \alpha_{n-1}Ay_m,$$
$$\vartheta_m = \|w_{m+1}\|/\tau_{m-1}, \quad c_m = 1/\sqrt{1 + \vartheta_m^2},$$
$$\tau_m = \tau_{m-1}\vartheta_m c_m, \quad \eta_m = c_m^2 \alpha_{n-1},$$
$$d_m = y_m + (\vartheta_{m-1}^2 \eta_{m-1}/\alpha_{n-1})d_{m-1},$$
$$x_m = x_{m-1} + \eta_m d_m.$$

If x_m has converged, stop.

3) *Compute*

$$\rho_n = \tilde{r}_0^T w_{2n+1},$$
$$\beta_n = \rho_n/\rho_{n-1},$$
$$y_{2n+1} = w_{2n+1} + \beta_n y_{2n},$$
$$v_n = Ay_{2n+1} + \beta_n(Ay_{2n} + \beta_n v_{n-1}).$$

We would like to point out that the iterates generated by the QMR Algorithm 3.1 and the TFQMR Algorithm 4.1 are different in general.

Another transpose-free QMR method was proposed by Chan, de Pillis, and Van der Vorst [5]. Their scheme is mathematically equivalent to the QMR Algorithm 3.1, when the latter is based on the classical Lanczos process without look-ahead. The method first uses a transpose-free squared version of the Lanczos algorithm (see, e.g., Gutknecht [28]) to generate the scalar tridiagonal Lanczos matrix H_n. The right Lanczos vectors v_n are then computed by running the standard Lanczos recurrence, and finally the QMR iterates are obtained as in Algorithm 3.1. Freund and Szeto [21] have derived yet another transpose-free QMR scheme, which is modeled after CGS and is based on squaring the residual polynomials of the standard QMR Algorithm 3.1. However, the algorithm given in [5] and the squared QMR approach both require three matrix-vector products with A at each iteration, and hence they are more expensive than CGS, Bi-CGSTAB, or TFQMR, which involve only two such products per step.

Finally, we remark that none of the transpose-free methods considered in this section, except for Freund and Zha's simplified QMR algorithm based on (4.1), addresses the problem of breakdowns. Indeed, in exact arithmetic, all these schemes break down every time a breakdown occurs in the BCG Algorithm 3.2. Practical look-ahead techniques for avoiding exact and near-breakdowns in these transpose-free methods still have to be developed.

5 Concluding Remarks

In this paper, we have covered only some of the recent advances in iterative methods for non-Hermitian linear systems. A more extensive survey of recent developments in this field can be found in [17].

The introduction of CGS in the 1980s spurred renewed interest in the nonsymmetric Lanczos algorithm, with most of the effort directed towards obtaining a method with better convergence properties than BCG or CGS. Several BCG-based algorithms were proposed, such as Bi-CGSTAB, introduced by Van der Vorst [53]. The quasi-minimal residual technique was introduced by Freund [13, 15] in the context of complex symmetric systems, then later coupled with a new variant of the look-ahead Lanczos approach to obtain a general non-Hermitian QMR algorithm [19]. Finally, several transpose-free algorithms based on QMR have been introduced recently, which trade the multiplication by A^T for one or more multiplications by A. However, their convergence properties are not well understood, and none of these algorithms have been combined with look-ahead techniques yet. In general, it seems that the transpose-free methods have more numerical problems than the corresponding methods that use A^T, and more research is needed into studying their behavior.

Finally, even though the field of iterative methods has made great progress in the last few years, it is still in its infancy, especially with regard to the packaged software available. Whereas there are well-established robust general-purpose solvers based on direct methods, the same cannot be said about solvers based on iterative methods. There are no established iterative packages of the same robustness and wide acceptance as, for example, the LINPACK library, and as a result many of the scientists who use iterative methods write their own specialized solvers. We feel that this situation needs to change, and we would like to encourage researchers to provide code for their methods.

References

[1] O. Axelsson, Conjugate gradient type methods for unsymmetric and inconsistent systems of linear equations, *Linear Algebra Appl.* **29** (1980), pp. 1–16.

[2] D.L. Boley, S. Elhay, G.H. Golub, and M.H. Gutknecht, Nonsymmetric Lanczos and finding orthogonal polynomials associated with indefinite weights, *Numer. Algorithms* **1** (1991), pp. 21–43.

[3] D.L. Boley and G.H. Golub, The nonsymmetric Lanczos algorithm and controllability, *Systems Control Lett.* **16** (1991), pp. 97–105.

[4] C. Brezinski, M. Redivo Zaglia, and H. Sadok, A breakdown-free Lanczos type algorithm for solving linear systems, Technical Report ANO–239, Université des Sciences et Techniques de Lille Flandres-Artois, Villeneuve d'Ascq, France, 1991.

[5] T.F. Chan, L. de Pillis, and H.A. Van der Vorst, A transpose-free squared Lanczos algorithm and application to solving nonsymmetric linear systems, Technical Report CAM 91-17, Department of Mathematics, University of California, Los Angeles, CA, 1991.

[6] P. Concus and G.H. Golub, A generalized conjugate gradient method for nonsymmetric systems of linear equations, in *Computing Methods in Applied Sciences and Engineering* (R. Glowinski and J.L. Lions, eds.), Lecture Notes in Economics and Mathematical Systems 134, Springer, Berlin, 1976, pp. 56–65.

[7] P. Concus, G.H. Golub, and D.P. O'Leary, A generalized conjugate gradient method for the numerical solution of elliptic partial differential equations, in *Sparse Matrix Computations* (J.R. Bunch and D.J. Rose, eds.), Academic Press, New York, 1976, pp. 309–332.

[8] J. Cullum and R.A. Willoughby, A practical procedure for computing eigenvalues of large sparse nonsymmetric matrices, in *Large Scale Eigenvalue Problems* (J. Cullum and R.A. Willoughby, eds.), North-Holland, Amsterdam, 1986, pp. 193–240.

[9] A. Draux, *Polynômes Orthogonaux Formels – Applications*, Lecture Notes in Mathematics 974, Springer, Berlin, 1983.

[10] V. Faber and T. Manteuffel, Necessary and sufficient conditions for the existence of a conjugate gradient method, *SIAM J. Numer. Anal.* **21** (1984), pp. 352–362.

[11] V. Faber and T. Manteuffel, Orthogonal error methods, *SIAM J. Numer. Anal.* **24** (1987), pp. 170–187.

[12] R. Fletcher, Conjugate gradient methods for indefinite systems, in *Numerical Analysis Dundee 1975* (G.A. Watson, ed.), Lecture Notes in Mathematics 506, Springer, Berlin, 1976, pp. 73–89.

[13] R.W. Freund, Conjugate gradient type methods for linear systems with complex symmetric coefficient matrices, RIACS Technical Report 89.54, NASA Ames Research Center, Moffett Field, CA, 1989.

[14] R.W. Freund, A transpose-free quasi-minimal residual algorithm for non-Hermitian linear systems, RIACS Technical Report 91.18, NASA Ames Research Center, Moffett Field, CA, 1991.

[15] R.W. Freund, Conjugate gradient-type methods for linear systems with complex symmetric coefficient matrices, *SIAM J. Sci. Stat. Comput.* **13** (1992), to appear.

[16] R.W. Freund, G.H. Golub, and M. Hochbruck, Krylov subspace methods for non-Hermitian p-cyclic matrices, Technical Report, RIACS, NASA Ames Research Center, Moffett Field, CA, 1992.

[17] R.W. Freund, G.H. Golub, and N.M. Nachtigal, Iterative solution of linear systems, RIACS Technical Report 91.21, NASA Ames Research Center, Moffett Field, 1991, to appear in Acta Numerica.

[18] R.W. Freund, M.H. Gutknecht and N.M. Nachtigal (1991b), An implementation of the look-ahead Lanczos algorithm for non-Hermitian matrices, Technical Report 91.09, RIACS, NASA Ames Research Center, Moffett Field, CA, 1991, *SIAM J. Sci. Stat. Comput.*, to appear.

[19] R.W. Freund and N.M. Nachtigal, QMR: a quasi-minimal residual method for non-Hermitian linear systems, *Numer. Math.* **60** (1991), pp. 315–339.

[20] R.W. Freund, N.M. Nachtigal, and T. Szeto, An implementation of the QMR method based on coupled two-term recurrences, Technical Report, RIACS, NASA Ames Research Center, Moffett Field, CA, 1992.

[21] R.W. Freund and T. Szeto, A quasi-minimal residual squared algorithm for non-Hermitian linear systems, Technical Report 91.26, RIACS, NASA Ames Research Center, Moffett Field, CA, 1991.

[22] R.W. Freund and H. Zha, A look-ahead algorithm for the solution of general Hankel systems, Technical Report 91.24, RIACS, NASA Ames Research Center, Moffett Field, CA, 1991.

[23] R.W. Freund and H. Zha, Simplifications of the nonsymmetric Lanczos process and a new algorithm for solving indefinite symmetric linear systems, Technical Report, RIACS, NASA Ames Research Center, Moffett Field, CA, 1992.

[24] P.E. Gill, W. Murray, D.B. Ponceleón, and M.A. Saunders, Preconditioners for indefinite systems arising in optimization, Technical Report SOL 90-8, Stanford University, Stanford, CA, 1990.

[25] G.H. Golub and R.S. Varga, Chebyshev semi-iterative methods, successive overrelaxation iterative methods, and second order Richardson iterative methods, *Numer. Math.* **3** (1961), pp. 147–168.

[26] W.B. Gragg, Matrix interpretations and applications of the continued fraction algorithm, *Rocky Mountain J. Math.* **4** (1974), pp. 213–225.

[27] W.B. Gragg and A. Lindquist, On the partial realization problem, *Linear Algebra Appl.* **50** (1983), pp. 277–319.

[28] M.H. Gutknecht, The unsymmetric Lanczos algorithms and their relations to Padé approximation, continued fractions, and the QD algorithm, in *Proceedings of the Copper Mountain Conference on Iterative Methods*, 1990.

[29] M.H. Gutknecht, A completed theory of the unsymmetric Lanczos process and related algorithms, Part II, IPS Research Report No. 90–16, ETH Zürich, Switzerland, 1990.

[30] M.H. Gutknecht, Variants of BiCGSTAB for matrices with complex spectrum, IPS Research Report No. 91–14, ETH Zürich, Switzerland, 1991.

[31] M.H. Gutknecht, A completed theory of the unsymmetric Lanczos process and related algorithms, Part I, *SIAM J. Matrix Anal. Appl.* **13** (1992), to appear.

[32] G. Heinig and K. Rost, Algebraic methods for Toeplitz-like matrices and operators, Birkhäuser, Basel, 1984.

[33] M.R. Hestenes and E. Stiefel, Methods of conjugate gradients for solving linear systems, *J. Res. Nat. Bur. Stand.* **49** (1952), pp. 409–436.

[34] W.D. Joubert, Generalized conjugate gradient and Lanczos methods for the solution of nonsymmetric systems of linear equations, Ph.D. Dissertation, University of Texas, Austin, TX, 1990.

[35] W.D. Joubert and D.M. Young, Necessary and sufficient conditions for the simplification of generalized conjugate-gradient algorithms, *Linear Algebra Appl.* **88/89** (1987), pp. 449–485.

[36] S.-Y. Kung, Multivariable and multidimensional systems: analysis and design, Ph.D. Dissertation, Stanford University, Stanford, CA, 1977.

[37] C. Lanczos, An iteration method for the solution of the eigenvalue problem of linear differential and integral operators, *J. Res. Natl. Bur. Stand.* **45** (1950), pp. 255–282.

[38] C. Lanczos, Solution of systems of linear equations by minimized iterations, *J. Res. Natl. Bur. Stand.* **49** (1952), pp. 33–53.

[39] T.A. Manteuffel, The Tchebychev iteration for nonsymmetric linear systems, *Numer. Math.* **28** (1977), pp. 307–327.

[40] J.A. Meijerink and H.A. van der Vorst, An iterative solution for linear systems of which the coefficient matrix is a symmetric M-matrix, *Math. Comp.* **31**, 1977, pp. 148–162.

[41] N.M. Nachtigal, A look-ahead variant of the Lanczos algorithm and its application to the quasi-minimal residual method for non-Hermitian linear systems, Ph.D. Dissertation, Massachusetts Institute of Technology, Cambridge, MA, 1991.

[42] C.C. Paige and M.A. Saunders (1975), 'Solution of sparse indefinite systems of linear equations, *SIAM J. Numer. Anal.* **12** (1975), pp. 617–629.

[43] P.N. Parlett, Reduction to tridiagonal form and minimal realizations, *SIAM J. Matrix Anal. Appl.* **13** (1992), to appear.

[44] B.N. Parlett, D.R. Taylor, and Z.A. Liu, A look-ahead Lanczos algorithm for unsymmetric matrices, *Math. Comp.* **44** (1985), pp. 105–124.

[45] J.K. Reid, On the method of conjugate gradients for the solution of large sparse systems of linear equations, in *Large Sparse Sets of Linear Equations* (J.K. Reid, ed.), Academic Press, New York, 1971, pp. 231–253.

[46] Y. Saad, The Lanczos biorthogonalization algorithm and other oblique projection methods for solving large unsymmetric systems, *SIAM J. Numer. Anal.* **19** (1982), pp. 485–506.

[47] Y. Saad, Krylov subspace methods on supercomputers, *SIAM J. Sci. Stat. Comput.* **10** (1989), pp. 1200–1232.

[48] Y. Saad and M.H. Schultz, Conjugate gradient-like algorithms for solving nonsymmetric linear systems, *Math. Comp.* **44** (1985), pp. 417–424.

[49] Y. Saad and M.H. Schultz, GMRES: a generalized minimal residual algorithm for solving nonsymmetric linear systems, *SIAM J. Sci. Stat. Comput.* **7** (1986), pp. 856–869.

[50] P. Sonneveld, CGS, a fast Lanczos-type solver for nonsymmetric linear systems, *SIAM J. Sci. Stat. Comput.* **10** (1989), pp. 36–52.

[51] J. Stoer, Solution of large linear systems of equations by conjugate gradient type methods, in *Mathematical Programming – The State of the Art* (A. Bachem, M. Grötschel and B. Korte, eds.), Springer, Berlin, 1983, pp. 540–565.

[52] D.R. Taylor, Analysis of the look ahead Lanczos algorithm, Ph.D. Dissertation, University of California, Berkeley, CA, 1982.

[53] H.A. Van der Vorst, Bi-CGSTAB: A fast and smoothly converging variant of Bi-CG for the solution of nonsymmetric linear systems, *SIAM J. Sci. Stat. Comput.* **13** (1992), to appear.

[54] P.K.W. Vinsome, Orthomin, an iterative method for solving sparse sets of simultaneous linear equations, in *Proceedings of the Fourth Symposium on Reservoir Simulation*, Society of Petroleum Engineers of AIME, Los Angeles, CA, 1976.

[55] V.V. Voevodin, The problem of a non-selfadjoint generalization of the conjugate gradient method has been closed, *USSR Comput. Math. and Math. Phys.* **23** (1983), pp. 143–144.

[56] O. Widlund, A Lanczos method for a class of nonsymmetric systems of linear equations, *SIAM J. Numer. Anal.* **15** (1978), pp. 801–812.

CONVERGENCE ACCELERATION ALGORITHMS

Mohamed Hafez

Department of Mechanical Engineering
University of California, Davis
Davis, CA 95616

For practical three dimensional problems, direct solvers are not feasible (for some time to come), and one must resort to iterative methods. Techniques to reduce the number of iterations and the work required to obtain accurate solutions have always been in demand. Recent developments of modified and new techniques suitable for parallel computations are still in progress. The papers in this session present the state of the art and offer recommendations for future research directions in this important field.

Emphasis is placed on multigrid (and multilevel) techniques as well as preconditioned conjugate gradient type algorithms. The remarkable success of multigrid CFD was presented by Jameson. Unfortunately, his paper was not available at the time of publication of this monograph and the reader is referred to the literature where some of this progress has been documented.

The first paper was given by B. van Leer, who found, with cooperation of P. Roe, an amazing preconditioning matrix for Euler equations, which reduces the spread of the characteristic speeds from a factor $(M+1)/\min(M, |M-1|)$ to a factor $1/\sqrt{1 - \min(M^2, M^{-2})}$. An application to an explicit upwind scheme for a two-dimensional transonic problem is shown, where with the local time-stepping strategy, the residual is reduced seven orders of magnitude in three thousand iterations, while the characteristic time stepping requires only one thousand iterations. Hopefully, the proposed preconditioning matrix will be useful for multigrid calculations as well; by accelerating the single grid scheme and by making the single grid scheme a more reliable smoother. At any rate, the effect of preconditioning is to cluster the eigenvalues of the discrete spatial operators which should be useful for most acceleration techniques.

Jameson argued that there is a minimum number of iterations required for the solution of a nonlinear problem and his multigrid, multidimensional calculations of Euler and Navier Stokes equations converge already in very few cycles!

Next, R. Temam presented a very interesting method for elliptic boundary value problems. The new "Incremental Unknowns" method is related to the nonlinear Galerkin method, he introduced recently to study turbulence and dynamical systems. In the application to elliptic problems, a nested grid is used where new variables are introduced on the fine grid and their relations to the original variables are enforced. For one-dimensional problems, it is shown that it is possible to construct such a relation (interpolation) consistent with the discrete form of the differential operator and hence the method becomes exact, leading to a cyclic reduction procedure, similar to the multigrid method. For two-dimensional problems, the method provides a systematic way to construct preconditioned discrete equations which can be solved very efficiently with conjugate gradient type algorithms. The preconditioning matrix has a multilevel structure and depends on the relation between the new and the original variables on the fine grids. The impressive results of the test cases indicate the potential of the new method for elliptic problems, particularly because of its simplicity. On the other hand, it remains to be seen, how sensitive the method is to the choice of the interpolants. The performance of the new method for more general problems, for example transonic flows with shocks, will determine its real merits. (It may be necessary to incorporate some features of local solutions of Euler equations i.e. a wave model, in the interpolation process).

In passing, there are other ways to combine multigrid and conjugate gradient type algorithms. One can use extrapolation techniques (which are equivalent to Bi-Conjugate Gradient method) to accelerate multigrid calculations, particularly in the presence of shock waves as demonstrated by the writer a few years ago. In general, for nonlinear problems, the interaction of the inner and outer iterations is a growing subject of great interest to CFD and it deserves more attention in the 90's.

Y. Saad studied preconditioning techniques for GMRES and their implementation on parallel computers. Incomplete LU decomposition (with more fill in) looses on parallelism since they are more sequential in nature. Saad recommends multicolor SSOR on reduced systems and an example is shown for an elliptic problem to demonstrate the effectiveness of the proposed strategy. Alternatives are also discussed, for example the use of direct solvers, for small blocks in a domain decomposition approach, may be preferred.

Finally, a review of recent advances in conjugate gradient type methods for nonsymmetric systems is given by Freund, Golub and Nachtical. The two main features of interest are the minimization property and the recursive relations for updating the algorithm parameters. Only for special classes of nonsymmetric matrices, conjugate gradient type algorithms with these two features are available. While the numerous algorithms have been developed with emphasis on the minimization property (e.g. Orthomin and GMRES), work per iteration and overall storage requirements grow linearly with the iteration number for all these algorithms. "Consequently, in practice one cannot afford to run the full version of these algorithms, and it is necessary to use restarts. For difficult problems, this often results in very slow convergence." The second category of Conjugate-gradient type methods share the second feature but not the first as in the Bi-Conjugate Gradient method. These algorithms can exhibit irregular convergence behavior and breakdowns may occur. Moreover, in addition to matrix-vector products, multiplications with the transpose of the system matrix is required. This disadvantage has been removed in the Conjugate Gradient Squared methods and several transpose-free schemes have been developed.

The paper discusses the so-called look-ahead variants of the Lanczos process which remedy the breakdown problem. Also, the quasi-minimal residual technique, QMR, introduced by Freund is coupled with a variant of the look-ahead Lanczos approach to obtain a general non-Hermitian QMR algorithm. Transpose-free algorithms based on QMR have also been introduced. However, their convergence properties are not well understood, and none of these algorithms have been combined with look-ahead techniques yet. The authors conclude with an important remark concerning software packages. "Whereas there are well-established robust general-purpose solvers based on direct methods, for example the LINPACK Library, the same cannot be said about solvers based on iterative methods."

Hopefully, this session will stimulate even more fruitful ideas and research in convergence acceleration techniques.

SPECTRAL AND HIGHER-ORDER METHODS

SPECTRAL METHODS
FOR VISCOUS, INCOMPRESSIBLE FLOWS

Claudio Canuto

Dipartimento di Matematica, Politecnico di Torino
10129 TORINO, Italy,
and Istituto di Analisi Numerica del C.N.R., 27100 PAVIA, Italy

ABSTRACT

Recent results and future trends in the numerical analysis and implementation of spectral methods for the incompressible Navier-Stokes equations are discussed. The issues of discrete representation of functions, Stokes solvers, temporal discretization and resolution of flow structures are addressed.

1. Introduction

Spectral methods have reached in the 80's a level of high performance, reliability and flexibility in the accurate simulation of incompressible flows at the laminar, transitional and turbulent regimes. The early success of spectral methods drove attention on them from an incresing number of numerical analysts. Proficuous and deep interactions between numerical analysts and spectral method practitioners have remarkably increased in the second half of the decade, with feed-back in both ways.

In this paper, we will address those aspects of spectral methods for the incompressible Navier-Stokes equations, which are more supported by a sound mathematical analysis. Clear mathematical insight can help in designing new algorithms, understanding the performance of existing algorithms and improving them, and, last but not least, debugging codes.

If one wants to design a spectral algorithm for incompressible problems, one has to make appropriate choices about the following basic issues: 1) discrete representation of functions, and mesh(es); 2) Stokes solvers; 3) temporal discretization; 4) resolution of structures. Considerable progress has been made in the 80's on each of these topics, and particularly on points 1) and 2). Further research in the

90's is expected to improve the efficiency of spectral methods in those problems for which they are currently in use, as well as extend their range of applicability. In the next sections, we will briefly address each of the previously mentioned basic issues.

2. Discrete representation of functions, and mesh(es)

The core of a spectral method (what is usually referred to as a *spectral brick*, or *spectral element*) is characterized by the following features: the computational domain is a tensor product of intervals; the trial (and test) functions are tensor products of one-dimensional truncated orthogonal expansions; a double representation of functions co-exists, namely, "Fourier" coefficients with respect to the orthogonal basis and grid-values at one or more tensor-product Gaussian grids.

In the early ages of spectral methods, the physical domain of interest was invariably covered by (the image of) one single spectral brick; increase in resolution came from truncating the orthogonal expansion at higher and higher levels. Unfortunately, this strategy could be applied to very simple geometries only. During the 80's, there has been a flourishing of efficient methods for decomposing the physical domain into (the images) of many spectral bricks, in order to satisfy geometric constraints and/or enhance local resolution. So, in recent years, the philosophy of spectral methods has become closer and closer to that of finite element methods. At the beginning of the 80's, conventional (single domain, highest order) spectral methods were deeply influenced by finite difference ideas; a wide gap existed between spectral methods and conventional (low order) finite element methods, only partially mitigated by the birth of early ideas on subdomain patching and overlapping from one side [Orszag (1980)], and by the p-version of the finite element method from the other side [Babuska, Szabo and Katz (1981)]. Presently, at the beginning of the 90's, the influence of finite difference ideas on spectral methods is confined to specific problems; conversely, the gap with finite elements has been remarkably reduced by the development of new domain decomposition strategies, such as variational or integral interface conditions, finite element subdomain matching, spectral element methods, and the h-p version of the finite element methods.

Subdomain matching is a crucial aspect of any spectral method based on domain decomposition (we refer to Ch.13 in Canuto, Hussaini, Quarteroni and Zang (1990) for a complete discussion on spectral domain decomposition methods). *Patching conditions* amount to collocating the equations that glue together contiguous subdomains, at selected interface grid points. These were the first matching conditions to be used, as they are quite simple at least for standard approximations. However, they are difficult to extend to nontrivial situations (such as non-aligned grids, and cross-points); furthermore, they may reduce the convergence rate of iterative solvers. *Integral conditions* are derived by flux-balance enforcement across interfaces. They have been proposed by Macaraeg and Streett (1986), who successfuly applied them to convection-dominated problems. *Variational conditions* are based on integration-by-parts arguments. They lead to a better conditioning of the corresponding algebraic systems, and allow for more flexible matchings. Furthermore, their theoretical analysis is easier than that for patching conditions. Among the methods based on variational matching conditions, let us mention the well-celebrated *Spectral Element methods* introduced by Patera (1984), and the more recently developed *Mortar* (and *Sliding*) *Element methods* (Bernardi, Maday and Patera (1989)).

The classical variational formulations presume that the weight function defining the orthogonal system be the Legendre weight: indeed, this is the only Jacobi weight which is not identically 0 or ∞ on the boundary. This prevents the use of Fast transforms methods in variable coefficient problems. In fact, spectral element methods presume a moderate degree of the polynomials in the expansion within each subdomain. A viable remedy is to enforce the subdomain matching via a *finite element preconditioning* [Deville and Mund (1885), Canuto and Pietra (1990)]: this approach allows spectral residuals to be computed at any set of collocation nodes in the subdomains, including Chebyshev nodes for which FFT is available. Hereafter, we will sketch some details about this method.

Subdomain matching via finite elements

Let us first recall the basic ideas of single-domain finite element preconditioning. Consider the mixed Dirichlet-Neumann boundary value problem

$$(2.1) \qquad \begin{cases} -\nabla \cdot \mathbf{F}(\nabla u) = f & \text{in } \Omega = (-1,1)^2, \\ \mathbf{F}(\nabla u) \cdot \mathbf{n} = g & \text{on } \Gamma \subset \partial\Omega, \\ u = 0 & \text{on } \partial\Omega - \Gamma, \end{cases}$$

where $\mathbf{F} : R^2 \to R^2$ is a smooth function such that $\mathbf{F}(\xi) \cdot \xi \geq \alpha|\xi|^2$ for all $\xi \in R^2$, and $\mathbf{n}$ is the outward normal to $\partial\Omega$.

If we define the residuals $R(u) = f - \nabla \cdot \mathbf{F}(\nabla u)$ in Ω, and $r(u) = g - \mathbf{F}(\nabla u) \cdot \mathbf{n}$ on $\Gamma \cap \partial\Omega$, we are led to introduce the preconditioned form of problem (2.1):

$$(2.2) \qquad \begin{cases} -\nabla \cdot \mathbf{F}(\nabla w) = R(u) & \text{in } \Omega, \\ \mathbf{F}(\nabla w) \cdot \mathbf{n} = r(u) & \text{on } \Gamma \subset \partial\Omega, \\ w = 0 & \text{on } \partial\Omega - \Gamma. \end{cases}$$

Note that $w \equiv 0$ if and only if u solves (2.1).

Now, the idea is to approximate (2.2) by computing the residuals on the left hand side by a spectral expansion, while solving the system on the right hand side by low order finite elements. More precisely, let us look for an approximation $u_N \in P_N(\Omega)$ of u by computing the previous residuals at the nodes of a tensor-product Gaussian grid $\mathcal{G}_N \subset \bar{\Omega}$. Next, let us introduce the space V_h of finite element functions on $\bar{\Omega}$, i.e., $V_h = V_h(\Omega) = \{v \in C^0(\bar{\Omega}) \mid v \text{ is bilinear between four neighboring nodes of } \mathcal{G}_N\}$, and the space T_h of the finite element functions on each side of Γ, i.e., $T_h = T_h(\Gamma) = \{v \in C^0_{side}(\bar{\Gamma}) \mid v \text{ is linear between two neighboring nodes of } \mathcal{G}_N \cap \Gamma\}$. Let $I_h : C^0(\bar{\Omega}) \to V_h$, resp., $J_h : C^0_{sides}(\bar{\Gamma}) \to T_h$, denote the finite element interpolation operator in $\bar{\Omega}$, resp. on $\mathcal{G}_N \cap \bar{\Gamma}$.

The discrete version of (2.2) will be as follows:

Given $u_N \in P_N(\Omega)$, *find* $w_h \in V_h$ *such that*

(2.3)
$$\int_\Omega \mathbf{F}(\nabla w_h) \cdot \nabla z_h = \int_\Omega I_h(R(u_N))z_h + \int_\Gamma J_h(r(u_N))z_h, \qquad \forall z_h \in V_h.$$

Note that

$$w_h \equiv 0 \iff \int_\Omega I_h(R(u_N))z_h + \int_\Gamma J_h(r(u_N))z_h = 0, \qquad \forall z_h \in V_h.$$

A suitable choice of test functions allows one to prove that the latter condition is in turn equivalent to enforcing $R(u_N) = 0$ at $\mathcal{G}_N \cap \Omega$ and a linear combination of $r(u_N) = 0$ at $\mathcal{G}_N \cap \bar{\Gamma}$.

If we write (2.3) in matrix form as $\mathcal{L}_h w_h = L_N u_N - d$, we are led to drive w_h to 0 by an iterative technique such as Richardson or conjugate gradient methods. The convergence properties of these schemes are related to the behavior of the eigenvalues of $\mathcal{L}_h^{-1} L_N$. Fraenken, Deville and Mund (1990) show that for a Dirichlet problem they lie in the interval [.69, 1.] of the real axis, whereas if finite differences are used as preconditioners the related eigenvalues lie in the interval $[1., \pi^2/4]$ (see Fig.1 in their paper). Furthermore, Canuto and Pietra (1990) produce evidence that the convergence behavior of Richardson iterations is not sensibly affected by the presence of Neumann boundary conditions (see Fig.6 in their paper).

Let us now go back to the problem of subdomain matching. For simplicity, we assume that the domain Ω is decomposed into two subdomains Ω^1 and Ω^2, with interface $\Gamma = \bar{\Omega}^1 \cap \bar{\Omega}^2$. The Dirichlet model problem

(2.4)
$$\begin{cases} -\nabla \cdot \mathbf{F}(\nabla u) = f & \text{in } \Omega \\ u = 0 & \text{on } \partial\Omega \end{cases}$$

is equivalent to

(2.5)
$$\begin{cases} -\nabla \cdot \mathbf{F}(\nabla u^k) = f & \text{in } \Omega^k \ (k = 1, 2) \\ u^k = 0 & \text{on } \partial\Omega^k \cap \partial\Omega \ (k = 1, 2) \\ u^1 = u^2 & \text{on } \Gamma \\ \mathbf{F}(\nabla u^1) \cdot \mathbf{n}^1 + \mathbf{F}(\nabla u^2) \cdot \mathbf{n}^2 = 0 & \text{on } \Gamma. \end{cases}$$

Thus, we are looking for two functions u^1 and u^2 (satisfying the boundary conditions and with equal value on Γ) which annihilate the interior residuals $R(u^k) = f - \nabla \cdot \mathbf{F}(\nabla u^k)$ in Ω^k $(k = 1, 2)$, and the interface residual $\mathcal{N}(u^1, u^2) = \mathbf{F}(\nabla u^1) \cdot \mathbf{n}^1 + \mathbf{F}(\nabla u^2) \cdot \mathbf{n}^2$ on Γ. On the discrete level, let $\mathcal{G}_N^k$ (k=1,2) be two grids in $\bar{\Omega}^k$ with "conformal" matching, and define the space $S_N = \{v \in C^0(\bar{\Omega}) \mid u_{|\Omega^k} \in P_N(\Omega^k), \ k = 1, 2\}$; furthermore, define the finite element space $V_h = \{w \in C^0(\bar{\Omega}) \mid w_{|\Omega^k} \in V_h(\Omega^k), \ k = 1, 2\}$ and the finite element interpolants $I_h^k : C^0(\bar{\Omega}^k) \to V_h(\Omega^k)$ on the grids $\mathcal{G}_N^k$ (k=1,2), and $J_h : C_{sides}^0(\bar{\Gamma}) \to T_h$ on $\mathcal{G}_N \cap \bar{\Gamma}$).

The discrete preconditioned form of (2.5) will be as follows: Given $u_N \in S_N$, find $w_h \in V_h$ such that

$$\int_\Omega \mathbf{F}(\nabla w_h) \cdot \nabla z_h = \int_{\Omega^1} I_h^1(R(u_N^1)) \, z_h + \int_{\Omega^2} I_h^2(R(u_N^2)) \, z_h$$

$$+ \int_\Gamma J_h(\mathcal{N}(u_N^1, u_N^2)) \, z_h, \qquad \forall z_h \in V_h.$$

Then, w_h is identically zero if and only if

$$\int_{\Omega^k} I_h^k(R(u_N^k)) \, z_h = 0, \qquad \forall z_h \in V_h(\Omega^k) \ (k = 1, 2),$$

$$\int_\Gamma J_h(\mathcal{N}(u_N^1, u_N^2)) \, z_h = 0, \qquad \forall z_h \in V_h.$$

It can be shown that these conditions are equivalent to

$$
\begin{cases}
R^k(u_N) = 0 & \text{on } \mathcal{G}_N \cap \Omega^k \ (k=1,2) \\
a \ linear \ combination \ of \ \mathcal{N}(u_N^1, u_N^2) = 0 & \text{on } \mathcal{G}_N \cap \Gamma.
\end{cases}
$$

In matrix form, the previous problem can be written as

$$(2.6) \quad S_h w_h = M_{h,1} \cdot R(u_N^1) + M_{h,2} \cdot R(u_N^2) + \mathcal{M}_{h,\Gamma}(\mathcal{N}(u_N^1, u_N^2)),$$

where S_h denotes the global stiffness matrix, and the M_h denote the mass matrices. Again w_h can be driven to zero by a suitable iterative scheme. Each iteration requires three independent stages: *i)* compute the spectral residuals $R(u_N^1), R(u_N^2), \mathcal{N}(u_N^1, u_N^2)$ at the grid nodes in Ω^1, Ω^2 and Γ; *ii)* multiply them by interior and interface mass-matrices to get the right-hand side of (2.6); *iii)* solve the finite element system. Note that the enforcement of the spectral collocation equations requires a simple modification of the right-hand side of a standard finite element method. The finite element system can be solved (in exact or approximate way) using the "best available" finite element solver, independently of the underlying spectral grid. According to Canuto and Pietra (1990), the discrete problem (2.6) enjoys good conditioning properties, even when cross points are present (see Figure 9 in their paper). Furthermore, the domain decomposition method here described leads to a consistent, spectrally accurate approximation of (2.4) (see Table II in the above reference).

3. Stokes solvers

In this section, we will discuss the approximation by spectral methods of the Stokes problem in primitive variables (velocity and pressure). Other formulations of the Stokes problem have been considered for spectral discretizations. Let us mention here the stream-function (and vorticity/stream-function) approach, intensively used by R.Peyret and his coworkers (see, e.g., Vanel, Peyret and Bontoux (1985), Ehrenstein (1986)) and theoretically investigated by Bernardi, Coppoletta, Girault and Maday (1989) and Bernardi, Girault and Maday (1990); and the so-called Galerkin approach, based on a divergence-free velocity basis, introduced by Leonard and Wray (1982) and analyzed by Pasquarelli, Quarteroni and Sacchi Landriani (1987). Here, we prefer to discuss the velocity/pressure formulation only, because we believe that in 90's the interest of numerical simulation will be more and more directed towards 3D problems in non-simple geometries, for which the other formulations are less applicable.

3.1 Compatibility conditions between discrete velocity and pressure

Let us consider the following Helmholtz-Stokes Problem with no-slip boundary conditions in a bounded domain $\Omega \subset R^2$:

$$(3.1) \qquad \begin{cases} \gamma u + \Delta u + \nabla p = f & \text{in } \Omega \\ \nabla \cdot u = 0 & \text{in } \Omega \\ u = 0 & \text{on } \partial\Omega. \end{cases}$$

The zeroth-order term in the momentum equation accounts for a possible discretization in time ($\gamma \geq 0$). We are looking for a solution $u \in V = \{(u_1, u_2) \mid u_i \in H_0^1(\Omega), \ i = 1, 2\}$ and $p \in Q = \{p \in L^2(\Omega) \mid \int_\Omega p \, dx = 0\}$. Introducing the bilinear forms $e(u, v) = \int_\Omega (\gamma uv + \nabla u \nabla v) : V \times V \to R$ and $d(v, q) = -\int_\Omega (\nabla \cdot v)q : V \times Q \to R$, and the linear form $(f, v) = \int_\Omega f v : V \to R$, we can formulate problem (3.1) in the variational form:

$$(3.2) \qquad \begin{cases} u \in V, \qquad p \in Q \\ e(u, v) + d(v, p) = (f, v), & \forall v \in V, \\ d(u, q) = 0, & \forall q \in Q. \end{cases}$$

Let us now define a discrete Stokes approximation of Legendre Galerkin or Collocation type. To this end, let us decompose Ω into spectral bricks, and set $\mathcal{P}_N = \{q \mid q \in P_N \text{ on each spectral brick in } \Omega\}$. All the discrete functions we are dealing with will be in $\mathcal{P}_N$. Let us introduce finite dimensional subspaces $V_N \subset V$ and $Q_N \subset Q$, and (in the case of collocation) approximations of the bilinear and linear forms $e_N(u, v) \cong e(u, v) : V_N \times V_N \to R$, $d_N(v, q) \cong d(v, q) : V_N \times Q_N \to R$ and $(f, v)_N \cong (f, v) : V_N \to R$ based on Gaussian quadratures. Finally, consider the following discrete approximation of (3.2):

$$(3.3) \qquad \begin{cases} u_N \in V_N, \qquad p_N \in Q_N \\ e_N(u_N, v) + d_N(v, p_N) = (f, v)_N, & \forall v \in V_N, \\ d_N(u, q) = 0, & \forall q \in Q_N. \end{cases}$$

This system can be written in matrix form as

$$(3.4) \qquad \begin{cases} E_N \underline{u} + G_N \underline{p} = \underline{f}_N \\ D_N \underline{u} = \underline{0} \end{cases}$$

where $E_N = E_N(\gamma)$ is the viscosity matrix, G_N is the (discrete) gradient matrix and D_N ($\equiv G_N^T$) is the (discrete) divergence matrix. Necessary and sufficient conditions for stability and consistency of the approximation (3.3) were given by Brezzi (1974); they are known as *Inf-Sup conditions*. Sufficient conditions are: *i)* there exists a constant $\alpha = \alpha_N$ such that

$$(3.5) \qquad e_N(v, v) \geq \alpha_N \|v\|_V^2, \qquad \forall v \in V_N;$$

ii) there exists a constant $\beta = \beta_N$ such that

$$(3.6) \qquad \sup_{v \in V_N} \frac{d_N(v, q)}{\|v\|_V} \geq \beta_N \|q\|_Q, \qquad \forall q \in Q_N.$$

Inf-Sup constants are of interest both to theoretical numerical analysts and to practitioners. Indeed, they influence the convergence of the discrete solution to the exact solutions of the Stokes problem as $N \to \infty$, as well as the conditioning of algorithms for solving the discrete Stokes problem for a fixed N. Two examples: *i)* the approximation error of the discrete problem is related to the best approximation error in the discrete subspaces via the inf-sup constants, as follows:

$$\|u - u_N\|_V \leq (1 + \frac{C}{\alpha_N}) \inf_{w_N \in K_N} \|u - w_N\|_V + \frac{C}{\alpha_N} \inf_{q_N \in Q_N} \|p - q_N\|_Q$$

$$\|p - p_N\|_Q \leq \frac{C}{\beta_N}(1 + \frac{C}{\alpha_N}) \inf_{w_N \in K_N} \|u - w_N\|_V$$
$$+ \frac{C}{\alpha_N \beta_N} \inf_{q_N \in Q_N} \|p - q_N\|_Q,$$

where $K_N = \{v \in V_N \mid d_N(v, q) = 0, \quad \forall q \in Q_N\}$. Thus, "small" inf-sup constants indicate that the corresponding discrete method is not optimal. *ii)* The eigenvalues of the discrete Uzawa operator $U_N = D_N E_N^{-1} G_N : Q_N \to Q'_N$ (by which the Stokes problem can be reduced to a scalar problem for the pressure) satisfy the estimate

$$\frac{\lambda_{max}}{\lambda_{min}} \simeq \frac{C}{\alpha_N \beta_N^2};$$

again, "small" inf-sup constants indicate a poorly conditioned operator, hence, a slow convergence of any iterative method used to solve the discrete Uzawa problem (see next sub-section for more detail).

The previous examples show how important is to investigate the asymptotic behavior of the inf-sup constants as $N \to \infty$. We anticipate that, for most of the existing schemes, $\alpha_N = \alpha$ is independent of N, whereas β_N does depend on N. The asymptotic behavior of β_N depends on the interaction of the pressure modes with the velocity field. More precisely, let us fix an element $q \in \mathcal{P}_N$, and let us try to satisfy

$$\beta_N(q) \overset{def}{=} \sup_{v \in V_N} \frac{d_N(v, q)}{\|v\|_V \|q\|_Q} > 0.$$

Note that $\beta_N = inf_{q \in Q_N} \beta_N(q)$. Three situations are possible:
i) $\beta_N(q) = 0$. Such pressure modes, known as *spurious modes*, must be rules out of Q_N in order to satisfy the inf-sup condition (3.6).
ii) $\beta_N(q) \cong C N^{-\sigma}$ as $N \to \infty$ for some $\sigma > 0$. These modes are termed *weakly spurious modes* after Vandeven (1989), who first investigated them. If Q_N contains such modes, then $\beta_N \to 0$ as $N \to \infty$.
iii) $\beta_N(q) = O(1)$ as $N \to \infty$. We will term these modes the *essential modes*; clearly, they are the "good" modes with respect to the inf-sup condition (3.6).

In conclusion, one can think of the pressure space as decomposed into spurious modes, weakly-spurious ones and essential ones, i.e. (with obvious notation),

$$\mathcal{P}_N = \mathcal{S}_N \oplus \mathcal{WS}_N \oplus \mathcal{E}_N.$$

The truly computed discrete pressure p_N will belong to a subspace $Q_N \subset \mathcal{P}_N$ which will satisfy $Q_N \cap \mathcal{S}_N = 0$ (in order to match the inf-sup condition), and, in most cases, $Q_N \subseteq \mathcal{WS}_N \oplus \mathcal{E}_N$. Note that the larger is Q_N, the better are the approximation results (e.g., exact divergence free velocity), at the expense of the existence of weakly-spurious modes; conversely, the smaller is Q_N the worse are the approximation results (e.g., non divergence free velocity), but the inf-sup constant β_N will be independent of N.

Before considering some examples, we want to point out that the discrete variational formulation (3.3) is the simplest framework in which to cast approximations of spectral type. However, it does not include several important cases, such as Chebyshev collocation schemes, tau methods and even Legendre collocation schemes in cylindrical coordinates (see Azaies and Vandeven (1991)). Basically, the reason is that in these cases the discrete gradient operator G_N is not the adjoint (up to a sign) of the discrete divergence operator D_N. In such cases, one has to resort to a generalized saddle-pont formulation, which has been studied by Nicolaides (1982) and Bernardi, Canuto and Maday (1988). The well-posedness of the new problem is ensured by *generalized inf-sup conditions*, which lead to a pressure decomposition similar to the one discussed above. The generalized saddle-point formulation is also useful in the analysis of approximations to the generalized Stokes problem for a compressible flow (see Bernardi, Laval, Metivet and Thomas (1990). In the coming years, we foresee a development of new, efficient spectral schemes for the compressible Navier-Stokes equations: the numerical analysis of these methods should take advantage of the tool represented by the generalized inf-sup conditions.

3.2 Examples (Legendre Collocation methods)

Let us illustrate the pressure decomposition for several popular collocation schemes. In the sequel, GL_N, resp., G_N, will denote the (N+1)-node Gauss-Lobatto, resp., Gauss, grid on the reference interval (-1,1).

A) <u>Channel Flow</u>

Let $\Omega = (0, 2\pi) \times (-1, 1)$ and assume periodicity in the x-direction. Look for $u_N \in (T_N \otimes P_N)^2$ and $p_N \in T_N \otimes P_N$, where $T_N = span\{e^{ikx} \mid |k| \leq N\}$; collocate the x-Fourier transform of momentum and continuity equations at the $F_N \otimes GL_N$ grid, where F_N is the (2N+1)-node Fourier grid on $(0, 2\pi)$. This method yields $\nabla \cdot u_N \equiv 0$ in Ω. Since $d_N(\hat{v}, \hat{q}) = \sum_{\xi \in GL_N} [\hat{q}'(\xi)\hat{v}(\xi) + ik\hat{q}(\xi)\hat{v}(\xi)] \omega_\xi$, and ξ are the zeroes of $L'_N(y)$ or $\xi = \pm 1$, there is one spurious mode, precisely $\mathcal{S}_N = span\{L_N(y)\}$. Concerning the weakly spurious modes, Vandeven (1989) proved that, if $q = e^{ikx}L_N(y)$, then

$$\sup_{v \in V_N} \frac{d_N(v, q)}{\|v\|\|v\|\|q\|_Q} \cong (\frac{k}{N})^{1/2}.$$

Hence, $\mathcal{WS}_N \subset \{e^{ikx}L_N(y) \mid 1 \leq |k| \leq N\}$ and $\beta_N = O(N^{-1/2})$. Note that, for a *fixed* N, it is not possible to decide which one of the modes $q = e^{ikx}L_N(y)$ is a weakly spurious mode, and which one is an essential mode. In other words, the concept of weakly-spurious mode is meaningful only in an asymptotic sense; this makes it difficult to filter out weakly-spurious modes in an "optimal way". A drastic remedy consists of removing all the candidates to become weakly-spurious modes, i.e., the pressure space can be restricted to $M_N =$

$T_N \otimes P_{N-1}$. This is equivalent to a *staggered grid* method [Malik, Zang and Hussaini (1985), Bernardi, Maday and Métivet (1987a)], in which the momentum equation is collocated at the $F_N \otimes GL_N$ grid, and the continuity equation is collocated at the $F_N \otimes G_{N-1}$ grid. We get $\beta_N = O(1)$ (i.e., no spurious modes), but the price to be paid is that now $\nabla \cdot u_N \not\equiv 0$ in Ω.

B) The (P_N, P_N)method

First, let us consider the square cavity $\Omega = (-1,1)^2$. Look for $u_N \in (P_N(\Omega))^2$ and $p_N \in P_N(\Omega)$; collocate the momentum and continuity equations at the $GL_N \otimes GL_N$ grid. The method yields $\nabla \cdot u_N \equiv 0$ in Ω. Bernardi, Maday and Métivet (1987) characterized the spurious modes as

$$\mathcal{S}_N = span\{L_N(x), L_N(y), L_N(x)L_N(y), L_N'(x)(1 \pm x)L_N'(y)(1 \pm y)\}.$$

Thus, $dim\ \mathcal{S}_N = 7$. Such a dimension holds also for a variety of boundary conditions on the velocity, and for different internal constraints [Azaies and Labrosse (1991)]. In 3D problems, $dim\ \mathcal{S}_N = 12N + 3$. To our knowledge, the weakly-spurious modes have not yet been characterized; however, Bernardi, Maday and Metivet (1987b) proved that $\beta_N = O(N^{-1})$.

The extension of this method to the case of a domain Ω decomposed into K subdomains was studied by Sacchi-Landriani and Vandeven (1989). Now, u_N and p_N are continuous in Ω and satisfy $u_N \in (P_N(\Omega^k))^2$ and $p_N \in P_N(\Omega^k)$ for $k = 1, .., K$. The momentum and continuity equations are collocated at the $GL_N \otimes GL_N$ grid in Ω^k, and a variational matching condition based on the continuity of the normal stress tensor is enforced on the interfaces. The inf-sup constant β_N behaves again as $O(N^{-1})$. However, the characterization of the spurious modes is quite involved: in the simplest case of M aligned bricks, one can prove that $dim\ \mathcal{S}_N = 4M + 3$; but, if the bricks are not aligned, the compatibility conditions become quite cumbersome. An implicit filtering of the spurious modes is possible by the finite element preconditioning (see the next subsection).

C) The (P_{N+1}, P_{N-1})method [Bernardi and Maday (1986)]

On the square cavity $\Omega = (-1,1)^2$, look for $(u_N)_1 \in P_N(x) \otimes P_{N+1}(y)$, $(u_N)_2 \in P_{N+1}(x) \otimes P_N(y)$, and $p_N \in P_{N-1}(\Omega)$. Setting $\widetilde{G}_{N+1} = G_{N-1} \cup \{\pm 1\}$, collocate the x-momentum equation at the $GL_N \otimes \widetilde{G}_{N+1}$ grid, the y-momentum equation at the $\widetilde{G}_{N+1} \otimes GL_N$

grid, and the continuity equation at the $GL_{N-1} \otimes GL_{N-1}$ grid. There are no spurious modes, but $\nabla \cdot u_N \not\equiv 0$ in Ω. Weakly spurious mode are present; indeed, $\beta_N = O(N^{-1})$. Unlike the previous schemes, now the ellipticity constant α_N is not independent of N, indeed $\alpha_N = O(N^{-1})$. The main draw-back of the method is that it requires expensive interpolation procedures among the three grids.

D) The (P_N, P_{N-2})method [Maday and Patera (1987)]

On the square cavity $\Omega = (-1, 1)^2$, look for $u_N \in (P_N(\Omega)^2)$ and $p_N \in P_{N-2}(\Omega)$; collocate the momentum equation at the $GL_N \otimes GL_N$ grid, and the continuity equation at the $GL_{N-2} \otimes GL_{N-2}$ grid. Recently, Azaies and Labrosse (1991) proposed a single-grid version of the method, in which the enforcement of the continuity equation requires only values at the $GL_N^\star \otimes GL_N^\star$ grid, where $GL_N^\star = GL_N - \{\pm 1\}$. There are no spurious modes, but $\nabla \cdot u_N \not\equiv 0$ in Ω. Weakly spurious mode are present, leading to an inf-sup constant β_N for which the bound $\beta_N \geq O(N^{-1/2})$ has been proven by Maday and Patera (1987); numerical computations by Azaies and Coppoletta (1991) indicate that $\beta_N \cong O(N^{-1/8})$.

The extension of the method to domains decomposed into subdomains is quite easy, via variational interface conditions. The absence of spurious modes does not require any pressure filtering. The (P_N, P_{N-2}) basis is precisely the core of the *Spectral Element method* for the Stokes problem.

E) Pressure stabilization via bubble functions [Canuto (1992)]

In order to satisfy the inf-sup condition, a dual action (with respect to restricting the pressure space) consists of enlarging the velocity space. Arnold, Brezzi and Fortin (1984) showed that any finite element scheme for the Stokes problem can be stabilized by adding enough functions with support on one element (*bubble functions*) to the velocity space. The same idea can be applied in the framework of spectral methods. Let V_N, Q_N a couple of velocity-pressure spaces which do not satisfy the inf-sup condition; let $\mathcal{B}_N$ be a space of C^2 velocities with local support on the quadrangle defined by four adjacent grid-points on the $GL_N \otimes GL_N$ grid, and set

$$\tilde{V}_N = V_N \oplus \mathcal{B}_N.$$

Then, under suitable assumptions, the couple $\tilde{V}_N$, Q_N does satisfy the inf-sup condition, i.e., no spurious pressure modes are present.

The bubble functions can be eliminated in the solution process (e.g., by local static condensation). Although the elements of $\mathcal{B}_N$ are not infinitely differentiable functions, the spectral accuracy preserved, since $\tilde{V}_N \subset V_N$, hence,

$$\|u - u_N\| \le C \inf_{w \in \tilde{V}_N} \|u - w\| \le C \inf_{w \in V_N} \|u - w\|.$$

3.3 Some algorithmic aspects

We will now briefly discuss how the presence of spurious and weakly spurious pressure modes can influence some of the algorithms used for the solution of the discrete Stokes system.

1) Uzawa's method

This algorithm, well-known in the framework of finite order methods, was first used in conjunction with spectral methods by Maday, Patera and Ronquist (1986). The discrete system (3.4) can be written as

$$\begin{cases} E_N \underline{u} = \underline{f}_N - G_N \underline{p} \\ U_N \underline{p} = D_N \underline{f}_N, \end{cases}$$

where $U_N = D_N E_N^{-1} G_N$ is the Uzawa operator. The second equation can be solved for the pressure by an iterative method (e.g., conjugate gradient), next the velocity is obtained from the first equation.

The rate of convergence of the iterative scheme depends on the condition number of the Uzawa operator. In the pure Stokes case (i.e., $\gamma = 0$ in (3.1)), $E_N = E_N(0) \cong D_N G_N$, hence $U_N \cong$ a zeroth order differential operator. So, one would expect $cond(U_N) \cong 1$. However, weakly spurious modes affect the conditioning, indeed, $\lambda_{max}(U_N) = O(\alpha_N^{-1})$ and $\lambda_{min}(U_N) = O(\beta_N^{-2})$. Thus, $cond(U_N) \to 0$ as $N \to \infty$, whenever weakly-spurious modes are present. If β_N decays slowly and N is not too large, a diagonal preconditioning suffices (Patera). In the Helmholtz-Stokes case (i.e., a large γ in (3.1)), $E_N(\gamma) \cong \gamma I$, hence, $U_N \cong \gamma^{-1} D_N G_N$ and $cond(U_N) \cong O(N^4)$. So, preconditioning U_N is a must. Streett and Hussaini (1989) propose a Multigrid/Conjugate Gradient algorithm, based on a Full Approximation (FAS) multigrid scheme, few conjugate gradient iterations as relaxation, and the direct inversion of the Uzawa operator on the coarsest grid. They report that this method is more expensive (per time step) than a splitting method, but it requires less time steps to achieve the same accuracy on a fixed time interval.

2) Influence Matrix method
    This method is the core of the well-known algorithm proposed by Kleiser and Schumann (1980) for the simulation of Poiseuille flows in a infinite slab. The condition $D_N \underline{u} = \underline{0}$ in (3.4) is replaced by the *equivalent* boundary value problem

$$\begin{cases} E_N D_N \underline{u} = \underline{0} \\ B_N D_N \underline{u} = \underline{0}, \end{cases}$$

where B_N is a suitable boundary operator. In turns, this is converted into the boundary value problem for the pressure

$$\begin{cases} D_N G_N \underline{p} = D_N \underline{f}_N \\ B_N \underline{p} = \varphi. \end{cases}$$

The pressure boundary data φ are chosen in order to enforce $B_N D_N \underline{u} = \underline{0}$, by solving a linear system of the form

$$C_N \varphi = Z_N \underline{f}_N.$$

C_N is termed the *influence* or *capacitance matrix*. If we define the lifting operator B_N^{-1} by $B_N^{-1} \varphi = p$, solution of

$$\begin{cases} D_N G_N \underline{p} = \underline{0} \\ B_N \underline{p} = \varphi, \end{cases}$$

then $C_N = B_N U_N B_N^{-1}$, where U_N is the Uzawa operator. The existence of spurious pressure modes for (3.3) may or may not yield a singular C_N. Indeed, if there exists some pressure boundary data $\varphi \neq 0$ for which $p = B_N^{-1} \varphi$ is a spurious mode, then $G_N B_N^{-1} \varphi = \underline{0}$, hence, $C_N \varphi = \underline{0}$. Two *examples*:
i) The P_N-method for the Channel Flow has one spurious mode $\{L_N(y)\}$. But the solutions $p_\pm$ to the problems

$$\begin{cases} p_\pm''(y) = 0 & y \in (-1, 1) \\ p_\pm(\pm 1) = 1, \quad p_\pm(\mp 1) = 0 \end{cases}$$

have no components along $\{L_N(y)\}$. Hence, the Kleiser-Schumann influence matrix is non-singular.
ii) The (P_N, P_N)-method on the square cavity has 7 spurious modes But the collocation solutions p to

$$\Delta p = 0 \qquad \text{on } (GL_N \otimes GL_N) \cap \Omega$$

have no components along $\{L_N(x),\ L_N(y),\ L_N(x)L_N(y)\}$. Hence, the corresponding influence matrix has 4 zero eigenvalues only.

In general, the singularity of C_N can be fixed by deriving extra conditions on the pressure, or replacing zero eigenvalues of C_N by non zero (e.g., unity) eigenvalues. In both cases, this amounts to restricting the pressure space Q_N. Further details on the influence matrix method can be found in Tuckerman (1989).

3) Finite Element Pseudo $-$ Collocation

The finite element preconditioner described in Sect. 2 takes the following form for the Stokes problem:

Given $u_N \in V_N$, $p_N \in Q_N$, *find* $w_h \in V_h$, $r_h \in Q_h$ *solutions of*

$$
\begin{cases}
\displaystyle\int_\Omega \nabla w_h \nabla v_h - \int_\Omega \nabla \cdot v_h\, r_h = \\
\hfill \forall v_h \in V_h, \\
\displaystyle\hspace{2em} = \int_\Omega I_h^{vel}(-\Delta u_N + \nabla p_N - f)\, v_h, \\
\displaystyle -\int_\Omega \nabla \cdot w_h\, q_h = \int_\Omega I_h^{press}(\nabla \cdot u_N)\, q_h \hfill \forall q_h \in Q_h.
\end{cases}
$$

Here, $V_N = (P_N(\Omega))^2$, $Q_N = P_N(\Omega)$, i.e., the spectral approximation is based on the (P_N, P_N)-method described in Sect.3.2. On the other hand, V_h (resp., Q_h) is a finite element velocity (resp., pressure) space based on the $GL_N \otimes GL_N$ grid, whereas I_h^{vel} (resp., I_h^{press}) are the corresponding finite element interpolation operators. At convergence of the iterative scheme, one gets a *Pseudo-Collocation* scheme

$$
\begin{cases}
\int_\Omega I_h^{vel}(-\Delta u_N + \nabla p_N - f)\, v_h = 0, & \forall v_h \in V_h, \\
\int_\Omega I_h^{press}(\nabla \cdot u_N)\, q_h = 0, & \forall q_h \in Q_h.
\end{cases}
$$

The precise algebraic form of these equations depends on the choice of the finite element spaces [Demaret and Deville (1989)]. For instance, using bilinear velocities and constant pressures yields the usual (P_N, P_N)-method, which, as we know, is affected by spurious modes; furthermore, the f.e.m. preconditioner itself is affected by the checkerboard spurious mode. Demaret and Deville recommend the use of biquadratic velocities and bilinear pressures: in this case, the f.e.m. preconditioner has no spurious modes, and the resulting pseudo-collocation spectral scheme is also free of spurious modes. Indeed, the spectral residual is now interpolated both at the nodes of the Gauss-Lobatto grid, and at the mid-nodes in between. Since the spurious modes of the (P_N, P_N)-method have non-vanishing gradients at the latter nodes, the iterative finite element residual calculation forces them to zero. Deville and Mund (1990) show that the eigenvalues of the preconditioned Stokes operator are bounded between 1. and a value < 2.1. Finally, the spectral domain decomposition can be handled via the finite element interface matching, as described in Sect.2 [Demaret and Deville (1991)].

4. Temporal Discretization

Most of the time advancing schemes used so far together with spectral discretizations of the incompressible Navier-Stokes equations have been of semi-implicit type: the linear Stokes problem is advanced implicitly, the non-linear convective part is advanced explicitly. Even in the calculation of steady states, the use of iterative schemes based on a semi-implicit time advancing has been the most popular choice. On the contrary, fully implicit methods, as well as Newton-like methods in the local convergence to steady state solutions, have been a viable choice in finite order methods (finite differences, finite elements). There are at least two reasons for the preference of semi-implicit schemes by the spectral community: 1) accuracy requirements in the simulation of truly evolutionary problems have imposed a time step well below the stability limit of explicit schemes for the convective part; 2) the existing techniques for solving linear systems of equations arising from spectral discretizations are relatively well developed for elliptic problems, but still unsatisfactory for hyperbolic ones.

Recently, attempts have been made to reduce the load unbalance between implicit Stokes and explicit convective calculations, in favor of a smaller global computational cost. For instance, Frohlich, Gerhold, Lacroix and Peyret (1991) consider a third order backward Euler time advancing for the Rayleigh-Benard Problem in a 2D periodic channel with high aspect ratio. At each time step, the implicit Helmholtz-Stokes + Convection operator is inverted by several iterations of the fixed-point algorithm

$$\gamma u^* + \mathcal{S}(u^*, p^*)^T = f - \mathcal{C}(u^k)$$
$$u^{k+1} = (1-\alpha)u^k + \alpha u^*$$

The authors report savings by a factor in the order of 3 over the corresponding semi-implicit algorithm.

Ho, Maday, Patera and Ronquist (1989) resort to the Lagrangian form of Navier-Stokes equations in order to integrate in time along characteristics by an implicit method; a high-order remeshing (char-acteristic-foot calculations) based on the explicit integration of a convective equation (invariant equation) is applied. The method allows for the use of a symmetric Stokes solver at each time step, even if the convective terms are significant. Consequently, a reduction of computational cost of the time integration by more than one order of magnitude over a semi-implicit scheme is observed, although temporal errors may affect the spatial accuracy.

Demaret and Deville (1991) use Newton's method to compute steady solutions of the Navier-Stokes equations; the linearized spectral equations to be solved implicitly are preconditioned by biquadratic (velocity) - bilinear (pressure) finite elements. The authors find the preconditioning properties nearly independent of the Reynolds number, for a sufficiently wide parameter interval. A preconditioner based on bilinear finite elements for velocity would have been very inefficient. Indeed, the information at the Gauss-Lobatto grid is not sufficient for preconditioning the first derivative operator [Macareag and Streett (1986), Funaro (1987)]. On the contrary, biquadratic finite elements take information also at the mid-nodes, thus yielding an effective preconditioning for first order terms [Deville and Mund (1990)].

Looking towards the future, it is likely that the interest in implicit time-advancing methods will considerably grow. Indeed, the increase in computer power will allow the direct numerical simulation of a larger portion of the spectrum of the exact solution. Now, as far as stability is concerned, higher frequency modes impose a more severe restriction on the explicit time step; on the other hand, as far as accuracy is concerned, their temporal evolution is slower than that of lower frequency modes, thus, they allow a larger implicit time step. Therefore, it is conceivable the future development of *mixed schemes*, which advance explicitly the lower portion of the spectrum, and implicitly the higher one. (Nonlinear Galerkin methods - see Sect.5 - are already an example of this approach.) Clearly, more efficient techniques for solving implicit equations will be required. For instance, techniques such as Krylov subspace methods for solving linear unsymmetric systems, or backtracking and trust-region methods for solving nonlinear systems, should receive more attention in the framework of spectral methods.

5. Resolution of structures

During the 80's, spectral methods have consolidated their theoretical grounds, as well as developed efficient algorithms for the simulation of certain physical problems. The future of spectral methods will greatly depend on the amount of *flexibility* that will be possible to incorporate in them. Although spectral methods will hardly become "general purpose" methods, yet their area of application will be widened by new means of localizing resolution (both in physical and in transform space), and of coupling them with other numerical methods. This will be a major challenge in the 90's.

Domain decomposition is the most natural way of enhancing resolution in regions of highly structured flow patterns. The achievements of *spectral elements* in this area are already quite impressive. Other nice applications of domain decomposition techniques can be found in Peyret (1990). However, a lack of flexibility associated with the (large) quadrilateral spectral bricks appears if classical (conformal) matching between subdomains is enforced: local refinement propagates along cartesian directions even in regions of limited structure (a similar situation occurs in finite difference mesh refinements). The problem can be alleviated by resorting to loosely coupled subdomains, i.e., by allowing for a weak, integral-type (nonconforming) matching. An example of this approach is the *Mortar Element Method* [Bernardi, Maday and Patera (1989)]. The task of glueing subdomains together is attributed to an auxiliary space of mortar functions, defined on the union of all internal interfaces. The trial functions are non-conforming across subdomains, in that pointwise continuity is (partially) replaced by more flexible integral conditions involving mortar functions. The stability, consistency and spectral convergence of the method can be proven. An efficient (parallel) implementation is possible, via global-skeleton / local-elemental evaluations.

An alternative approach to localizing resolution is *self-adaptive gridding*. Promising results have been obtained in this direction (see, e.g., Bayliss, Gottlieb, Matkowsky and Minkoff (1989), Guillard and Peyret (1989), Augenbaum (1989)): the grid mapping is adapted in order to minimize suitable Sobolev norms of the discrete solution. The extension of these techniques to more complex situations (2 or 3 space dimensions, multiple fronts) still has to be carried out. More generally, effort should be put in designing efficient a-posteriori error estimators, that could be coupled with a mesh generation and/or domain decomposition procedure in order to obtain self-adaptive localization.

As previously pointed out, in the near future spectral methods may well become one of the constitutive components of *blended numerical methods*, which exploit the best features of each component. The coupling of spectral and finite difference methods for hyperbolic problem with discontinuous solutions is under investigation since several years. In particular, Gottlieb and coworkers are now trying to incorporate ENO ideas into spectral methods. On the other hand, the non-conforming, integral-type subdomain matching (*mortar*) previously mentioned, allows for a mathematically founded, optimal coupling between spectral bricks and other numerical discretizations based on *integral* discretizations (finite elements, finite volumes). Recent, although still partial, results in this direction have been obtained by Bernardi, Maday and Sacchi Landriani (1989), Debit and Maday (1990).

Any efficient coupling strategy should be based on the automatic (self-adaptive) detection of the regions where it is preferable to use spectral methods or, conversely, local methods. For viscous flows at high Reynolds numbers, the superior accuracy of the more expensive spectral methods can be exploited in the viscosity-dominated region, whereas less accurate but cheaper local methods usually suffice in the inviscid region. Brezzi, Canuto and Russo (1989) have recently developed a *self-adaptive* formulation of the viscous/inviscid coupling (see also Arina and Canuto (1990)). The Navier-Stokes equations are modified in order to retain the viscous terms only wherever they are "large enough". An automatic domain decomposition is obtained, with smooth (C^1) inviscid/viscous matching. The method can be extended to formulate the self-adaptive coupling of different mathematical models (e.g., parabolized Navier-Stokes equations, Euler equations and potential equations - see Achdou and Pironneau (1991)).

Finally, the high accuracy of spectral methods makes them attractive as back-ground methods for modern subgrid-resolution (or improved grid-resolution) techniques. For instance, the recently developed *Nonlinear Galerkin methods* [Marion and Temam (1989)] are based on the decomposition of the effective spectrum in two (or more) parts $u = y + z$, plus an appropriate law $z = \varphi(y)$ between large and small eddies, suggested by the existence of an exact or approximate invariant manifold. In the field of large eddy simulation, among the variuos subgrid-scale stress models, Germano (1991) proposed a *dynamical Eddy Viscosity model* based on information on two scales (or filters) and extrapolation from the resolved to the unresolved portion of the spectrum. Both methods, although conceptually different, find a natural realization in the framework of spectral methods.

References

Achdou, Y., Pironneau, O. (1991): The χ-method for the Navier-Stokes equations. C.R. Acad.Sci.Paris, **188**, and Report Math. Dept. University of Houston.

Arnold, D., Brezzi, F., Fortin, M. (1984): A stable finite element for the Stokes equations. Calcolo **21**, 337-344.

Augenbaum, J.M. (1989): An adaptive pseudospectral method for discontinuous problems. Appl.Numer.Math.**5**, 459-480.

Azaies, M., Coppoletta, G. (1991): Calcul de la pression dans le probleme de Stokes par une methode spectrale de "quasi-collocation" a grille unique sans modes parasites, to appear in Annales Magrebines de l'Ingegneur.

Azaies, M., Labrosse, G. (1991): A multidimensional periodic nonperiodic steady Stokes solver for primitive variables, by spectral collocation. Submitted to J.Comput.Phys.

Azaies, M., Vandeven, H. (1991): Resolution numerique des equations de Stokes pour des ecoulements axisymetriques confines en geometrie cylindrique annulaire. Report LIMSI 91-5, Orsay, France.

Babuska, I., Szabo, B.A., Katz, I.N. (1981): The p-version of the finite element method. SIAM J.Numer.Anal.**18**, 515-545.

Bayliss, A., Gottlieb, D., Matkowsky, B.J., Minkoff, M. (1989): An adaptive pseudospectral method for reaction-diffusion problems. J.Comput.Phys. **81**, 421-443.

Bernardi, C., Canuto, C., Maday, Y. (1988): Generalized inf-sup conditions for Chebyshev spectral approximations of the Stokes problem. SIAM J. Numer. Anal.**25**, 1237-1271.

Bernardi, C., Coppoletta, G., Girault, V., Maday, Y. (1990): "Spectral methods for the Stokes problem in stream-function formulation", pp. 229-236 in Canuto and Quarteroni, eds. (1990).

Bernardi, C., Girault, V., Maday, Y. (1990): "Mixed spectral element approximation of the Navier-Sokes equations in the streamfunction and vorticity formulation", Report R90045, Université P.&M.Curie, Paris.

Bernardi, C., Laval, F., Metivet, B., Thomas, B. (1990): "Equations de Navier-Stokes avec masse volumique variable: formulation bien poseé du probleme de Stokes", Report HI-72/6690 EDF Clamart, France.

Bernardi, C., Maday, Y. (1988): Analysis of a staggered grid algorithm for the Stokes equation. Int.J.Numer.Meth.Fluids **8**, 537-557.

Bernardi, C., Maday, Y., Métivet, B. (1987a): Spectral approximation of the periodic/nonperiodic Navier-Stokes equations. Numer.Math.**51**, 655-700.

Bernardi, C., Maday, Y., Métivet, B. (1987b): Calcul de la pression dans la résolution spectrale du probleme de Stokes. La Rech.Aéro., 1-21.

Bernardi, C., Maday, Y., Patera, A.T. (1989): "A new non-conforming approach to domain decomposition: the mortar element method", in *Nonlinear Partial Differential Equations and their Applications*, H.Brezis and J.L.Lions, eds. (Pitman, London).

Bernardi, C., Maday, Y., Sacchi Landriani, G. (1989): Nonconforming matching conditions for coupling spectral and finite element methods. Appl.Numer.Math. **6**, 65-84.

Brezzi, F. (1974): On the existence, uniqueness and approximation of saddle-point problems arising from Lagrangian multipliers.

RAIRO Numer. Anal.**8**, 129-151.

Canuto, C. (1992): "Spectral Pressure Stabilization via Bubble Functions", in preparation.

Canuto, C., Hussaini, M.H., Quarteroni, A., Zang, T.A. (1990): *Spectral Methods in Fluid Dynamics* (Springer-Verlag, Berlin New York), 2nd printing.

Canuto C., Pietra, P. (1990): Boundary and interface conditions within a finite element preconditioner for spectral methods. J.Comput.Phys.**91**, 310-343.

Canuto, C., Quarteroni, A., eds. (1990): *Spectral and High Order Methods or Partial differential Equations* (North-Holland, Amsterdam)

Debit, N., Maday, Y. (1990): The coupling of spectral and finite element method for the approximation of the Stokes problem. Report R90013, Université P.&M.Curie, Paris.

Demaret, P., Deville, M. (1989): Chebyshev pseudospectral solution of the Stokes equations using finite element preconditioning. J.Comput.Phys.**83**, 463-484.

Demaret, P., Deville, M. (1991): Chebyshev collocation solutions of the Navier-stokes equations using multi-domain decomposition and finite element preconditioning. J.Comput.Phys.**95**.

Deville, M., Mund, E. (1985): Chebyshev pseudospectral solution of second order elliptic equations with finite element preconditioning. J.Comput.Phys.**60**, 517-533.

Deville, M., Mund, E. (1991): "Fourier analysis of finite element preconditioned collocation schemes", to appear in SIAM J. Sci.Stat.Comput.

Ehrenstein, U. (1986): *Methodes Spectrales de Resolution des Equations de Stokes et de Navier-Stokes. Applications a des Ecoulements de Convection Double*, Ph.D. thesis, University of Nice.

Franken, P., Deville, M., Mund, E. (1990): "On the spectrum of the iteration operator associated to the finite element preconditioning of Chebyshev collocation calculations", pp.295-304 in Canuto and Quarteroni, eds. (1990).

Frohlich, J., Gerhold. T., Lacroix, J.M., Peyret, R. (1991): Fully implicit spectral methods for convection. Report 301-1991, University of Nice.

Funaro, D. (1987): A preconditioning matrix for the Chebyshev differencing operator. SIAM J.Numer.Anal. **24**, 1024-1031.

Germano, M. (1991): An algebraic property of the turbulent stress tensor and its possible use in subgrid modelling. Proc. 8th Symposium on Turbulent Shear Flows, Munich, September 1991, **2**, 19, 1-6.

Guillard, H., Peyret, R. (1988): On the use of spectral methods for the numerical solution of stiff problems. Comput.Methods Appl.Mech.Engrg.**66**, 17-43.

Ho, L.-W., Maday, Y., Patera, A.T., Ronquist, E.M. (1990): A high order Lagrangian-decoupling method for the incompressible Navier-Stokes equations, pp.65-90 in Canuto and Quarteroni, eds. (1990).

Kleiser, L., Schumann, U. (1980): "Treatment of incompressibility and boundary conditions in 3-D numerical spectral simulations of plane channel flows", pp.165-173 in *Proc. 3rd GAMM Conf. Numerical Methods in Fluid Mechanics*, E.H.Hirschel, ed. (Vieweg, Braunschweig)

Leonard, A., Wray, A. (1982): "New numerical method for the simulation of three-dimensional flow in a pipe", pp.335-342 in *Proc. 8th Int. Conf. Numerical Methods in Fluid Dynamics*, E.Krause, ed. (Springer-Verlag)

Macaraeg, M., Streett, C.L. (1986): Improvements in spectral collocation through a multiple domain technique. Appl.Numer. Math.**2**, 95-108.

Maday, Y., Patera, A.T. (1989): "Spectral Element Methods for the Incompressible Navier-Stokes Equations", pp.71-143 in *State of the Art Surveys in Computational Mechanics*, A.Noor and J.T.Oden, eds. (ASME, New York).

Maday, Y., Patera, A.T., Ronquist, E.M. (1986): A well-posed optimal spectral element approximation for the Stokes problem, to appear in SIAM J.Numer.Anal.

Malik, M.R., Zang, T.A., Hussaini, M.Y. (1985): A spectral collocation method for the Navier-Stokes equations. J.Comput.Phys. **61**, 64-88.

Marion, M., Temam, R. (1989): Nonlinear Galerkin methods. SIAM J.Numer.Anal. **29**, 1139-1157.

Nicolaides, R.A. (1982): Existence, uniqueness and approximation for generalized saddle point problems. SIAM J.Numer.Anal. **19**, 349-357.

Orszag, S.A. (1980): Spectral methods for problems in complex geometries. J.Comput.Phys.**37**, 70-92.

Pasquarelli, F., Quarteroni, A., Sacchi-Landraini, G. (1987): Spectral approximation of the Stokes problem by divergence-free functions. J.Sci. Comput.**2**, 195-226.

Patera, A.T. (1984): A spectral element method for fluid dynamics: laminar flow in a channel expansion. J.Comput.Phys.**54**, 468-488.

Peyret, R. (1990): The Chebyshev multidomain approach to stiff problems in fluid mechanics, pp.129-146 in Canuto and Quarteroni, eds. (1990).

Sacchi-Landriani, G., Vandeven, H. (1989): A multidomain spectral collocation method for the Stokes problem. ICASE Report 89-42, submitted to Numer.Math.

Streett, C.L., Hussaini, M.Y. (1991): A numerical simulation of the appearance of chaos in finite-length Taylor-Couette flow. Appl.

Numer.Math.**7**, 41-71.

Tuckerman, L.S. (1989): Divergence-free velocity fields in non-periodic geometries. J.Comput.Phys.**80**, 403-441.

Vandeven, H. (1990): Compatibilité des espaces discrets pour l'approximation par méthodes spectrales du probleme de Stokes periodique-nonperiodique. MMAN **23**, 649-688.

Vanel, J.M., Peyret R., Bontoux, P. (1985): "A pseudospectral solution of vorticity- streamfunction equations using the influence matrix technique", pp.477-488 in *Numer. Methods for Fluid Dynamics*, K.W.Morton and M.J. Baines, eds. (Clarendon Press)

ISSUES IN THE APPLICATION
OF HIGH ORDER SCHEMES

David Gottlieb[1]

Division of Applied Mathematics
Brown University
Providence, RI 02912

ABSTRACT

We argue, in this paper, that the type of simulations to be carried out in the next decade will entail the use of high order schemes.

A discussion of some issues in the application of those schemes to time dependent problems are discussed. In particular we will review spectral shock capturing techniques and the asymptotic behavior of high order compact schemes.

1 Introduction

The aim of this work is to discuss some issues in the application of high order finite difference schemes and spectral methods, to the numerical solution of the Navier Stokes equations. The term "high order" is of course vague, we will use it for schemes with formal order of accuracy of (its *space* discretization) more than three.

The discussion will be motivated by the type of problems that will be, in the author's belief, in the forefront of the research in the next decade. We will attempt to identify some common features in the physical problems to be numerically simulated and try to assess the merits of high order schemes concerning these features.

It is our prediction that future applications of computational method in fluid mechanics will involve fluid problems sharing common features:

[1]Research was supported by the National Aeronautics and Space Administration under NASA Contract No. NAS1-18605 while the author was in residence at the Institute for Computer Applications in Science and Engineering (ICASE), NASA Langley Research Center, Hampton, VA 23665. Research was also supported by AFOSR grant 90-0093 and DARPA grant N00014-91-J-4016.

- The problems will be truly time dependent. The interest will not be focused upon the steady-state solution of the time dependent system of equations, but rather on the time development of the solution over a long period of time.

- Understanding of fine details of the flow field, rather than its gross features, will be required. Thus , the engineers' goals will be more ambitious than just finding , say , the lift and the drag of an airfoil.

It is interesting that both properties listed above point to the direction of *high order schemes.*

Kreiss and Oliger [13] analyzed the phase error of difference schemes applied to a simple wave equation. They found out that the number of grid points N_k , required to resolve a wave , when a scheme of order of accuracy 2k is used, is given by the formula

$$N_k \sim A_k j^{\frac{1}{2k}}$$

Where j is the number of periods in time. Thus, it is important, for a meaningful long time integration, to use a higher order scheme. Moreover, they also showed that A_k decreases fast enough with the order of accuracy k that it compensates for the extra work, needed in the application of a high order scheme, even if we are interested in short time integration.

The model analysis of Kreiss and Oliger seems to hold for more complicated problems. By now it is widely recognized that high order schemes are needed if either fine details of the flow are required or a long term integration is performed. Thus the conclusion is that high order schemes will be used more and more in applications in the next decade.

Two examples to illustrate the above are in order.

The first example has to do with the numerical simulation of a low Mach number, compressible, viscous flow past a circular cylinder at a moderately low Reynolds number. The unsteady wake generated by the cylinder has been of great interest to computational fluid dynamicists as well as to theoretical and experimental aerodynamicists. The Reynolds number range between 40 and 1000 has been of particular interest because it spans the transition from steady flow to an unsteady wake flow dominated by the period shedding of vortices from the cylinder. The shedding frequency of this vortices is known theoretically and *all* numerical methods, applied to this problem, obtain this frequency within reasonable accuracy. However, Srinivasan [20] measured (in a wind tunnel experiment) more than one distinct frequency in the shedding regime. In addition to the vortex shedding frequency he found clearly, discernible lower frequencies and concluded that this was a feature of the initial stage of transition to turbulence. A numerical simulation, using the second order accurate MacCormack scheme did find a secondary frequency, very nearly the same found in the wind-tunnel experiment. See picture 1 for the pressure signal.

A more recent simulation using a spectral method (Chebyshev-Fourier) found no secondary frequencies. Spectral methods are ,of course, methods with high order accuracy, and as any other high order method, are very sensitive to the kind of outflow boundary conditions treatment. In particular they tolerate only those treatments based on the *characteristic variables.* Low order methods are more tolerant and allow high variety of boundary conditions. When the far field boundary conditions from the spectral code, were incorporated into the second order code, the secondary frequency disappeared. The spurious frequency was generated by a boundary condition that was tolerated by the low order (and robust) MacCormack scheme but not by the high order scheme used. See picture 2.

The above example illustrates the type of surprises one might encounter when using low order schemes for long time integrations. Stability, in the classical sense of Lax , is not enough anymore, one has to be careful not to have spurious phenomena. It also illustrates the fact that the gross features of the flow (in this case the shedding frequency) were obtained even with low order schemes. It is only when some delicate features of the flow were sought for that an unphysical solution had been observed. The fact that the lower order scheme produced a spurious solution does not contradict the fact that the scheme is stable and converges *as the number of grid-points increases.* The spurious solution will eventually disappear if further grid refinements will be carried out. However for realistic grids it is impossible to distinguish between a physical and a spurious solutions. *a high order scheme is less likely to produce a spurious solution.*

In the next example (see [18]) the problem of a shock wave interacting with density wave is considered. This is a situation in which one is interested not only in the accurate simulation of the shock wave, but also , it is very important to get the behavior of the density wave very accurately. In figure 2 and 3 we present the density profile obtained- by the second order Muscle scheme with 400 grid points - and by a spectral method with 257 points.These results are compared to a third order ENO scheme with 1201 points (this is taken as the analytic solution). *The second order scheme completely misses the fine structure of the flow behind the shock wave though it does get the shock very accurately.* Here , again, resolving delicate features of the flow necessitates a high order scheme.

The paper does not attempt to give an definitive overview of the performance of high order schemes. Instead, we will discuss some issues in their applications to long term integrations of the Navier -Stokes equations. High order methods are less robust than low order once, and their application demands more sophistication and better knowledge of theory. The issues that we will discuss are related to the sensitivity of high order methods. We will restrict ourselves to spectral methods and high order compact finite difference schemes.

One can identify two main stumbling blocks in the application of spectral methods to hyperbolic systems.

1. Treatment of Shock waves

2. Application to Complex Geometries.

Both items follow from the formal high accuracy of spectral methods. By no means these are limited to spectral methods, in fact when one uses higher order finite difference schemes one encounters similar problems.

In the last few years significant progress had been made in those areas.

Complex geometries are being dealt with by a combination of domain decomposition spectral techniques and mappings. Domain decomposition techniques not only overcome difficulties resulting from complex geometries but also seem to be amenable to parallel computing. We will not deal here with this issue since it merits a paper by itself.

We would rather review the state of the art of shock capturing spectral techniques and try to predict future development. This is done in Section 2.

Compact high order schemes are now being seriously considered as an alternative to spectral methods. In Section 3 we discuss issues in the applications of fourth-and sixth order compact schemes for long time integration of hyperbolic systems.

2 Spectral Shock Capturing Methods

Spectral methods are based on the *global* expansions of the unknowns in term of Fourier series or Chebyshev polynomials. This is the reason why these expansions converge exponentially fast for analytic functions, but also why the rate of convergence deteriorates *everywhere* in the presence of discontinuities (e.g. shock waves) in the flow field. This phenomenon **(known as the Gibbs phenomenon)** was found a hundred years ago, and seems to rule out the possibility of applying spectral methods (or any linear, high order finite difference schemes) to shocked flows. This is a fundamental difficulty that must be resolved before attempting to apply spectral methods to shock waves.

Despite the Gibbs phenomenon ,it should be noted that spectral methods are being successfully applied to simulate problems that include shocks. Of course, it should be stressed that one *should not* use spectral methods if one is interested primarily in the shock, and not in other details of the flow. Spectral methods are beneficial when fine details of the flow have to be resolved. We refer back to the example in the introduction.

As an another example example, we show in picture 5 (taken from a recent paper by W.S. Don) the flow field obtained by a spectral simulation of an interaction between a shock wave and a hydrogen bubble. The main issue here is not the resolution of the shock, but rather the final form of the bubble.Thus this problem requires higher accuracy in the flow. It clearly demonstrates that spectral methods can and should be used for such problems.

In the following we will briefly review recent developments in the theory and application of shock capturing spectral techniques.

We will divide the review into three subsections. First, we will review the basic issues concerning the approximation theory. This is important since in scientific computations we compute approximations to *projections* of the solutions. It is important to find out how accurate are the projections themselves. Next we will review linear hyperbolic problems, and show that in this case a stable spectral method gives the projection of the solution within spectral accuracy. Finally a brief review of spectral methods for nonlinear hyperbolic equations will be given .

2.1 Approximation Theory - Gibbs Phenomenon

The fundamental question can be very simply phrased:

Is it possible to recover, with high accuracy, pointwise values of a piecewise continuous function from its first N Fourier coefficients?

This is the most basic question in the application of spectral methods to shock wave calculations , since the best we can hope for - say in the Fourier-Spectral method applied to PDE's - is to approximate well the first Fourier coefficients of the solution to the equations. In other words, we find only an approximation to the projection of the solution . Thus the question is whether this information is enough to construct a highly accurate approximation to the solution.

There are many variants of this question. First, In the commonly used Pseudospectral-Fourier method we attempt to find the *point values* of a projection of the solution and not the Fourier coefficients. Moreover, for initial-boundary value problems one uses polynomial methods rather than Fourier methods. The fundamental question can be rephrased using the appropriate Chebyshev or Legendre coefficients. However, answering satisfactory the above posed question is the key to the success of spectral methods.

To further simplify the discussion, and also to quantify it, consider an analytic but non-periodic function $f(x)$ defined in $[-1,1]$. Notice that $f(x)$ has a discontinuity at the boundary $x = \pm 1$ if it is extended periodically with period 2. The Fourier coefficients of $f(x)$ are defined by

$$\hat{f}(k) = \frac{1}{2} \int_{-1}^{1} f(x)e^{-ik\pi x} dx \qquad (2.1)$$

Assume that the first $2N+1$ Fourier coefficients $\hat{f}(k)$, $|k| \leq N$, are known but the function $f(x)$ is not. A classical result is that if $f(x)$ is analytic and periodic than the Fourier series converges exponentially fast. Our objective is to recover the nonperiodic function $f(x)$ for $-1 \leq x \leq 1$ with exponential accuracy in the maximum norm.

The traditional Fourier partial sum using the first $2N + 1$ modes

$$f_N(x) = \sum_{k=-N}^{N} \hat{f}(k)e^{ik\pi x} \qquad (2.2)$$

does a poor job: it produces a first order approximation to $f(x)$ with an error $O(\frac{1}{N})$ away from the boundary $x = \pm 1$, and shows $O(1)$ spurious oscillations near the boundary $x = \pm 1$ known as the Gibbs phenomenon. Thus there is no convergence in the maximum norm.

The first important observation is that high order information is contained in the Fourier modes $\hat{f}(k)$, in a weak sense. It can be shown that *integrals* of $f_N(x)\psi(x)$ approximate , with exponential accuracy, integrals of $f(x)\psi(x)$ if $\psi(x)$ is analytic and periodic.

This fact is being utilized when *filters* are used. Filters in the Fourier space are obtained by multiplying the given Fourier coefficient by $\hat{\sigma}_k^N$ to form:

$$f_N^\sigma(x) = \sum_{k=-N}^{N} \hat{\sigma}_k^N \hat{f}(k)e^{ik\pi x} \qquad (2.3)$$

In the past it had been known that one can get exponential accuracy away from the boundary $x = \pm 1$ if $\hat{\sigma}_k^N$ are chosen as suitable *real* numbers [7], [13], [15], .

The notion of a filter is not well understood and deserves some clarification. We follow the work by Vandeven [23] and define a filter by:

definition: A Filter of order p is defined by

$$\hat{\sigma}_k^N = \hat{\sigma}(\frac{k}{N})$$

where $\hat{\sigma}$ is a smooth function in its argument i.e.

$$\hat{\sigma}(x) \in C^p$$

and $\hat{\sigma}(x)$ is tangent to the function $f = 1$ at $x = 0$ and to the zero function as $x = 1$ i.e.

$$
\begin{aligned}
\hat{\sigma}(0) &= 1 \\
\hat{\sigma}^l(0) &= 0 \qquad 1 \le l \le p - 1 \\
\hat{\sigma}^l(1) &= 0 \qquad 0 \le l \le p - 1
\end{aligned}
$$

With this definition one can prove:

THEOREM:

For any $-1 \le x \le 1$ we have

$$|f(x) - f_N^\sigma(x)| \le C_p N^{1-p} d(x)^{1-p} |||F|||$$

The quantity $d(x)$ is the distance from the discontinuity.

Vandeven found optimal filters, in some sense, and showed that by taking p proportional to N exponential accuracy can be recovered *away from the discontinuity.*

To better understand the role of filters it is worthwhile to look at the graph of the function $\sigma(x)$ given by

$$\sigma(x) = \sum_{k=-N}^{N} \hat{\sigma}(\frac{k}{N}) e^{ik\pi x}$$

In Picture 6 we draw the function for an eights order filter, it is evident that $\sigma(x)$ is a function with narrow support, in the sense that the function is small (though not vanishing) outside a small interval. This is a crucial observation since the filtered function $f_N^\sigma(x)$ is the convolution of $f_N(x)$ with the function $\sigma(x)$

$$f_N^\sigma(x) = \int_{-1}^{1} f_N(x - \xi)\sigma(\xi)d\xi$$

The meaning of the above equation is that the numbers $\hat{\sigma}_k^N \hat{f}(k)$ are the Fourier coefficients of a **localized** version of the original function $f(x)$. Thus away from the discontinuity the localized function has a rapidly convergent Fourier series.

A similar idea had been utilized in the physical space by [7]. In this approach the function $\sigma(x)$ is constructed in the physical space as a product of an approximation to the delta function and a localizer, the Fourier coefficients of this σ are the filters in the Fourier space.

The only difference between the two types of filters is that one has a finite Fourier series expansion and it is almost local in space, where the other one is compact in space but has an infinite Fourier expansion.

The problem with this approach is that the accuracy can not be recovered in the neighborhood of the shock, see [23] [11]. The reason for it is that real $\hat{\sigma}_k^N$ corresponds to an <u>even</u> function. Thus, the interval of smoothness around the point of reconstruction has to be symmetric, points close to the shock can not be recovered.

In [4] the authors showed that by taking complex $\hat{\sigma}$ one can get one side accurate reconstruction. For those one-sided filters one can use two different approximations in $-1 \leq x \leq 0$ and in $0 < x \leq 1$, right-sided for the former and left-sided for the latter, to obtain exponential convergence globally.

In a recent paper [9] the authors adopted a different point of view. The basic idea is to realize that the problem with the Fourier approximation is the non-periodicity of the function and the fact that the functions $e^{\pi i k x}$ are the solutions of a *regular* Sturm-Liouville (S-L) problem. In [6] it is shown that expanding an analytic, non-periodic function $f(x)$ by the eigenfunctions of a *singular* Sturm-Liouville problem yields rapid convergence. For example, a Chebyshev or Legendre expansion of $f(x)$ converge exponentially. Thus if the first $2N + 1$ Fourier coefficients can provide enough information to reconstruct the coefficients of an expansion based on a singular S-L problem, one might recover the accuracy. Unfortunately, one can <u>not</u> recover the coefficients of the Chebyshev or the Legendre expansion within high enough accuracy.

In the above mentioned paper it is shown that from the first $2N + 1$ Fourier coefficients of an analytic but non-periodic function, one can get the first $m \sim N$ coefficients in the *Gegenbauer* expansion based on the Gegenbauer polynomials $C_n^\lambda(x)$, provided that the parameter λ, appearing in the weight function $(1 - x^2)^{\lambda - \frac{1}{2}}$, grows with the number of Fourier modes- N. This yields exponential accuracy in the maximum norm.

The proof consists of two separate and independent steps: The first step is to show that from the Fourier partial sum of the first $2N + 1$ Fourier modes, of an arbitrary L_2 function $f(x)$, it is possible to recover the partial sum of the first m terms in the Gegenbauer expansion of the same function to exponential accuracy (in the maximum norm) by letting the parameter λ and the number of terms

m in the Gegenbauer expansion grow linearly with N. In this step $f(x)$ needs not be smooth. Any L_2 function will do. In the second step they prove the exponential convergence, in the maximum norm, of the Gegenbauer expansion of an analytic function when λ grows linearly with m.

Thus it is shown that one can construct an exponentially convergent approximation to an analytic, non-periodic function, from its first $2N + 1$ Fourier coefficients.

We quoted the above theorem is some detail because it shows how to overcome the Gibbs Phenomenon for any piecewise analytic (or C^p) function. One can show that from the first N Fourier coefficients of a piecewise analytic function in $[-1, 1]$ one can get an exponentially convergent *Gegenbauer* expansion at *any subinterval* $[a, b]$ which is free of discontinuities.

This is the first step towards the removal of Gibbs oscillations from every eigenfunction expansions. Further work along these lines should be devoted to Chebyshev methods and not only Fourier methods. Moreover one has to extend the above procedure to collocation methods where the information given is the first N Fourier modes of the interpolant, rather than that of the function itself given in the Galerkin methods.

However we feel that the method discussed in [9] provides the main clue for removing the Gibbs phenomenon.

2.2   Hyperbolic system of linear equations

We have seen, in the last Section , that once the Fourier coefficients of a given function are obtained with high accuracy, then it is possible to reconstruct a spectrally accurate approximation to the point values of the function.

It has been shown that this is indeed the case in the Spectral-Fourier approximation to a system of *linear* hyperbolic equations, even in several space dimensions.

In fact, by using the Green's identity , one can show that when a linear system of hyperbolic equations, with *discontinuous initial conditions* is being approximated by **stable** Fourier Galerkin *or* Fourier collocation methods then the *integrals* of the discontinuous solutions multiplied by smooth functions , are obtained within high accuracy. Thus using the results of the last Section, spectrally accurate *pointwise* approximations to the point value of the solution can be obtained.

The above can be summarized (in one dimension) in the following

Theorem

Let $u(x,t)$ be the periodic solution of the system of linear hyperbolic equations of the form

$$
\begin{aligned}
u_t(x,t) &= A(x,t)u_x(x,t) \\
u(x,0) &= u_0(x)
\end{aligned}
$$

Where $A(x,t)$ is a $p \times p$ matrix , and $u(x,t)$ is a function valued , p component vector. We assume periodic boundary conditions in [-1,1].

Let $u_N(x,t)$ be the Spectral Fourier approximation to $u(x,t)$ Then there exists a function $\Phi(x,y,t)$ such that the pointwise error

$$
\int_{-1}^{1} u_N(x,t)\Phi(x,y,t)dx - u(y)
$$

tends *exponentially* to zero.

The theorem is easily extended to several space dimensions, however it does not carry over to the case of the Pseudospectral (collocation) Fourier method applied to linear systems. In this case the information furnished initially is the point values of the initial conditions. It might be argued that, since only point values are used then the exact location of the shock is not known, it can be everywhere between two grid points. This may lead to deterioration of the accuracy in large regions.

One remedy is to start with the exact Fourier coefficients of the solution. This approach poses problems in the nonlinear case , the problem should be approached from different angle. For example using ideas of subcell spectral resolution one can alleviate the problem. It is readily verified that if the discontinuity is *exactly* in the middle between two grid points then there is no deterioration in accuracy. By using the correct initial location of the shock one can show that the pseudospectral method contains information on the exact location of the shock and thus the accuracy in the pseudospectral (collocation) method is the same as in the spectral (Galerkin) method. There is still no counterpart to the above theorem in the case of Chebyshev methods applied to linear initial-boundary hyperbolic problems. Numerical experiments, however, indicate that the situation here is no worse then in the periodic case. It seems that some more sophisticated theoretical tools should be developed in order to produce a similar proof.

2.3 Nonlinear Systems of Hyperbolic equations.

The attempts to adopt spectral methods to the numerical solution of non linear hyperbolic problems with discontinuous solutions can be classified into four different categories. We wish here to review briefly the four different methods and to try to assess the merit of the different methods as well as to justify the way in which we chose to follow.

The main problem in the The most robust method is to switch automatically to a lower order, finite difference, scheme. This is done usually in the context of cell averaged methods, developed in [16] for Fourier methods and in [3] for Chebyshev methods. In this approach the spectral flux is replaced by a lower order, finite difference flux, in the neighborhood of the shock to produce a method that avoids oscillations. In particular the combination of spectral (or the spectral element) methods with FCT methods seem to be a robust way of simulating shock waves. We refer the reader to the review by Zalesak in this volume for details. The validity of this approach is limited by the fact that the overall accuracy deteriorates *globally*. Thus spectral accuracy is lost even away from the shock . However it seems that for engineering purposes the method performs very satisfactory. A yet unpublished report by Karniadakis indicates that by combining this approach with the Spectral - Element methods, one can simulate complex shock interactions in complex geometries.

The second approach is the Spectral Viscosity (SV) approximation adopted by Tadmor and coworkers.([22]). Tadmor considered the scalar nonlinear equation of the form

$$\frac{\partial}{\partial t}u(x,t) + \frac{\partial}{\partial x}f(u(x,t)) = 0.$$

The equation is modified to include, in the right hand side, an approximation to *the high modes only* of the second derivative of the solution $u(x,t)$. In particular, in the periodic case one adds the term

$$\epsilon_N \frac{\partial}{\partial x} Q_N * \frac{\partial}{\partial x} u_N(x,t)$$

The viscosity kernel $Q_N(x,t)$ is activated only on the high frequencies. Its form is

$$Q_N(x,t) = \sum_{|k|=l_N}^{N} \hat{Q}_k(t) \hat{u}_k(t) e^{ikx}$$

Where

$$\hat{Q}_k(t) \geq 1. - (\frac{l_N}{|k|})^q$$

The parameters l_N and q are appropriately chosen.

With this dissipation term added to the equation Tadmor was able to prove that the SV method is Total-Variation Bounded, moreover he could show convergence of the numerical approximation to the correct entropy solution.

We believe that the theory of Tadmor is a manifistation of the fact a *stable* spectral method contains a highly accurate information on the solution, and that with a suitable post processing one can obtained a spectrally accurate approximation to the correct entropy solution. It seems that these results hold for em every filtering technique that keeps the spectral method stable. We see the value of Tadmor's work in it's beautiful theoretical achievements rather than in the algorithmic aspects of it.

The third method , adopted by Cai Gottlieb and Shu is based on the spectral ENO approach. In this approach, the equation is solved in its cell averaged form .

The solution procedure consists of two stages. In the first stage the point values u_j^n are reconstructed from the cell averages $\bar{u}_j^n$ by a total variation bounded method, based on locating and subtracting the discontinuity at every step.

The second stage consists of advancing the solution in time with the use of the total variation bounded Runga-Kutta methods developed by Shu.

This technique is not limited to Fourier methods, but can be used also in the case of initial-boundary value problems solved with the use of Chebyshev techniques.

The method of reconstructing the point values from the cell averages is evidently nonlinear, since it is based on the solution itself. Thus this method requires a nonlinear numerical treatment at every step. This can be quite expensive. Also as in the case of all methods based on the cell averaged form, extensions to several space dimensions are not trivial.

The most efficient of all the current methods , in our opinion, is the fourth method. In this technique one uses an extremely weak filter, at every time step, just enough to keep the method stable.

One way to keep the method stable is to solve at each time the equation

$$\frac{\partial u}{\partial t} = \frac{\partial f(u)}{\partial x} + (-1)^s \frac{1}{N^{2s-1}} \frac{\partial^{2s}}{\partial x^{2s}}$$

Where s depends on the number of grid point N. This has the flavor of Tadmor's suggestion, however it is exactly a low pass filter.

At the end of the run, an accurate point values of the solution, are obtained at every sub-interval free of discontinuities by finding the *Gegenbauer* expansion for this sub-interval. This is based on the theorem (see [9]) that this is a spectrally accurate procedure. Thus, if the spectral calculations contain spectrally accurate information (and Tadmor's theory seems to support this conjecture) then we have spectral accuracy in *any interval where the solution is smooth.*

In figure 7 we bring several numerical simulations of complicated interactions of shock waves and complex structured flows reported on by W. S. Don.

The problem is an one dimensional interaction of a Mach 3 shock with a sinusoidal density wave. This problem has become a standard problem for checking the accuracy of a given scheme.

Since low order schemes miss the fine structure behind the shock, it is important to use an high order scheme for this problem. The spectral method described above resolved the shock wave within one grid point and the complicated flow field is well resolved.

3 Stable and Asymptotically Stable Compact Schemes

Recently, higher-order compact schemes have seen increasing use in the DNS (Direct Numerical Simulations) of the Navier-Stokes equations. Although they do not have the spatial resolution of Spectral methods, they offer significant increases in accuracy over conventional second order methods. They can be used on any smooth grid, and do not have an overly restrictive CFL dependence as compared with the $O(N^{-2})$ CFL dependence observed in Chebyshev Spectral methods on finite domains. In addition, they are generally more robust and less costly than Spectral methods. The issue of the relative cost of higher-order schemes (accuracy weighted against physical and numerical cost) is a far more complex issue, depending ultimately on what features of the solution are sought and how accurately they must be resolved. In any event, the further development of the underlying stability theory of these schemes is important.

It turns out that this schemes are very sensitive to boundary treatments. In particular *all* of the boundary conditions , currently used , allow non physical time growth of the solution. Recently, the stability characteristics of various compact fourth- and sixth-order spatial operators were assessed in reference [5], using the theory of Gustafsson, Kreiss and Sundstrom (G-K-S) for the semi-discrete Initial Boundary Value Problem (IBVP). The results were then generalized to the fully discrete case with Runge-Kutta time advancement using a recently developed theory by Kreiss[14]. In all cases, favorable comparisons were obtained between G-K-S theory, eigenvalue determination, and numerical simulation. The conventional definition of stability is then sharpened to include only those spatial discretizations that are asymptotically stable (bounded, Left Half-Plane eigenvalues). It is shown that many of the higher-order schemes which are G-K-S stable are not asymptotically stable. It was concluded that in practical calculations, only those schemes which satisfied both definitions of stability were of any great usefulness.

It was shown in the above work of Carpenter et al. that conventional (optimal) finite difference closures at the boundaries of order greater than four are not G-K-S (or asymptotically) stable. Since fifth-order boundary closures possessing both stability properties were needed for sixth-order inner schemes, an alternate method for closing the boundaries was sought. The solution was to parametrize the fifth-order difference formula at several points at each end of the spatial domain, thereby creating adjustable coefficients in the spatial operator. The asymptotic properties of the operator were established by the numerical determination of the eigenvalue spectrum, and the parameters were then adjusted until the desired spectrum was obtained. The resulting scheme was then tested for G-K-S stability, and if stable, satisfied both the desired criteria for a numerical discretization.

Several technical difficulties were encountered in trying to determine stable formulations in this manner. In general, a large number of free parameters were needed to find a combination which resulted in a stable formulation. This results from trying to achieve a high-order discretizations at the inflow boundary where the stencils are dramatically downwind, and mostly unstable. Although a stable closure condition was found for the sixth-order compact scheme, $(5^2, 5^2 - 6 - 5^2, 5^2)$ it was apparent that if schemes of higher accuracy were to be obtained, a systematic procedure was required to constrain the parameter space over which the search was performed. Another difficulty was that the numerical eigenvalue determination did not yield the exact eigenvalues of the spatial operator, but rather depended on numerical round off and the condition number of the resulting spatial operator. This was not found to be a significant problem for the schemes determined in the study, but it was found that many of the high-order schemes were not well conditioned.

The fundamental difficulty with determining a spatial operator based on the results from an eigenvalue analysis, is that it uses as a basis for the method the the spatial matrix resulting from discretization of the scalar wave equation $U_t + aU_x = 0$. **While G-K-S stability of a discretization on the scalar wave equation implies G-K-S stability on a system of hyperbolic equations, (if the boundary conditions are imposed in characteristic form)[8], the same is not in general true for asymptotic stability.** Therefore, there is no guarantee that the numerical scheme determined in this manner will be stable for an arbitrary hyperbolic problem. . An obvious remedy for the analysis presented by Carpenter et al.[5] would to have used the system not the scalar eigenvalue determination as a basis for devising stable closure formula. This would further constrain an already difficult search procedure to isolate the parameters at the boundaries which would produce an strictly asymptotically stable scheme. An altogether different procedure must be used if an arbitrarily high-order scheme is sought.

The approach of devising suitable boundary closures and then testing them with various stability techniques (such as finding the norm) is entirely the wrong approach when dealing with high-order methods. Very seldom are high-order boundary closures stable, making them difficult to isolate. An alternative approach is to begin with a norm which satisfies all the stability criteria for the hyperbolic system, and look for the boundary closure forms which will match the norm exactly. This method was used recently by Strand[21] to isolate stable boundary closure schemes for the explicit central fourth- and sixth-order schemes. The norm used was an energy norm mimicking the norm for the differential equations. Further research should be devoted to BC for high order schemes in order to make sure that the results obtained are reliable.

References

[1] S. Abarbanel and D. Gottlieb, "Information content in spectral calculations", in Progress and Supercomputing in Computational Fluid Dynamics, edited by E. Murman and S. Abarbanel, (1984), Birkhauser, pp. 345-356.

[2] W. Cai, "Spectral methods for shock wave calculations", Ph.D. thesis, Brown University 1989.

[3] W. Cai, D. Gottlieb and A. Harten, "Cell averaging Chebyshev methods for hyperbolic problems", Comput. and Math. with Appli. 1990.

[4] W. Cai, D. Gottlieb and C. -W. Shu, "Essentially non-oscillatory spectral Fourier methods for shock wave calculations", Math. Comput., V52, (1989), pp. 389-410.

[5] Mark H. Carpenter, David Gottlieb and Saul Abarbanel, "The stability of numerical boundary treatments for compact high-order finite difference schemes", ICASE report 91-71, september 1991.

[6] D. Gottlieb and S. Orszag, "Numerical analysis of spectral methods: theory and applications", SIAM-CBMS, Philadephia, 1977.

[7] D. Gottlieb and E. Tadmor, "Recovering pointwise values of discontinuous data", in Progress and Supercomputing in Computational Fluid Dynamics, edited

[8] Gottlieb D., Gunzburger M., and Turkel E. " ON numerical boundary treatment of hyperbolic systems for finite difference and finite element methods" SIAM Journal on Numerical Analysis 19,4 (1982).

[9] Gottlieb D. Shu, C.W. , Solomonoff A., Vandeven H. "On the Gibbs Phenomenon I" to appear in Applied Numerical Mathematics.

[10] A. Harten, B. Engquist, S. Osher and S. Chakravarthy, "Uniformly high order accurate essentially non-oscillatory schemes, III", J. Comput. Phys., V71, (1987), pp. 231-303.

[11] D. A. Kopriva, "A practical assessment of spectral accuracy for hyperbolic problems with discontinuities", J. Sci. Comput., V2, (1987), pp. 249-262.

[12] H. Kreiss and J. Oliger, "Comparison of accurate methods for the integration of hyperbolic problems", Tellus, V24, (1972), pp. 199-215.

[13] H. Kreiss and J. Oliger, "Stability of the Fourier method", SIAM J. Numer. Anal., V16, (1979), pp. 421-433.

[14] Kriess H.O. and Wu L. "On the stability definitions of difference approximations for Initial Boundary valur problems", Comm. Pure Appl. Math. to appear.

[15] A. Madja, J. McDonough and S. Osher, "The Fourier method for nonsmooth initial data", Math. Comput., V32, (1978), pp. 1041-1081.

[16] B. McDonald, "Flux corrected pseudospectral methods for scalar hyperbolic conservation laws", J. Comput. Phys., V82, (1989), pp. 413-428.

[17] S. Osher and C. -W. Shu, "High order essentially non-oscillatory schemes for Hamilton-Jacobi equations", SIAM J. Numer. Anal., V28, (1991), pp. 907-922.

[18] C.-W. Shu and S. Osher, "Efficient implementation of essentially non-oscillatory shock capturing schemes", J. Comput. Phys., V77, (1988), pp. 439-471.

[19] C.-W. Shu and S. Osher, "Efficient implementation of essentially non-oscillatory shock capturing schemes, II", J. Comput. Phys., V83, (1989), pp. 32-78.

[20] K. R. Sreenivasan, "Transition and turbulent wakes and chaotic dynamical systems", in S.H. Davis and J.L. Lumley, eds, Frontiers in Fluid Mechanics (Springer, New York, 1985), pp. 41-67.

[21] B. Strand- Ph.D thesis Uppsala University, to appear.

[22] E. Tadmor, "Convergence of spectral methods for nonlinear conservation laws", SIAM J. Numer. Anal., V26, (1989), pp. 30-44.

[23] H. Vandeven, "Family of spectral filters for discontinuous problems", J. Sci. Comput., V6, (1991), pp. 159-192.

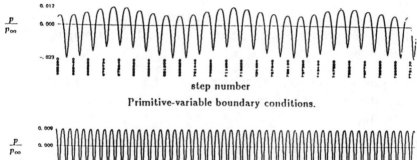

step number

Primitive-variable boundary conditions.

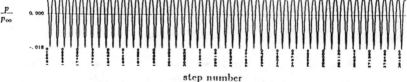

step number

Characteristic-variable boundary conditions.

(a.) Time history of pressure.

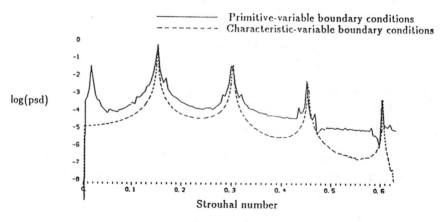

Strouhal number

(b.) Frequency spectrum of pressure.

FIGURE 1 Effect of boundary conditions on pressure in wake region at location 10 cylinder diameters downstream of cylinder and one diameter above wake centerline. Finite-difference code. psd = power spectral density.

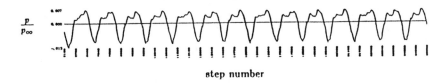

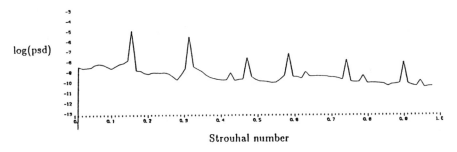

(a.) Time history of pressure.

(b.) Frequency spectrum of pressure.

FIGURE 2 Computed pressure in wake region at location 10 cylinder diameters downstream of cylinder and one diameter above wake centerline. Fully-spectral code. psd = power spectral density.

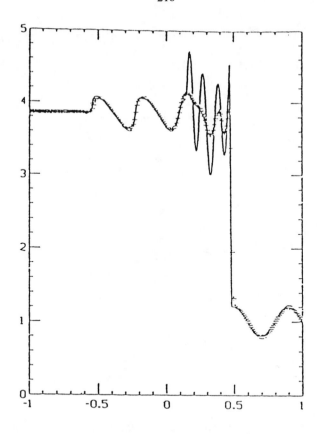

FIGURE 3 2nd order MUSCL TVD

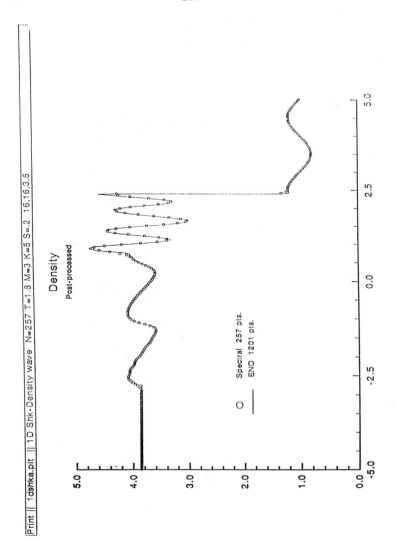

FIGURE 4

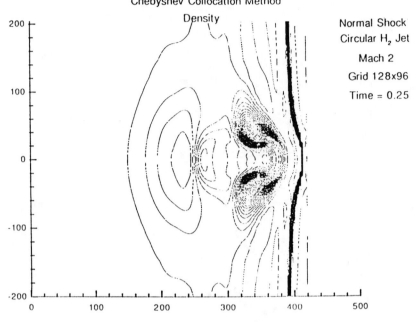

FIGURE 5

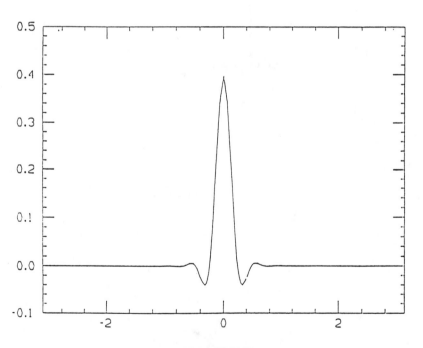

FIGURE 6

ESSENTIALLY NONOSCILLATORY POSTPROCESSING FILTERING METHODS

F. Lafon[1]

C.E.A - CEL-V BP 27
94195 Villeneuve Saint Georges, France

S. Osher[2]

Department of Mathematics, University of California
Los Angeles, CA 90024-1555

ABSTRACT

High order accurate centered flux approximations used in the computation of numerical solutions to nonlinear partial differential equations produce large oscillations in regions of sharp transitions. In this paper, we present a new class of filtering methods denoted by ENO-LS (Essentially Nonoscillatory Least Squares) which constructs an upgraded filtered solution that is close to the physically correct weak solution of the original evolution equation. Our method relies on the evaluation of a least squares polynomial approximation to oscillatory data using a set of points which is determined via the ENO framework.

Numerical results are given in one and two space dimensions for both scalar and systems of hyperbolic conservation laws. Computational running time, efficiency and robustness of the method are illustrated in various examples such as Riemann initial data for both Burgers' and Euler's equations of gas dynamics. In all standard cases the filtered solution appears to converge numerically to the correct solution of the original problem. Some interesting results based on nonstandard central difference schemes, which exactly preserve entropy, and have been recently shown generally not to be weakly convergent to a solution of the conservation law, are also obtained using our filters.

[1]Research supported by ONR Grant N00014-86-K-0691 and NSF Grant DMS 88-11863.

1 Introduction

Numerical improvements in the computation of high order accurate numerical solutions to nonlinear hyperbolic conservation laws have been recently obtained. Hence, following Total Variation Diminishing (TVD) schemes, the Essentially Nonoscillatory (ENO) method has been introduced and proved to be very efficient in the computation of high accurate numerical solutions for several types of physical problems including Computational Fluid Dynamics (CFD) problems or front propagation using the Hamilton-Jacobi framework. However, these high accurate methods use a lot of computational time. For that reason, filtering methods were developed, beginning in the late eighties. The first one, described by B. Engquist and B. Sjogreen in [1], uses simple TVD and conservation properties to correct nonphysical spurious oscillations from one time step to another. The correction step consists in pushing numerical data points up or down to an acceptable level while preserving conservation. In [5], we presented a new class of filtering methods of any order of accuracy. Our method relies on switching fluxes at locations in which spurious oscillations are detected. This method was observed to be very efficient and its cost relatively low since high order ENO fluxes are evaluated at a few points only- a central difference method is used most often.

In this paper, we investigate some interesting computational properties of centered schemes after numerical oscillations have developed and propagated for some time. We define a new class of filtering methods that can be applied to highly oscillatory numerical solutions. This relies on the construction of an ENO stencil of points ([2, 7, 8]) which is fitted with high degree polynomials from a least squares procedure. Our numerical filter is capable of smoothing oscillations having large amplitude and high frequency, but without removing sharp singularities which are crucial components of these solutions. Furthermore, the filtered solution seems to retain the oscillatory solution properties of the unfiltered schemes in some special "entropy preserving" cases as defined in [3] by J. Goodman and P. D. Lax, and in [6] by J. Liu and D. Levermore. We investigate some numerical examples using several centered approximations in order to illustrate the former property. The main conclusion indicates that for standard central differences, our filtered solution always converges accurately to the strong limit, whereas the predicted oscillatory behavior is retained even after using our filter in the examples of [3, 6].

Our main test problems will be inviscid Burgers' equation and the inviscid Euler equations of gas dynamics. We first consider Burgers' equation:

$$U_t + \left(\frac{U^2}{2}\right)_x = 0, \tag{1}$$

with smooth initial condition $U(x,0) = U_0(x)$, $U_0 \in C^\infty(0,1)$, and periodic boundary conditions. We will discuss the main properties of the numerical solution obtained from some schemes based on the approximate fluxes:

$$F_{j+\frac{1}{2}} = \tfrac{1}{2}(U_{j+1}^2 + U_j^2), \tag{2}$$

$$F_{j+\frac{1}{2}} = \tfrac{7}{24}(U_{j+1}^2 + U_j^2) - \tfrac{1}{24}(U_{j+2}^2 + U_{j-1}^2), \tag{3}$$

$$F_{j+\frac{1}{2}} = \tfrac{37}{120}(U_{j+1}^2 + U_j^2)) - \tfrac{8}{120}(U_{j+2}^2 + U_{j-1}^2) + \tfrac{1}{120}(U_{j+3}^2 + U_{j-2}^2) \tag{4}$$

$$F_{j+\frac{1}{2}} = \tfrac{1}{6}(U_{j+1}^2 + U_j U_{j+1} + U_j^2), \tag{5}$$

$$F_{j+\frac{1}{2}} = \tfrac{1}{2} U_j U_{j+1}. \tag{6}$$

The fluxes (2,3,4) are just standard central differencing of second, fourth and sixth order of accuracy, respectively; while (5,6) are the interesting examples analyzed in [6] and in [3], respectively. The oscillatory solution obtained from any of these fluxes is then corrected by the ENO-LS method, preserving the formal order of accuracy.

The Euler equations of gas dynamics:

$$\begin{aligned} \mathbf{U}_t + (\mathbf{F}(\mathbf{U}))_x &= 0, \\ \mathbf{U}(x,0) &= \mathbf{U}_0(x), \end{aligned} \tag{7}$$

are to be solved for $t > 0$ and x in some interval Ω with appropriate boundary conditions, where

$$\mathbf{F}(\mathbf{U}) = v\mathbf{U} + (0, p, vp)^T$$

and $\mathbf{U} = (\rho, q, p)$, ρ is density, q is momentum, v is velocity, and p is the pressure. In this work, we use conventional second, fourth and sixth order central differencing with ENO-LS post processing applied to Euler's equations. See [4] for an analysis of the oscillatory Von Neumann-Richtmyer scheme approximating Euler's equation, [9].

2   The ENO-LS Filter, Algorithms

The ENO-LS method mimics the construction of ENO polynomials but without involving the evolution equations. In short, we follow the algorithm just below:

Algorithm 2.1   • *1.) Compute N times the numerical solution of*

$$U_t + A(U)_x = \quad 0$$
$$U(x,0) = \quad U_0(x)$$

i.e we let $V_j^0 = U_0(x_j)$, and compute $V_j^N = I(V^0, N\Delta t)$, for all $j = 1, ..., n$, where I is the solution semigroup operator that transforms the initial pointwise data V_j^0 to V_j^N after N iteration time steps.

• *2.) Filter the numerical solution computed in step 1.) by an iteration procedure similar to Jacobi or Gauss-Seidel elliptic solvers; first let:*

$$W_j^0 = \quad V_j^N$$

for all $j = 0, ..., n$, then make use of primitive variables:

$$U_{j+\frac{1}{2}}^m = \sum_{i=0}^{j} W_i^m \Delta x_i, \tag{8}$$

and finally construct a sequence $\{U_{j+\frac{1}{2}}^m\}_{m=0,...,M}$ so that

$$U_{j+\frac{1}{2}}^{m+1} = \quad E(U^m, U^{m+1}), \tag{9}$$

where M is defined from the stopping criterion:

$$\|W^{m+1} - W^m\| \leq \varepsilon,$$

for $m = 0, ..., M - 1$, where ε is a small parameter of order Δx^α and

$$W_j^m = \frac{U_{j+\frac{1}{2}}^m - U_{j-\frac{1}{2}}^m}{\Delta x_j}. \tag{10}$$

Finally, let $\bar{V}_j^N = W_j^M$, for $j = 1, ..., n$.

- *3.) Go to step 1.) unless $t = t_{max}$.*

We notice that relation (10) and the use of primitive variables (8) implies the conservation property of the sequence $\{W_j\}_{j=1,...,n}$, i.e the resulting finite difference scheme is always in conservation form. Moreover, the number of correction steps M can be initially fixed so that the ratio $\frac{N}{M}$ is as large as desired. The operator E is a non trivial linear combination of $U^m_{j+\frac{1}{2}}, U^{m+1}_{j+\frac{1}{2}}$, for some j, as in point Jacobi or Gauss Seidel method. In the Jacobi procedure we have:

$$U^{m+1}_{j+\frac{1}{2}} = E(U^m_{\frac{1}{2}}, U^m_{\frac{3}{2}}, ..., U^m_{j-\frac{1}{2}}, U^m_{j+\frac{1}{2}}, ..., U^m_{n-\frac{1}{2}}),$$

and either

$$
\begin{cases}
U^{m+1}_{j+\frac{1}{2}} = E(U^{m+1}_{\frac{1}{2}}, ..., U^{m+1}_{j-\frac{1}{2}}, U^m_{j+\frac{1}{2}}, ..., U^m_{n-\frac{1}{2}}) & \text{if } j \text{ varies from 0 to} \\
 & n, \text{ or} \\
U^{m+1}_{j+\frac{1}{2}} = E(U^m_{\frac{1}{2}}, ..., U^m_{j-\frac{1}{2}}, U^{m+1}_{j+\frac{1}{2}}, ..., U^{m+1}_{n-\frac{1}{2}}) & \text{in the reverse} \\
 & \text{direction;}
\end{cases}
$$

for the Gauss-Seidel method.

An important property of the ENO-LS algorithm comes from the fact that the corrected solution V_j^N satisfies a conservation equation. To see that, we first use the relation (10) which can be rewriten as:

$$
W_j^{m+1} = W_j^m + \frac{E^{j+\frac{1}{2}}(U_{j+\frac{1}{2}-r_+}, ..., U_{j+\frac{1}{2}+s_+}) - U^m_{j+\frac{1}{2}}}{\Delta x_j}
$$

$$
- \frac{E^{j-\frac{1}{2}}(U_{j-\frac{1}{2}-r_-}, ..., U_{j-\frac{1}{2}+s_-}) - U^m_{j-\frac{1}{2}}}{\Delta x_j}
$$

(11)

and then discuss the construction of the least squares process. In (11), the pair of integers $(r_\pm, s_\pm)$ limit the width of the stencil used in the evaluation of the least squares polynomial. The appropriate stencil is defined as in the ENO algorithms (refer to [7], [8]). We briefly indicate the main steps leading to such a polynomial: We first compute the divided differences table of W^m and define the ENO stencil of points in the region which is the smoothest for successive space derivatives of W^m. We denote the ENO stencil in the set $(x_{j-r}, ..., x_{j+s})$, where $r + s = p - 1$, and p is the number of data points that we want to take into account in the evaluation of the least squares polynomial. Hence if we denote this polynomial by

$$P^{j+\frac{1}{2}}(x) = \sum_{i=0}^{q} Y_i \varphi_i(x), \tag{12}$$

where $(\varphi_0, ..., \varphi_q)$ are the basis functions of some polynomial space of degree q, then the unknown coefficients $\mathbf{Y} = (Y_0, ..., Y_q)$ are solutions of the linear system:

$$C^{j+\frac{1}{2}}\mathbf{Y} = \mathbf{F}^{j+\frac{1}{2}},$$

where $C^{j+\frac{1}{2}}$ is a $(q+1) \times (q+1)$ square matrix, $\mathbf{F}^{j+\frac{1}{2}}$ is a $q+1$ column vector, and both can be computed from the basis functions:

$$C_{k,l}^{j+\frac{1}{2}} = \sum_{i=-r}^{s} \varphi_k(x_{j+\frac{1}{2}+i}) \varphi_l(x_{j+\frac{1}{2}+i}),$$

$$F_k^{j+\frac{1}{2}} = \sum_{i=-r}^{s} U_{j+\frac{1}{2}+i} \varphi_k(x_{j+\frac{1}{2}+i}).$$

The updated value $U_{j+\frac{1}{2}}^{m+1}$ follows from letting $x = x_{j+\frac{1}{2}}$ on the previously constructed LS polynomial:

$$U_{j+\frac{1}{2}}^{m+1} = P^{j+\frac{1}{2}}(x_{j+\frac{1}{2}}) = \sum_{i=0}^{q} \{(C^{j+\frac{1}{2}})^{-1} F^{j+\frac{1}{2}}\}_i \varphi_i(x_{j+\frac{1}{2}}).$$

The global conservation feature follows provided that we write:

$$U_{j+\frac{1}{2}}^{m+1} = \sum_{t=-r}^{s} \alpha_{j+\frac{1}{2}+t}^{j+\frac{1}{2}} U_{j+\frac{1}{2}+t}^{m+l},$$

where $l = 0$ or 1 depending on whether the Jacobi or the Gauss-Seidel method is used, and

$$\alpha_{j+\frac{1}{2}+t}^{j+\frac{1}{2}} = \sum_{i=0}^{q} \sum_{k=0}^{q} D_{i,k}^{j+\frac{1}{2}} \varphi_i(x_{j+\frac{1}{2}}) \varphi_k(x_{j+\frac{1}{2}+t}),$$

where $D^{j+\frac{1}{2}} = (C^{j+\frac{1}{2}})^{-1}$.

Note first that the coefficients $\{\alpha_{j+\frac{1}{2}+t}^{j+\frac{1}{2}}\}_{t=-r,...,s}$ form a sequence of bounded real numbers, and second, that the basis functions $(\varphi_0, ...\varphi_q)$ can be appropriately chosen as a sequence of Chebyshev polynomials or some other set of orthogonal polynomials. The last choice permits us to compute directly the coefficients $\alpha^{j+\frac{1}{2}}$ without inverting the matrix $C^{j+\frac{1}{2}}$:

$$\alpha^{j+\frac{1}{2}}_{j+\frac{1}{2}+t} = \sum_{i=0}^{q} \frac{\varphi_i(x_{j+\frac{1}{2}})\varphi_i(x_{j+\frac{1}{2}+t})}{C^{j+\frac{1}{2}}_{i,i}}.$$

This yields a fast algorithm since matrix inversions are no longer needed.

We now present an extension of algorithm 2.1 to two space dimensions. The simplest possible extension would be to apply the previous algorithm in two sweeps. The first one will freeze one coordinate and correct the oscillatory solution with respect to the other free variable. The second one will simply reverse the role of each variable. This method, while simple, has difficulties near curved shocks. We shall use instead a fully two dimensional ENO-LS filtering algorithm. The latter will provide the construction of least squares polynomials in two space dimensions using a set of points which is chosen as the intersection of one dimensional ENO stencils of data points in each separate direction. The algorithm below describes our procedure.

Algorithm 2.2   • 1.) *Compute N times the numerical solution of*

$$U_t + A(U)_x + B(U)_y = 0$$
$$U(x, y, 0) = U_0(x, y)$$

using a very simple numerical method and then let $V^0_{i,j} = U_0(x_j, y_i)$, and $V^N_{i,j} = I(V^0, N\Delta t)$, for all $j = 1, ..., n_1$, and $i = 1, ..., n_2$, where I is the solution semigroup operator that we have already encountered in previous algorithm.

• 2.) *Let $W^0_{i,j} = V^N_{i,j}$, and filter M times the primitive variable*

$$U^m_{i+\frac{1}{2},j+\frac{1}{2}} = \sum_{l,k=0}^{i,j} W^m_{l,k}\Delta x_k \Delta y_l, \tag{13}$$

by a Jacobi or point Gauss-Seidel iteration procedure, i.e perform:

$$U^{m+1}_{i+\frac{1}{2},j+\frac{1}{2}} = E(U^m, U^{m+1}),$$

for positive integers m. Iterate as long as $\|W^{m+1} - W^m\| \le \varepsilon$, $m = 0, ..., M - 1$, and finally let the filtered solution: $V_{i,j}^N = W_{i,j}^M$, for $j = 1, ..., n_1$, $i = 1, ..., n_2$.

The construction of the two dimensional least squares polynomial is as follows:

- *2.1) At the location $(x_{i+\frac{1}{2}}, y_{j+\frac{1}{2}})$, compute the ENO stencils of points*

  $\{(x_{i-r_x}, y_j), ..., (x_{i+1+s_x}, y_j)\}$, and $\{(x_i, y_{j-r_y}), ..., (x_i, y_{j+1+s_y})\}$, *for $r_x + s_x + 1 = n_1$ and $r_y + s_y + 1 = n_2$, where n_1 and n_2 are the preset maximum number of points along the x and y axis. These sets of points form two segments crossing at (x_i, y_j). Then, define the x and y ENO stencils of points along these y and x segments, respectively. The two dimensional ENO stencil is taken as the intersection of the union of the predefined x and y one dimensional ENO stencils (refer to figure (1)). The least squares polynomial is then simply defined on this two dimensional ENO stencil and is set to $P^{(i+\frac{1}{2},j+\frac{1}{2})}(x,y) = \sum_{k=0}^q Y_k \varphi_k(x,y)$. Again, the unknowns coefficients Y_k, $k = 1, ..., p$ are computed by solving the linear system $C^{(i+\frac{1}{2},j+\frac{1}{2})}\mathbf{Y} = \mathbf{F}^{(i+\frac{1}{2},j+\frac{1}{2})}$.*

- *2.2) Let $U_{i+\frac{1}{2},j+\frac{1}{2}}^{m+1} = P^{(i+\frac{1}{2},j+\frac{1}{2})}(x_{i+\frac{1}{2}}, y_{j+\frac{1}{2}})$.*

- *2.3) Recover the conservative solution*

$$W_{i,j}^{m+1} = \frac{U_{i+\frac{1}{2},j+\frac{1}{2}}^{m+1} + U_{i-\frac{1}{2},j-\frac{1}{2}}^{m+1} - U_{i+\frac{1}{2},j-\frac{1}{2}}^{m+1} - U_{i-\frac{1}{2},j+\frac{1}{2}}^{m+1}}{\Delta x \Delta y}, \quad (14)$$

for $m = 0, ..., M - 1$.

- *3.) Go to step 1.) unless $t = t_{max}$.*

An example of two dimensional ENO stencil is given in figure 1. The main interesting feature of such construction is based on the localization of the least oscillatory part of the solution within the much larger rectangle $[x_{j-r_x}, x_{j+s_x}] \times [y_{i-r_y}, y_{i+s_y}]$.

In section 4, we investigate the two dimensional Burgers' equation and study numerical propagation of a shock along the radial axis. With centered fluxes, some spurious numerical oscillations propagate in the direction of the flow; however the two dimensional ENO-LS

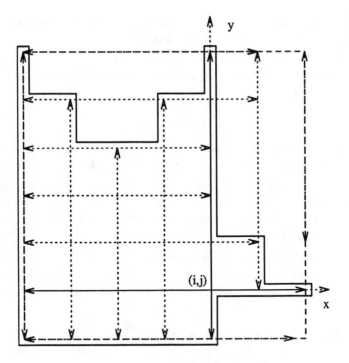

Figure 1: 2D ENO-LS stencil.

filtering method is able to filter all these oscillations while still giving the correct location of the curved shock wave.

Next, we consider hyperbolic systems of conservation laws exemplified by the inviscid Eulers' equations of compressible gas dynamics:

$$U_t + F(U)_x = 0,$$
$$U(x,0) = U_0(x)$$

for which there exists a complete set of real eigenvectors and eigenvalues; i.e $\nabla F(U) = P^{-1}\Lambda P$, where Λ is a diagonal matrix with entries $\lambda_1 \leq \lambda_2 ... \leq \lambda_d$, and P is a matrix whose columns define a complete set of eigenvectors of the system. Note that the eigenvalues can be multiple, which is the case for the Euler equations of gas dynamics in two space dimensions. Using the eigenvector decomposition, a field by field approach is used, i.e the fluxes will be corrected in each fields (our filter sometimes failed to remove oscillations near strong shocks when applied to the conserved quantities). Briefly, we proceed as follows:

Algorithm 2.3 (Field by field ENO-LS method)   • *2-1.) Let* $\mathbf{W}^m = (W_1^m, ..., W_q^m)^T$, *and define the primitive variable* $\mathbf{V}_{j+\frac{1}{2}}^m = \sum_{i=1}^j \mathbf{W}_i^m \Delta x$.

• *2-2.) Compute the left and right eigenvectors* $\mathbf{L}_{j+\frac{1}{2}}^{m,(k)}$, $\mathbf{R}_{j+\frac{1}{2}}^{m,(k)}$, *for* $k = 1, ..., d$ *(d is the number of equations) of the Roe matrix (see e.g. [2]) $A(\mathbf{W}_j^m, \mathbf{W}_{j+1}^m)$, and decompose the primitive vector* $\mathbf{V}_{j+\frac{1}{2}}^m$ *along each characteristic field:*

$$\sigma_{j+\frac{1}{2}}^{m,(k)} = \mathbf{L}_{j+\frac{1}{2}}^{m,(k)} \cdot \mathbf{V}_{j+\frac{1}{2}+l(k)}^m$$
$$\mathbf{Q}_{j+\frac{1}{2}}^m = \sum_{k=1}^d \sigma_{j+\frac{1}{2}+l(k)}^{m,(k)} \mathbf{R}_{j+\frac{1}{2}}^{m,(k)},$$

for which we have frozen the index j in order to get the same decomposition for all neighboring points $x_{j+\frac{1}{2}+l(k)}$, *for* $l(k) = -r(k), ..., +s(k)$ *involved in the calculation of the least squares polynomial which is constructed in step 2-3.).*

• *2-3.) Select an ENO stencil of p points, i.e* $\{x_{j-r(k)}, ..., x_{j+s(k)}\}$, *for* $r(k) + s(k) + 1 = p$; *and then define the least squares polynomial* $P^{(j+\frac{1}{2})}$ *of degree q so that:*

$$\sigma_{j+\frac{1}{2}}^{m+1,(k)} = P^{(j+\frac{1}{2})}(\sigma_{j+\frac{1}{2}-r(k)}^{m,(k)}, ..., \sigma_{j+\frac{1}{2}+s(k)}^{m,(k)}; x_{j+\frac{1}{2}}).$$

- *2-4.) Transform back the filtered field by field solution to the primitive vector:*

$$\mathbf{V}_{j+\frac{1}{2}}^{m+1} = \sum_{k=1}^{d} \sigma_{j+\frac{1}{2}}^{m+1,(k)} \mathbf{R}_{j+\frac{1}{2}}^{m,(k)},$$

and finally recover the desired physical variables:

$$\mathbf{W}_{j}^{m+1} = \frac{\mathbf{V}_{j+\frac{1}{2}}^{m+1} - \mathbf{V}_{j-\frac{1}{2}}^{m+1}}{\Delta x}. \tag{15}$$

- *2-5.) Iterate until the stopping criterion*

$$\|\mathbf{W}^{m+1} - \mathbf{W}^{m}\| \leq \varepsilon,$$

for $m = 0, ..., M - 1$, is reached; and finally let $\bar{\mathbf{U}}_{j}^{N} = \mathbf{W}_{j}^{M}$.

Note that this algorithm can be extended to two space dimensions by correcting separately oscillatory fields involved in the x and y fluxes, respectively. Moreover, this algorithm does not make use of the evolution equations but does use the eigenfunction decomposition in order to track efficiently the propagation of spurious oscillations.

To conclude this section, we indicate that we mainly supposed that the numerical oscillations always propagate with the flow speed along local characteristic fields and that the amplitudes of such oscillations are not too large so that the oscillatory solution does not become unbounded, e.g., negative density and/or pressure is not allowed. In all our numerical experiments, we had to turn the filter on not only to recover an acceptable final solution but also to reduce the amplitude and frequency of spurious oscillations during the calculations. Hence, we usually preset the value of the ratio $\frac{N}{M}$ in the numerical experiments just after singularities have developed.

3 Numerical Convergence Study

In this section, we investigate the numerical convergence of the approximate solution computed via the ENO-LS algorithm given an initial oscillatory solution which has been evaluated from one of the standard centered fluxes (2), (3), (4); or from one of the "entropy conserving" fluxes (5) or (6). As test problem, we study first the numerical evolution of Burgers' equation (1) in one space dimension:

$$U_t + (\frac{U^2}{2})_x = 0,$$

with initial condition $U(x,0) = \sin 2\pi x$, in the domain $[0,1]$, and extend the solution by periodicity outside 0 and 1. A shock wave develops at $t = \frac{2}{\pi}$ at the point $x = \frac{1}{4}$.

We first display in figure (2) the solution using 100 cells and a CFL condition of $\frac{1}{2}$ for a second order centered difference (CD) scheme using the flux (2), and then correct that solution by applying 7 iterations of the ENO-LS algorithm. Note first that the exact solution is perfectly recovered with exact location of the singularity because of the use of primitive variables in algorithm 2.1; second, these results are obtained for the set of coefficients $(p,q) = (13,2)$, which implies that the corrected solution is only second order accurate in smooth transition areas; third, the upper right plot displays the corrected solution when both (3,2)CD and (3,2)ENO schemes are sequentially used. In that case, we correct by 5 ENO iterations after every 25 CD steps. This method for filtering oscillations is also very efficient for one dimensional problems (refer to table 1); however, many more ENO iterations are needed in two dimensions. Basically, 10 ENO iterations are needed after every 10 CD steps in order to recover an accurate solution for the two dimensional Burgers' example 4.3, which is a quite expensive technique compared to the overall cost induced by the ENO-LS method.

We shall discuss the results obtained with larger values of q in section 4 in which we investigate numerical order of accuracy of the filtered solutions from the ENO-LS algorithm. Also, a similar study is investigated as the number of evaluation points p increases.

In figure (3), we plot the numerical solution before and after the filter steps when $n = 1000$. Again, convergence to the physical solution is reached within 10 Gauss Seidel iterations. Note that this number increases by a factor of nearly 2 when Jacobi iterations are used.

In figures (4) and (5), we show the filtered solutions which were initially computed using the standard (3,4)CD and (3,6)CD schemes (fluxes (3) and 4)), respectively. Note that the corrected solution is well reconstructed in smooth regions while a smearing of the shock over about 10 cells is observed. This smearing can be primarily

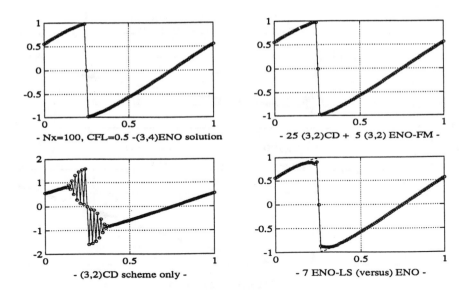

Figure 2: (3,2)CD scheme + ENO-LS method - 100 cells.

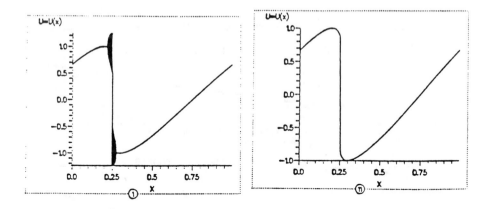

Figure 3: (3,2)CD scheme + ENO-LS method - 1000 cells.

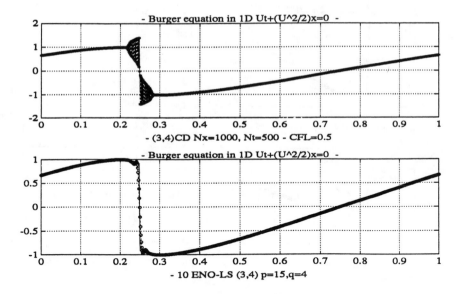

Figure 4: (3,4)CD scheme + 10 ENO-LS method - 1000 cells.

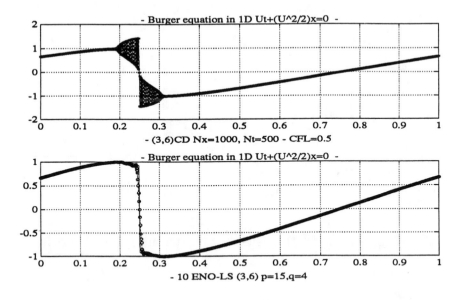

Figure 5: (3,6)CD scheme + 10 ENO-LS method - 1000 cells.

explained from the high (fourth) degree least squares polynomials ($q = 4$) used in those experiments.

Our second numerical test is devoted to showing that the filtered solution computed via the numerical flux (5):

$$F_{j+\frac{1}{2}} = \frac{1}{6}(U_{j+1}^2 + U_j U_{j+1} + U_j^2).$$

does not converge to the physical solution. It was observed in [6] that this scheme preserves mass and entropy (which in this case is taken to be $\frac{1}{2}U^2$). It was also shown there that the numerical solution does not converge to a weak solution of the original problem. Basically, as the mesh size tends to zero, some spurious oscillations are still visible near the shock and cannot be removed. We ran the last two previous experiments but used the approximate flux (5). We plot the numerical results in the figures (6) and (7) when $n = 100$ and $n = 1000$, respectively. For 100 cells (figure (6)), the oscillations near the shock are all smoothed out; however, the location of the shock is smeared over about 10 cells. This results were obtained for the same set of coefficients $(p, q) = (13, 2)$. Indeed, the solution is not well reconstructed. As the number of cells increases, the filtered solution is well reconstructed except near the shock. Numerical oscillations are still visible and cannot be removed. Note that the results visualized in figure (7) are given after 16 Gauss Seidel iterations which is much more than we needed when standard centered differences are taken.

Our final Burgers' equation test deals with a similar convergence failure property to the correct physical solution when the flux approximation (6) is implemented. The numerical flux is:

$$F_{j+\frac{1}{2}} = \frac{1}{2}U_j U_{j+1}.$$

This scheme again conserves both mass and entropy (this time the entropy is taken to be $\log U_j$), and it was shown in [3] that the numerical solution does not converge to a weak limit of the original problem as the stepsize Δx tends to zero. In the numerical experiments, we noticed that the amplitude of the oscillations grew very fast and was rapidly becoming unbounded. The results are plotted in the figures (8) and (9) for $n = 100$ and $n = 1000$, respectively. Note that, in the case $n = 100$, the filtered solution is again not very well reconstructed, and for 1000 cells, some oscillations are still visible after 20 ENO-LS iterations on the right of the shock. Moreover

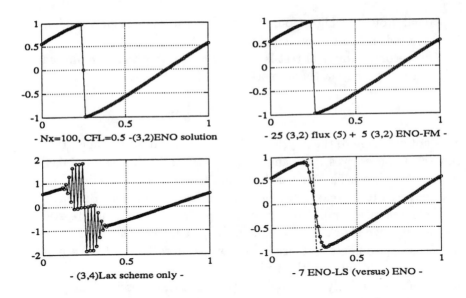

Figure 6: Flux (5) + ENO-LS method - 100 cells.

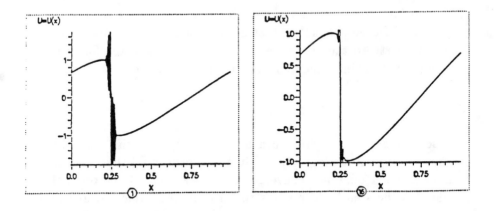

Figure 7: Flux (5) + ENO-LS method - 1000 cells.

these oscillations could not be removed, even after many additional filtering iterations.

Thus our techniques have been observed to construct converging sequences of filtered numerical solutions towards the expected solution given initial oscillatory data that has been computed from a central difference flux approximation. Our hope is now to show that our method is not only robust but is also fast. This is the topic of last section.

4　Time Efficiency of ENO-LS Algorithms

In this section, we want to test the ENO-LS filtering method for several test problems involving nonlinear hyperbolic systems of conservation laws. We will focus our attention on comparing precisely the CPU times of the ENO-LS method versus more classical and filtered methods. Among them, we will consider the straightforward central difference (CD) method, the expensive ENO (Essentially Nonoscillatory) technique [7, 8], and our FM scheme (Filtering Method) [5]. We will run three examples for 1D and 2D Burgers' equation, and for Eulers' equations of compressible gas dynamics.

4.1　1D Burgers'

We first compare the time efficiency of several numerical schemes for the 1D Burgers' equation (1). In table (1), we show the average computational time per iteration for 100 mesh points for the CD, ENO, and FM schemes with several correction angles (see [5]), and for various combinations, performing alternatively some centered difference and some ENO steps. The notation CD+ENO (40,5,20) simply means that the calculation starts with 40 CD steps, followed by 5 full ENO iterations, and back to the centered scheme; the last 20 steps of the calculations are finally performed by the ENO method in order to recover an acceptable solution. Note that all coefficients displayed in this table are tuned so that the numerical solution is high order accurate in smooth regions and no spurious oscillations are detected.

The contents of this table need a few comments. First of all, the fastest algorithm is the one based on central differences. This is indeed not surprising since only one Fortran instruction is needed in

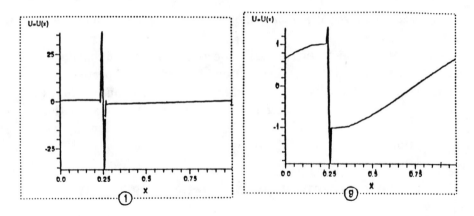

Figure 8: Flux (6) + ENO-LS method - 100 cells.

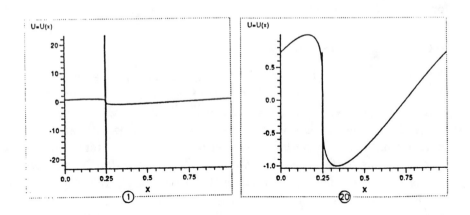

Figure 9: Flux (6) + ENO-LS method - 1000 cells.

Type of Scheme	Order of Accuracy	CPU time ×10⁻²	# of corrections, comments
CD	(3,2)	0.56	CD = Centered Differences
CD	(3,4)	0.72	–
ENO-RF	(3,2)	1.11	RF = Roe Fix, Entropy fix
ENO-RF	(3,4)	1.68	–
FM	(3,2)	0.76	FM = Filtering Method, 3% Corrections
FM	(3,4)	1.07	4% Corrections
FM	(3,2)	1.40	100% Corrections = full ENO
FM	(3,4)	2.04	100% Corrections
CD+ENO (40,5,20)	(3,2)	0.74	40CD, 5ENO, 20 final ENO
CD+ENO (40,1,10)	(3,2)	0.58	40CD, 1ENO, 10 final ENO
CD+ENO (20,7,10)	(3,4)	0.99	20CD, 7ENO, 10 final ENO
ENO-LS (7,2)	(2)	1.84	$(p,q) = (7,2)$, LU Inversion
ENO-LS (7,2)	(2)	0.65	Orthogonal polynomials
ENO-LS (7,3)	(3)	2.5	LU Inversion
ENO-LS (7,3)	(3)	0.89	Orthogonal polynomials

Table 1: CPU times of CD, ENO, FM and ENO-LS methods.

the coding of the approximate flux. Second, postprocessing a numerical solution computed from the CD scheme by an ENO method can be very efficient for lower orders. For a second order method, only one ENO iteration after every 40 CD steps has to be implemented in order to reduce sufficiently the amplitude of oscillations. Note however that for the fourth order method, 7 ENO iterations were needed every after only 20 CD steps. In fact, if these spurious oscillations are not regularly cut off, the final ENO iterations may not recover the correct numerical solution. Third, the ENO-LS filtering method is the most costly when full LU inversion of the $C^{j+\frac{1}{2}}$ matrix is performed. However, when orthogonal basis functions are introduced, the ENO-LS method is competitive with respect to the fast CD scheme. Note moreover that ENO-LS correction steps have to be performed at a few times only. Finally, after running many experiments, we noticed that if the ratio $\frac{p}{q}$ becomes too large, then shocks have a tendency to spread over a large number of cells. Again, there is a compromise that needs to be reached for fast convergence; this depends on the large value of the ratio $\frac{p}{q}$, and the approximation of the shape near shocks for which p needs to be slightly smaller. Note that in most of experiments, the ratio $\frac{p}{q} \in [4,6]$ was optimal.

Now, we want to measure the order of accuracy of the ENO-LS solution. To do so, we measure from computations the L^1 and L^∞ errors in the slabs $[0.10, 0.24]$ and $[0.26, 0.30]$. Numerical orders are

Type of Scheme	(q,p)	# of iterations	L^1 and L^∞ orders.
CD (3,2)	(2,5)	12	1.81 and 1.63
CD (3,2)	(2,9)	9	1.75 and 1.50
CD (3,2)	(2,13)	7	2.19 and 1.9
CD (3,4)	(3,5)	15	1.59 and 1.40
CD (3,4)	(3,9)	12	2.41 and 2.33
CD (3,4)	(3,13)	7	2.80 and 2.79

Table 2: Local L^1 and L^∞ order of accuracy.

shown in table 2.

Several comments about these results are now discussed. First of all, as the number of evaluation points increases, the better the quality of the results and the faster convergence is reached. Indeed, we have to pay the price of higher computations which are required to construct the coefficients involved in the $C^{j+\frac{1}{2}}$ matrix and in the vector $F^{j+\frac{1}{2}}$. On the other hand, faster calculations can be obtained provided that the values of p and q differ only slightly, i.e $p = q + l$, $l = 1, 2, 3, ...$, but local accuracy becomes obviously poor. Again, the optimal ratio seems to belong to the interval $[4, 6]$.

4.2 Example 2. Euler Equations of Gas Dynamics

The second test problem is devoted to the Euler equations of gas dynamics in one space dimension. We consider the initial condition given in example 8 of [8], that is:

$$\rho = 3.857143, q = 2.629369, p = 10.3333333 \text{ when } x < -4$$
$$\rho = 1 + \varepsilon \sin 5x, q = 0, p = 1. \text{ when } x \geq -4$$

When $\varepsilon = 0$, a pure Mach 3 shock is moving to the right from the initial discontinuity $x = 4$. When $\varepsilon = 0.2$, we not only have a Mach 3 shock propagating to the right, but have as well a succession of weaker rarefaction and shock waves propagating to the left. Numerical results for the ENO and FM methods of high order of accuracy can be found in [8] and in [5], respectively. We ran the same problem using 240 CD iterations and then plotted the results in figure (10). Next, we correct this highly oscillatory numerical solution by performing 7 ENO-LS correction steps. In this experiment, we use $(p, q) = (13, 3)$. The filtered solution is visualized in figure (11). Note that the pressure, velocity, and entropy are quite well reconstructed, whereas the density is not perfectly recovered near the strong shock

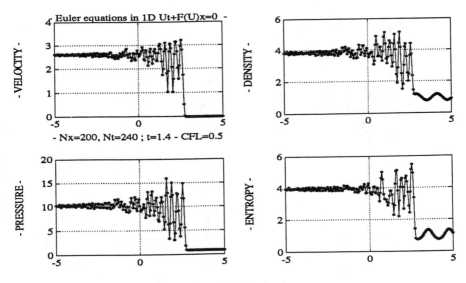

Figure 10: (3,2)CD scheme.

for which some physical oscillations should remain (refer to [10, 11]). However, we should note the remarkable improvement obtained from the solutions displayed in figures (10) and (11).

In figures (12) and (13), we use in sequence 50 CD steps and only one ENO iteration for second and fourth order methods before correcting the final oscillatory solution. The oscillatory solution is now postprocessed by 3 ENO-LS iterations. The filtered numerical results now look fairly similar to those shown in [8, 5].

4.3   Example 3. 2D Burgers' Equation

The last example is devoted to the two dimensional Burgers' equation to be solved in the square domain $[-1, 1] \times [-1, 1]$ with initial condition $U_0(x, y) = \sin 2\pi r$, where $r = \sqrt{x^2 + y^2}$, for $r \leq \frac{1}{4}$, and $U_0(x, y) = 0$ outside the disc $r = \frac{1}{4}$. In figure (14), we visualize the solutions obtained by the (3,2)CD scheme, followed by 4 iterations of ENO-LS correction steps. In that experiment, we use $(p_x, p_y, q) = (6, 6, 2)$, so that the local rectangle in which the

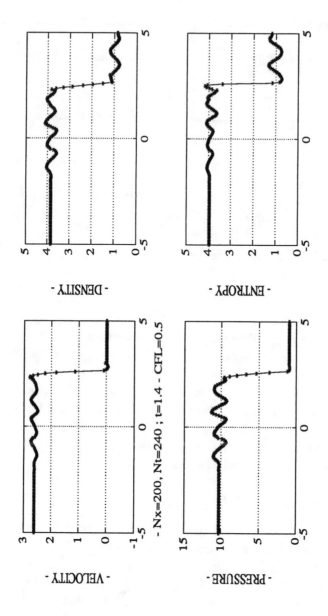

Figure 11: (3,2)CD scheme + 7 ENO-LS method (p,q)=(13,3).

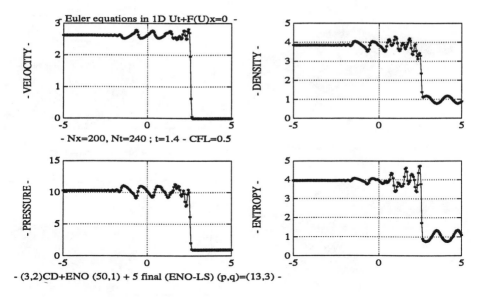

Figure 12: (3,2)(CD,ENO) schemes + 3 ENO-LS method.

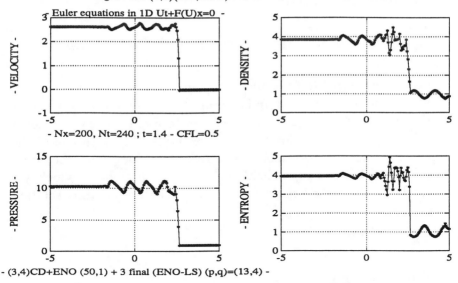

Figure 13: (3,4)(CD,ENO) schemes + 3 ENO-LS method.

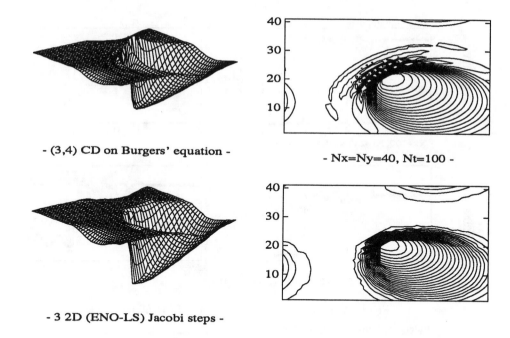

- (3,4) CD on Burgers' equation -

- Nx=Ny=40, Nt=100 -

- 3 2D (ENO-LS) Jacobi steps -

Figure 14: (3,2)CD scheme + 2D ENO-LS method.

two dimensional ENO-LS stencil of points is taken contains 36 mesh points. Note that a speed up factor of nearly 1.5 is obtained by using the two dimensional ENO-LS method instead of the one dimensional splitting version. In this numerical example, the amplitude of the numerical oscillations which propagate radially away from the initial shock location was approximately $\frac{1}{3}$ of the strength of the shock, yet the ENO-LS method recovered the nonoscillatory accurate solution quite well.

References

[1] B. Engquist, P. Lotstedt, and B. Sjogreen, Math. Comput., 52, 509 (1989).

[2] A. Harten, B. Engquist, S. Osher, and S. Chakravarthy, J. Comput. Phys., 71, 231 (1987).

[3] J. Goodman and P. D. Lax, Comm. Pure Appl. Math., 41, 591 (1988).

[4] T. Hou and P. D. Lax, Comm. Pure Appl. Math., 44, 1 (1991).

[5] F. Lafon and S. Osher, "High order filtering methods for approximating hyperbolic systems of conservation laws," ICASE Report, 90-25, 1990, J. Comput. Phys., to appear (1992).

[6] D. Levermore and J. Liu, to appear.

[7] C-W. Shu and S. Osher, J. Comput. Phys., 77, 439 (1988).

[8] C-W. Shu and S. Osher, J. Comput. Phys., 83, 32 (1989).

[9] J. Von Neumann and R. D. Richtmyer, J. Appl. Phys., 21, 380 (1950).

[10] T. Zang, M. Hussaini, and D. Bushnell, AIAA J., 22, 13 (1984).

[11] T. Zang, M. Kopriva, and M. Hussaini, "Pseudospectral Calculation of Shock Turbulence Interactions," ICASE Report 83-14, 1983.

SOME NOVEL ASPECTS OF SPECTRAL METHODS

George Em Karniadakis and Steven A. Orszag

Applied and Computational Mathematics
Princeton University
Princeton, New Jersey 08544

1 Introduction

The papers presented in this section of the book show how extensively spectral methods have changed since the development over
twenty years ago of transform methods enabled their efficient computational implementation. Both Fourier and polynomial spectral
methods are used routinely today in a wide range of applications
including weather simulation, wave dynamics, electromagnetics, and
turbulence simulation. In particular, spectral methods have been
used almost exclusively over the last decade in simulating turbulent
incompressible flows [6]. There are other areas however where the
potential of spectral methods has not been yet been fully explored,
primarily due to concerns that their performance, both in computational efficiency and convergence rate, may be degraded. This lower-
ing of convergence rate may come from irregularities in the computational domain, presence of discontinuities, inappropriate boundary
conditions, etc. We now believe that the integration of good ideas
from a variety of numerically-base disciplines may enable the application of spectral techniques to problem classes thought previously
to be inappropriate for them. While standard global spectral methods may indeed be inefficient in applications such as compressible
turbulence, plasma dynamics, nonlinear optics etc., simple established ideas from other discretization techniques can be employed
to make spectral methods also a useful simulation tool in these applications. There are several issues, for example, to be addressed
for a successful implementation of spectral methods to compressible
flow simulations. Treatment of shocks, recovery of spectral accuracy,
robustness, boundary conditions, and application to complex geometries are perhaps the most important ones. There are similar issues
involved in solving other problems, like micro-fabrication processes
such as ion etching where the presence of face corners produce discontinuous solutions and optical microlithography where small-scale

geometric irregularities need to be treated properly.

There has been an increasing trend over the last decade towards numerical schemes that are based on a hybrid construction of standard discretization algorithms, i.e. finite difference, finite elements, and spectral methods. A typical example of such a confluence of numerical algorithms is the spectral element method which is based on two weighted-residual techniques: finite element and spectral methods [8], [7]. This hybrid construction provides the flexibility for solving problems in complex geometries with spectral accuracy. More recently, hybrid spectral schemes have been formulated that employ heterogeneous discretizations in subdomains in order to effectively exploit the complementary merits of low order/high order methods [5]. Applications include, for example, computational domains with random boundaries.

Recently, there has also been significant progress in extending the use of spectral methods in shock wave calculations [2], [3], [4], [9] by using non-oscillatory constructs and flux limiters, techniques which are traditionally used in low order methods. It has been possible for example to simulate complex multi-dimensional interactions of vortex/shock wave interactions and detonation and deflagration processes in combustion.

In this paper, we briefly review hybrid spectral numerical methods which are used for solving two different classes of problems. In the first category we consider *elliptic* problems on complex geometry domains, and in the second we consider *hyperbolic* problems with discontinuous solutions or initial conditions. The paper is organized as follows: In Section 2 we describe a multi-domain spectral method and a hybrid spectral/low-order method and give examples from flow simulations. In Section 3 we review two different shock capturing schemes which have been successfully incorporated in spectral methods. The first is based on essentially non-oscillatory spectral reconstruction procedures and the second on the flux corrected transport (FCT) algorithm; examples from both methods will be presented.

2   Spectral Methods for Complex Geometries

In the spectral-element method, a complex domain is decomposed into simpler domains; in each of the simpler domains, a full spectral representation is used to represent the dependent variables, while the solution in the global region is obtained by a variational technique in terms of the elemental solutions. The accuracy of this spectral-element method may be summarized as follows. The convergence rate of the method depends only on the smoothness of the underlying solution, with infinite-order convergence obtained for analytic solutions. In contrast to finite-difference or h-type finite-element solutions in which doubling the spatial resolution decreases the error by a fixed factor (4 for a second-order scheme), the error using spectral elements decreases exponentially fast. Furthermore, a wide variety of studies have shown that to achieve engineering accuracy of several percent the spectral-element method requires at least an order-of-magnitude less resolution in two-dimensional flows than do difference or h-type element methods; once engineering accuracy has been achieved, further increase of the spatial resolution leads to exponentially small errors because of the high-order character of the method. The spectral-element discretization of the convection operator results in minimal numerical diffusion and dispersion, a fact which is of critical importance in the solution of difficult laminar or turbulent flows. Also, the spectral-element method offers good resolution of boundary layers due to both the intrinsic clustering of high-order polynomial interpolation and the flexibility of an elemental decomposition. Also, in contrast to difference or h-type element methods, errors due to geometrical patching or grid-resolution changes are small with the method capable of handling rapid changes in resolution across interfaces. In Figure 1, we illustrate the power of the spectral element method to solve complex fluid flow problems. In this figure, we plot the streamwise velocity contours in a direct numerical simulation of turbulent flow in a channel with a riblet wall at a Reynolds number of 3000.

Another class of hybrid spectral methods employ heterogeneous discretizations appropriate for simulating flows over walls of arbitrary roughness. The two main components of the algorithm are a high-order scheme (spectral element method) and a low-order scheme (finite-difference or finite-element method). The use of the finite-difference discretization is essential in geometries with random boundaries, where all discretization techniques based on mappings fail. The use of low-order finite elements can also be useful in a wider class of applications including, for example, flows in unbounded domains, flows over surfaces with distributed roughness elements, etc. A new general iterative relaxation procedure is applied to allow coupling of the two fundamentally different discretizations. In particular, the first component of the hybrid algorithm (spectral element method) is applied to the outer large-scale domain Ω_1, where the effective local Reynolds number is large and thus the spectral-like dispersive properties of the method are effectively utilized. In the near-wall region where an almost laminar flow prevails the second component (a low-order finite difference method) is applied providing sufficient resolution to simulate the viscous flow and account for the small-scale irregularity of the domain Ω_2. As regards time discretization, a high-order splitting scheme is employed that reduces the problem into solving a series of coupled hyperbolic and elliptic problems. Continuity of the solution along the spectral element-finite difference interface (boundary Γ is then imposed by requiring continuity of the elliptic components; the latter is accomplished using the iterative "Zanolli" patching procedure and appropriately chosen relaxation parameters [5].

The "Zanolli" patching procedure has been practiced in the past only in the context of similar discretizations on both domains (i.e. spectral collocation. It basically consists of solving a Dirichlet elliptic problem in domain Ω_1 and subsequently providing a pointwise flux (Neumann) condition for the solution of the corresponding elliptic problem in domain Ω_2; this procedure is then repeated until continuity of the two solutions at the interface is achieved. Convergence to the exact solution is typically obtained after three to five iterations depending on the problem size and the value of the relaxation parameter. In the current work, we have modified this patching procedure to first accommodate dissimilar discretization schemes across the two domains, and second to allow for a parallel implementation; the latter can be achieved by appropriately modifying the flux condition of the Neumann elliptic problem.

Several examples of the performance of this algorithm for spectral collocation are given in [5]. Here we demonstrate its effectiveness for 2-D problems with non-conforming discretizations by solving the Helmholtz equation in a complex domain (see Figure 2). The global domain is subdivided into two approximately equally sized subdomains and discretized using spectral elements in Ω_1 and finite differences in Ω_2. In Figure 2 we plot the solution for this complex domain with a random boundary. If sufficient resolution is placed in the low-order domain it has been shown that spectral convergence is still recovered in the high-order domain [5].

3   Spectral Methods for Shock Waves

There are three main issues associated to be addressed in formulating spectral methods for solving systems of hyperbolic conservation laws: conservative discretization procedure, monotonicity, and boundary and interface conditions in multi-domains. Conservation properties are honored if a cell-average procedure is followed where the quantities of interest are computed as cell averages (defined by successive collocation points) while their fluxes are computed at the end of the cells. This cell averaging procedure can be performed spectrally and it involves all collocation points of each subdomain [4]. Point values for the flux evaluation are recovered similarly through a spectral reconstruction procedure [3], [4]. Monotonicity can be imposed indirectly either by the use of appropriate non-oscillatory constructs or by the use of limiters in a diffusion-antidiffusion (FCT) algorithm which we describe below. Boundary and interface boundary conditions on the conservative variables should be imposed using characteristic decomposition. However since only reconstructed values are needed for the computation of fluxes boundary and interface conditions can be applied directly to fluxes [9].

Next, we briefly review the non-oscillatory spectral element method presented in [10] and an FCT based spectral method developed in [4], [9].

3.1   Non-oscillatory method

The main difficulty in applying spectral methods to discontinuous problems is the Gibbs phenomenon. If a discontinuous function is approximated by a spectral expansion (Chebyshev, Fourier etc.), the approximation is only $O(1/N)$ accurate in smooth regions and contains $O(1)$ oscillations near the discontinuity. When spectral methods are applied to partial differential equations with discontinuous solutions, the Gibbs phenomenon may also lead to numerical instability.

An interesting approach to constructing a non-oscillatory spectral approximation to a discontinuous function has been recently proposed in [2]. Let $u(x)$ be a piecewise C^∞ function with a jump discontinuity at point x_s of strength $[u]_{x_s}$. The key idea in [2] was to augment the Fourier spectral space with a saw-tooth function. It was shown that the approximation using the augmented spectral space will be non-oscillatory if the saw-tooth function approximates the strength and the location of the discontinuity with *second order* accuracy. In addition, a method for estimating the discontinuity parameters with specified accuracy based on the spectral expansion coefficients was suggested. More recently, it was pointed in [3] that a *first order* accurate approximation of discontinuity magnitude also leads to non-oscillatory behavior. Numerical experiments with discontinuous solutions using the Chebyshev spectral space augmented by a step function and cell averaging approach were reported in [1]. This approach was extended in [10] for the spectral element discretization. In addition, a subcell resolution method was used for estimation of the discontinuity parameters with an accuracy required for obtaining a non-oscillatory approximation.

3.2   FCT method

The spectral element-FCT method was formulated for scalar conservation laws in [4] and for systems in [9]. This method consists of two stages: a transport - diffusive stage and an antidiffusive or corrective stage. In the first stage, a first order positive type scheme is implemented, while in the second a "limited" correction due to the spectral element discretization is made.

The main steps of the proposed algorithm are as follows:

- Step 1: Evaluate the field of cell averages corresponding to the initial condition.

- Step 2: Compute the transportive fluxes corresponding to the low order scheme. The low order positive type scheme used here is Roe's scheme based on the cell-averaged values. The low order flux $\mathbf{F}_{I+}$ is defined as follows:

$$\mathbf{F}_{I+} = \frac{\mathbf{f}(\bar{\mathbf{u}}_{J+1}) + \mathbf{f}(\bar{\mathbf{u}}_J)}{2} - R \cdot |D| \cdot R^{-1} \frac{(\bar{\mathbf{u}}_{J+1} - \bar{\mathbf{u}}_J)}{2} \qquad (1)$$

  where R is the Jacobian matrix consisting of the right-eigenvectors of the Euler system linearized around the Roe-averaged state between $\bar{\mathbf{u}}_{J+1}$ and $\bar{\mathbf{u}}_J$.

- Step 3: Advance (explicitly) cell averages in time using low order fluxes to obtain the low order transported and diffusive solution $\bar{\mathbf{u}}_J^{td}$. This is done using the third order Adams-Bashforth scheme.

- Step 4: Compute the transportive fluxes $\mathbf{f}_I$ corresponding to the spectral element discretization.

- Step 5: Compute the antidiffusive fluxes $\mathbf{A}_I = \mathbf{f}_I - \mathbf{F}_I$ and limit them to obtain $\mathbf{A}_I^c$. It is important that the limiter be applied to the characteristic antidiffusive fluxes and not the componentwise fluxes.

- Step 6: Update (explicitly) the cell averages on the new time level using the limited antidiffusive fluxes (using the third order Adams-Bashforth scheme) $\bar{\mathbf{u}}_J^{n+1}$.

- Step 7: Reconstruct point values from the cell averages at the new time level.

- Step 8: If the target time is not achieved go to Step 2.

3.3 Numerical results

In this section we present results of several numerical experiments including Euler system with the non-oscillatory method and one- and two-dimensional solutions of the Euler system with the FCT method.

First, we present simulations using the spectral-non-oscillatory method for the test problem considered in [3]. This problem models the interaction between a moving shock and sinusoidal density disturbances. In Figure 3 we display the density profile obtained with resolution corresponding to $K = 22$ elements and $N = 10$ (corresponding to 199 grid points) at time $t = 1.8$. Low-order schemes (e.g. MUSCL) captures the shock quite accurately as well as the lower frequency oscillation but under-resolves dramatically the high-frequency, high-amplitude oscillation.

The first example for the spectral-FCT method is a numerical solution of Lax's problem (see Figure 4) for $K = 2$ and $N = 75$. The high order spectral element discretization is responsible for a remarkably sharp representation of discontinuities.

The last example we present is a two-dimensional simulation of supersonic flow over a forward-facing step, which is a standard benchmark test in the literature. The domain extends from $x = 0$ to $x = 3$ and from $y = 0$ to $y = 1$ with the origin located at the lower leftmost corner. Discretization is made with 42 elements of 13×9 collocations points each. The initial conditions as well as the boundary conditions at the inflow at $x = 0$ are $\rho = 1.4$, $u = 3.0$, $v = 0.0$, and $P = 1$ throughout the domain, which corresponds to a Mach 3 flow. The rest of the boundary conditions are: supersonic outflow at $x = 3$ and reflection boundary conditions along both the bottom and the top of the domain. A density contour plot with 30 levels (at time $t = 4.0$) is shown in Figure 5.

4 Discussion

Spectral methods have developed to the point where they have become the method of choice for a wide variety of difficult problems requiring accurate and efficient solutions. In this paper, we have focused on two new hybrid spectral element methods for the solution of systems of conservation laws.

The first method is based on non-oscillatory approximation concepts. The results show that this approach leads to a stable method. The new method is also capable of resolving very accurately fine structures arising from interactions of shocks with continuous disturbances. The generalization of this method in the case of multiple discontinuities is straightforward. However, the disadvantage of the method is that it requires to identify a discontinuity and to treat it in a special way (subcell resolution), and that it is not capable of treating rarefaction waves.

Another method we formulated here is a spectral element-FCT method. Unlike the previous one, this method is very robust. The numerical results demonstrate a clear superiority of high order shock capturing methods in resolving discontinuities (representing them by very sharp layers). However, this method is not as accurate in the smooth regions as the previous one.

Acknowledgements

The work presented here is based largely on the Princeton Ph.D. theses of R. D. Henderson and I. G. Giannakouros. This work was supported by the AFOSR, DARPA, NSF and ONR.

References

[1] W. Cai, D. Gottlieb, and A. Harten. Cell averaging Chebyshev methods for hyperbolic problems. Report No. 90-72, ICASE, 1990.

[2] W. Cai, D. Gottlieb, and C. W. Shu. Non-oscillatory spectral Fourier methods for shock wave calculations. *Math. Comp.*, 52:389–410, 1989.

[3] W. Cai and C. W. Shu. Uniform high order spectral methods for one and two dimensional Euler equations. Submitted for publication.

[4] J. Giannakouros and G. E. Karniadakis. Spectral element-FCT method for scalar hyperbolic conservation laws. *Int. J. Num. Meth. Fluids*, 14:707, 1992.

[5] R. D. Henderson and G. E. Karniadakis. Hybrid spectral element-low order methods for incompressible flows. *J. Sc. Comp.*, 6:79–100, 1991.

[6] M. Y. Hussaini and T. A. Zang. Spectral methods in fluid dynamics. *Ann. Rev. Fluid Mech.*, 19:339–367, 1987.

[7] G. E. Karniadakis. Spectral element simulations of laminar and turbulent flows in complex geometries. *Appl. Num. Math.*, 6:85, 1989.

[8] A. T. Patera. A spectral element method for Fluid Dynamics; Laminar flow in a channel expansion. *J. Comp. Phys.*, 54:468, 1984.

[9] D. Sidilkover, J. Giannakouros, and G. E. Karniadakis. Hybrid spectral element methods for hyperbolic conservation laws. In *Proc. 9th GAMM Conf. on Num. Meth. in Fluid Mech., Vieweg Verlag*, 1991.

[10] D. Sidilkover and G. E. Karniadakis. Non-oscillatory spectral element Chebyshev method for shock wave calculations. *J. Comp. Phys.*, to appear, 1992.

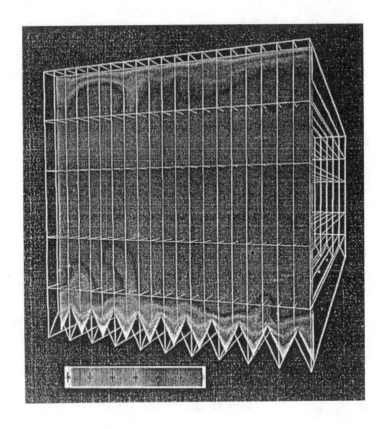

Figure 1: Instantaneous streamwise velocity contours on a z-plane in flow through a channel with a riblet wall at Re=3500.

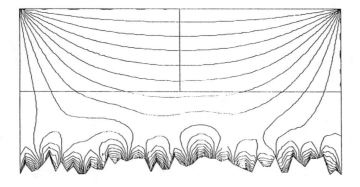

Figure 2: Simulated pressure distribution over a rough wall. The lower rectangle represents the boundary of the finite difference grid while the two upper rectangles are spectral elements. The patching procedure results in a smooth and continuous pressure distribution across the interface between the two domains.

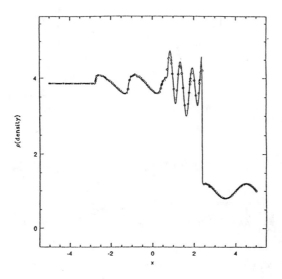

Figure 3: Spectral element non-oscillatory method. Moving shock interacting with sinusoidal density disturbances K=22, N=10 (199 grid points) at time t=1.8. The solid line represents the solution obtained by the third order ENO finite difference method with 1200 grid points (courtesy of Wai-Sun Don and David Gottlieb, Brown University).

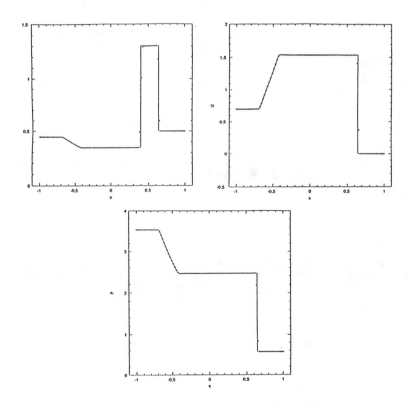

Figure 4: Spectral element-FCT method. Solution of the Lax's problem with K=2, N=75 at time t=0.26. (a) density, (b) velocity, (c) pressure (the solid line represents the exact solution).

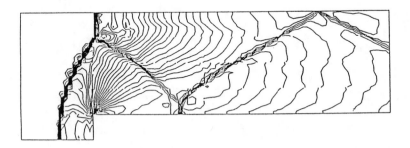

Figure 5: Supersonic flow impinging on a forward-facing step. Contours for 30 density levels at a time t=4.0. (Mach number of the incoming flow: 3.0).

MULTI-RESOLUTION AND SUBCELL RESOLUTION SCHEMES

WAVELET METHODS IN COMPUTATIONAL FLUID DYNAMICS

R. Glowinski[1], *J. Periaux*[2], *M. Ravachol*[2], *T. W. Pan*[3], *R. O. Wells*[4], *X. Zhou*[4]

[1]Department of Mathematics
University of Houston
Texas 77204-3476

and

INRIA, 78153 Le Chesnay, France

[2]Dassault Aviation
Saint-Cloud, France

[3]Department of Mathematics
University of Houston
Texas 77204-3476

[4]Department of Mathematics
Rice University
Houston, Texas 77251

ABSTRACT

We discuss in this paper the numerical solution of boundary value problems for partial differential equations by methods relying on compactly supported wavelet approximations. After defining compactly supported wavelets and stating their main properties we discuss their application to boundary value problems for partial differential equations, giving a particular attention to the treatment of the boundary conditions. Finally, we discuss application of wavelets to the solution of the Navier-Stokes equations for incompressible viscous fluids.

Introduction

Wavelets is a generic term denoting various mathematical objects which have been introduced mostly during the last decade. It is a topic in full evolution at the moment leading to a very active research effort, in the Mathematical and Electrical Engineering communities particularly. The limitations that we see concerning wavelet popularity in the Computation Fluid Dynamics community are essentially the following two:

(i) To enter the wavelet world requires a serious mathematical investment. Wavelets are considerably more complicated than finite element, finite difference, finite volume and spectral approximations.

(ii) A substantial effort has to be done to solve boundary value problems with complicated boundaries and/or boundary conditions. Indeed this difficulty is not specific to wavelets since we already encountered it for spectral approximations.

The content of this article is the following:

In Section 1 we define *compactly supported wavelets* à la Daubechies and state their basic properties. In Section 2 we briefly discuss some further properties of wavelet approximations which may be of interest for Scientific Computing. In Section 3 we briefly discuss the wavelet solution of boundary value problems for partial differential equations; a particular attention is given to the treatment of boundary conditions. Finally, we discuss in Section 4 the wavelet solution of the Navier-Stokes equations for incompressible viscous fluids.

1. Definition and Basic Properties of Compactly Supported Wavelets

1.1 Generalities

An inconvenient of traditional Fourier methods is that the sine and cosine functions have an unbounded support and even worst they do not vanish at infinity. On the other hand their spectra are very local consisting of a finite sum of Dirac measures. Conversely, if one uses approximations based on finite sum of Dirac measures such as (in one space dimension)

$$\sum_{j \in J} a_j \delta(x - x_j), \tag{1.1}$$

the spectrum of the basis "functions" $\delta(x - x_j)$, namely $s \to \exp(2i\pi x_j s)$ does not vanish at infinity in the frequency domain.

On the basis of the above observations it is then reasonable to look for orthonormal basis of $L^2(\mathbb{R})$, constructed from a *unique* generating function φ (the *scaling function*), via translation, dilation and linear combinations. We also want φ to be localized in x (space variable) and s (Fourier variable).

Example: Suppose that $\varphi(x) = e^{-x^2}$; concerning the Fourier transform $\hat{\varphi}$ of φ, we have then $\hat{\varphi}(s) = e^{-\pi^2 s^2}$; both functions vanish quickly for large values of their respective argument. Indeed wavelet families based on the above function φ are discussed in [1], [2]. □

More recently I. Daubechies has introduced in [3] wavelet families based on scaling functions φ such that:

(i) φ has a *bounded* support,

(ii) $\hat{\varphi}(s) \to 0$ *quickly* when $|s| \to +\infty$.

In the following parts of this paper the only wavelets that we shall consider are the Daubechies' ones. For an introduction to wavelets, intended for applied mathematicians and engineers we highly recommend [4].

1.2 The Daubechies Wavelets

Following [3] we require the *scaling function* φ to satisfy (N being a *positive integer*)

$$\varphi(x) = \sum_{k=0}^{2N-1} a_k \varphi(2x - k), \forall x \in \mathbb{R}, \tag{1.2}$$

$$\int_{\mathbb{R}} \varphi(x)dx = 1, \tag{1.3}$$

$$\int_{\mathbb{R}} \varphi(x - l)\varphi(x - m)dx = 0, \forall l, m \in \mathbb{Z}, l \neq m. \tag{1.4}$$

Relations (1.2) - (1.4) clearly imply

$$\sum_{k=0}^{2N-1} a_k = 2, \tag{1.5}$$

$$\sum_{k=0}^{2N-1} a_k a_{k-2m} = 0, \quad \forall m \in \mathbb{Z}, m \neq 0. \tag{1.6}$$

If the above relations hold, then the set

$$\bigcup_{j=0}^{+\infty} \bigcup_{l \in \mathbb{Z}} \varphi_{jl} \ \left(with \ \varphi_{jl}(x) = 2^{j/2}\varphi(2^j x - l)\right) \tag{1.7}$$

is an *orthogonal* basis of $L^2(\mathbb{R})$; also, the fact that the set of the coefficients, i.e. $\{a_k\}_{k=0}^{2N-1}$, is *finite* implies that φ has a *compact support*.

Example: Take $N = 1$ and $a_o = a_1 = 1$. We then have φ defined by

$$\varphi(x) = 1 \ if \ 0 \leq x \leq 1, \ \varphi(x) = 0 \ elsewhere. \tag{1.8}$$

The corresponding family (1.7) (namely the Haar functions family; cf. [5]) is an orthogonal basis of $\mathbb{R}$; its elements however are *discontinuous*. □

To "force" the *smoothness* of the scaling function φ we may require, for example, the monomials $1, x, \ldots x^{N-1}$ to be linear combination of the $\varphi(x - l)$; this implies the additional relations

$$\sum_{k=0}^{2N-1} (-1)^k k^m a_k = 0, \ m = 0, 1, \ldots, N - 1. \tag{1.9}$$

From the coefficients a_k we can use the *Fast Fourier Transform* (FFT) to construct φ via $\hat{\varphi}$ (see, e.g., [6] for this construction). Indeed, since $\text{supp}(\varphi) = [0, 2N - 1]$, it suffices to know φ at $0, 1, \ldots, 2N - 1$ and use the scaling relation (1.2) to compute φ at the *dyadic values* of x (in practice everywhere since the set of the dyadic numbers is *dense* in $\mathbb{R}$).

Once the coefficients a_k are known, we define the *wavelet function* ψ by

$$\psi(x) = \sum_{k=2-2N}^{1} (-1)^k a_{1-k} \varphi(2x - k), \tag{1.10}$$

and then the functions ψ_{jl} by

$$\psi_{jl}(x) = 2^{j/2}\psi(2^j x - l). \tag{1.11}$$

Let us denote by V_n (resp. W_n) the closure of the vector space span by $\{\varphi_{nl}\}_{l\in\mathbb{Z}}$ (resp. $\{\psi_{nl}\}_{l\in\mathbb{Z}}$); we have then the following properties

$$V_n \subset V_{n+1}, \tag{1.12}$$

$$\text{closure } (\bigcup_n V_n) = L^2(\mathbb{R}), \tag{1.13}$$

$$\{\varphi_{nl}\}_{l\in\mathbb{Z}} \text{ is an orthogonal basis for } V_n, \tag{1.14}$$

$$\{\psi_{nl}\}_{l\in\mathbb{Z}} \text{ is an orthogonal basis for } W_n, \tag{1.15}$$

$$\text{the orthogonal of } W_n \text{ in } V_{n+1} \text{ is } V_n. \tag{1.16}$$

The functions φ_{nl} and ψ_{nl} have *compact* supports; they also verify

$$\int_{\mathbb{R}} \varphi_{nl}(x)dx = 2^{-n/2}, \quad \int_{\mathbb{R}} \psi_{nl}(x)dx = 0. \tag{1.17}$$

Finally we have for $L^2(\mathbb{R})$ the following decomposition properties

$$L^2(\mathbb{R}) = V_n \oplus (\bigoplus_{j\geq n} W_j) = V_o \oplus (\bigoplus_{j\geq 0} W_j). \tag{1.18}$$

The potential of wavelets for multiscale analysis is related to relation (1.18).

2. Further Properties of Wavelets and Generalizations

From the following decomposition property

$$V_{n+1} = V_n \oplus W_n \tag{2.1}$$

we can expect wavelets to be well suited to multilevel solution methodologies and to the implementation of methods such as nonlinear Galerkin's (see [7]).

Remark 2.1: Several authors (see, e.g., [8]) have introduced *hierarchical finite element* bases leading to decomposition properties close to those mentioned above. These bases are well suited to adaptive mesh refinement. □

The wavelet functions considered so far are single variable functions. Concerning generalization to $\mathbb{R}^d$, with $d \geq 2$, we see at the moment the two following options:

(i) Use tensor products of one variable wavelet function spaces.

(ii) Use nonfactorable wavelets; such object exist, unfortunately they are not easy to handle (the support of the scaling function can be a region of $\mathbb{R}^d$ with a fractral boundary; see [9] for more details).

Finally, it can be shown that if f is sufficiently smooth then one has the following approximation property

$$\|f - P_n(f)\|_{H^m(\mathbf{R})} = 0(2^{-n(N-m)}),$$

where $P_n : L^2(\mathbb{R}) \rightarrow V_n$ is the orthogonal projector from $L^2(\mathbb{R})$ into V_n.

On Figure 2.1, we have shown the graphs of the scaling and wavelet functions corresponding to $N = 3$. These graphs suggest that φ and ψ are close to piecewise linear functions; in fact, it can be shown that for $N = 3$, φ and ψ are C^1 functions. We observe the asymmetry of the graphs of φ and ψ.

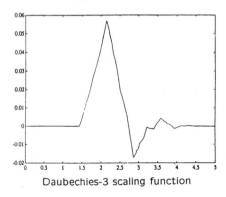
Daubechies-3 scaling function

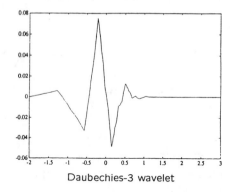
Daubechies-3 wavelet

Figure 2.1

Remark 2.2: Independently of the hierarchical structure of wavelet spaces, their double *localization* property (in the physical and Fourier spaces) make wavelets interesting for the solution of partial differential equations. □

3. Application to the Solution of Boundary Value Problems for Partial Differential Equations

3.1 Generalities

Wavelet based *Galerkin* methods have been applied, in [6], to the solution of boundary and initial/boundary value problems for linear and nonlinear elliptic, parabolic and hyperbolic equations in *one space variable*, with Dirichlet, Neumann and periodic boundary conditions. Indeed, as shown in [6] and [10], wavelet based Galerkin methods have proved quite efficient to solve the Burgers equation

$$\frac{\partial u}{\partial t} + u\frac{\partial u}{\partial x} - \nu\frac{\partial^2 u}{\partial x^2} = 0, \tag{3.1}$$

with $\nu \geq 0$.

Concerning multidimensional applications, let us make the following statement: *Everything you can do with (standard) spectral methods for periodic boundary conditions you can do better, for the same cost, with wavelet methods.* This (strong) statement is supported by the calculations, and comparisons to spectral methods, done by J. Weiss (see ref. [11]) for the incompressible Euler and Navier-Stokes equations, with periodic boundary conditions; wavelet based methods give in particular an excellent resolution of *shear layers*.

Concerning more general domains and/or boundary conditions the following approaches can be considered:

(i) Use boundary fitted wavelets as in [12]; their practical implementation seems quite complicated.

(ii) Use geometrical transformations to transform the multidimensional region in a box shaped region (or in a patch of such boxes). Indeed non periodic boundary conditions are complicated to implement even for a box shaped domain.

(iii) Extend to wavelet approximations the *domain imbedding methods* discussed in, e.g., [13] - [15].

Some domain imbedding methods will be discussed in Sections 3.2 (for Dirichlet boundary conditions) and 3.3 (for Neumann boundary conditions); we have chosen to discuss the solution of Dirichlet and Neumann problems since these problems are basic to many issues of Mechanics and Physics.

3.2 Domain Embedding Treatment of Dirichlet Boundary Conditions

3.2.1 Problem formulation

For simplicity, we shall concentrate on a simple *Dirichlet problem*, namely

$$\alpha u - \nabla^2 u = f \ in \ \omega, \ u = g \ on \ \gamma, \tag{3.2}$$

where, in (3.2), α is a positive number, ω is a bounded open connected region of $\mathbb{R}^d$ ($d \geq 1$), γ is the boundary of ω and $\nabla = \left\{ \frac{\partial}{\partial x_i} \right\}_{i=1}^d$; both functions f and g are given. We imbed ω in Ω, where Ω is a box shaped domain, as shown in Figure 3.1, below, where n denotes the unit outward normal vector at γ.

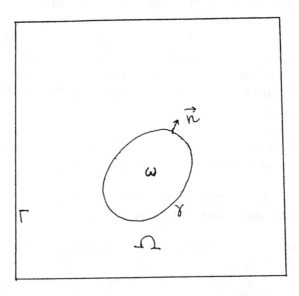

Figure 3.1

Two classical ways to implement the domain imbedding concept are *penalty* and the use of a *Lagrange multiplier*, respectively.

Both approaches rely on the fact that problem (3.2) is equivalent to

$$\begin{cases} Find\ \tilde{u} \in W_g\ such\ that \\[2mm] \int_\Omega (\alpha\tilde{u}v + \nabla\tilde{u}\cdot\nabla v)dx = \int_\Omega \tilde{f}vdx,\ \forall v \in W_o, \end{cases} \quad (3.3)$$

where, in (3.3), $\tilde{f}$ is an extension of f to Ω, and where W_g and W_o are defined by

$$W_g = \{v|v \in V, v = g\ on\ \gamma\}, \quad W_o = \{v|v \in V, v = 0\ on\ \gamma\}; \quad (3.4)$$

in (3.4), V is either $H^1(\Omega)$ or $H_o^1(\Omega)$, or $H_P^1(\Omega)$ (the space of the functions of $H^1(\Omega)$, periodic at Γ). If $g \in H^{1/2}(\gamma)$, then W_g is a closed nonempty affine subspace of V, implying that problem (3.3) has a *unique* solution; this solution clearly coincides with u on ω, justifying therefore the imbedding approach.

3.2.2 A penalty method

From now on we drop the $\sim$ in $\tilde{u}$ and $\tilde{f}$. We approximate then problem (3.3) by

$$\begin{cases} Find\ u_\epsilon \in V\ such\ that,\ \forall v \in V,\ we\ have \\[2mm] \int_\Omega (\alpha u_\epsilon v + \nabla u_\epsilon \cdot \nabla v)dx + \frac{1}{\epsilon}\int_\gamma u_\epsilon v d\gamma = \int_\Omega fvdx + \frac{1}{\epsilon}\int_\gamma gvd\gamma, \end{cases} \quad (3.5)$$

where, in (3.5), ϵ is a positive parameter. It can be easily shown (see, e.g., Appendix 1 in [16]) that

$$\lim_{\epsilon \to 0} \|u_\epsilon - u\|_{H^1(\Omega)} = 0, \quad (3.6)$$

where u is the solution of (3.3).

From a practical point of view we shall have to replace V by a finite dimensional subspace of it; also, in order to handle the integrals over γ, we have found convenient to proceed as follows:

Suppose that we want to evaluate $\int_\gamma gvd\gamma$, then

(i) Introduce G defined over Ω such that $g = G \cdot n$ on γ.

(ii) Observe that from the *divergence theorem* we have

$$\int_\gamma gvd\gamma = \int_\gamma G \cdot nvd\gamma = \int_\omega \nabla \cdot (vG)dx = \int_\Omega \chi_\omega \nabla \cdot (vG)dx,$$

(3.7)

where χ_ω is the *characteristic function* of ω, i.e., $\chi_\omega(x) = 1$ if $x \epsilon \omega$, $\chi_\omega(x) = 0$ elsewhere.

(iii) Approximate χ_ω by χ_ω^a, where χ_ω^a is *differentiable* over Ω and substitute to χ_ω in the last integral of (3.7). We have then (assuming that $\chi_\omega^a|_\Gamma = 0$ which is reasonable)

$$\int_\gamma gvd\gamma \sim \int_\Omega \chi_\omega^a \nabla \cdot (vG)dx = -\int_\Omega vG \cdot \nabla \chi_\omega^a dx. \qquad (3.8)$$

Concerning the approximate characteristic function χ_ω^a, numerical experiments done at Rice University suggest the following rule of thumb: *To approximate χ_ω by χ_ω^a use a "grid" four times finer than the one which will be used to compute u.*

3.2.3 A Lagrange multiplier method

Following [17], [18], we associate to the relation $u = g$ on γ a *Lagrange multiplier* λ defined over γ. Solving problem (3.3) is then equivalent to find a pair $\{\tilde{u}, \lambda\}$ such that

$$\begin{cases} \int_\Omega (\alpha\tilde{u}v + \nabla\tilde{u} \cdot \nabla v)dx = \int_\Omega \tilde{f}vdx + \int_\gamma \lambda vd\gamma, \ \forall v \in V, \\ \\ \int_\gamma (\tilde{u} - g)\mu d\gamma = 0, \ \forall \mu \in \Lambda; \tilde{u} \in V, \ \lambda \in \Lambda. \end{cases}$$

(3.9)

For sufficiently smooth data, we can take $\Lambda = L^2(\gamma)$; however, the *natural choice* for Λ is $H^{-1/2}(\gamma)$. Incidentally the multiplier λ in (3.9) is nothing but the jump of the normal derivative $\frac{\partial \tilde{u}}{\partial n}$ at γ.

The *conjugate gradient* solution of problem (3.9), together with the description of a finite element implementation is discussed in [17], [18]; numerical experiments show the validity of this approach; its wavelet implementation is currently taking place.

Remark 3.1: In principle one can combine the penalty and Lagrange multiplier methods via an *Augmented Lagrangian* approach, as shown in, e.g., [19], [20]. This new methodology still has to be investigated when applied to the solution of problem (3.3) (for multidimensional problems, at least, since we have already applied it in [6] for the wavelet solution of elliptic boundary value problems in one variable).

3.3 Domain Embedding Treatment of Neumann Boundary Conditions

3.3.1 Problem formulation

We keep the notation of Section 3.3. The problem that we consider is defined by

$$\alpha u - \nabla^2 u = f \ in \ \omega, \quad \frac{\partial u}{\partial n} = g \ on \ \gamma. \qquad (3.10)$$

Problem (3.10) has the following variational formulation:

$$\begin{cases} u \in H^1(\omega); \\ \int_\omega (\alpha u v + \nabla u \cdot \nabla v) dx = \int_\omega f v dx + \int_\gamma g v d\gamma, \ \forall v \in H^1(\omega). \end{cases}$$
$$(3.11)$$

The embedding methods to be described here are variants of those used for solving the Dirichlet problem; they are based on *regularization* and *Lagrange multiplier*, respectively.

Both methods rely on the following result:

With Ω, Γ, and V as in Section 3.2.1, consider the variational problem defined by

$$\begin{cases} \tilde{u} \in W, \\ \int_\Omega (\alpha \tilde{u} v + \nabla \tilde{u} \cdot \nabla v) dx = 0, \ \forall v \in W_o, \end{cases} \qquad (3.12)$$

where

$$W = \{v | v \in V, \int_\omega (\alpha v w + \nabla v \cdot \nabla w) dx - \int_\omega f w dx - \int_\omega g w d\gamma = 0, \ \forall w \epsilon V\},$$
$$(3.13)_1$$

and

$$W_o = \{v | v \in V, \int_\omega (\alpha v w + \nabla v \cdot \nabla w) dx = 0, \ \forall w \epsilon V\}; \qquad (3.13)_2$$

problem (3.12) has clearly a unique solution and it coincides with the solution u of (3.10), (3.11) over ω. In fact, $\tilde{u}$ is the extension of u over Ω which *minimizes* $\int_\Omega (\alpha v^2 + |\nabla v|^2) dx$.

3.3.2 A regularization method

Let ε be a *positive* number and consider the following variational problem

$$
\begin{cases}
u_\varepsilon \in V; \forall v \in V \ \text{one has} \\[2mm]
\varepsilon \int_\Omega (\alpha u_\varepsilon v + \nabla u_\varepsilon \cdot \nabla v) dx + \int_\omega (\alpha u_\varepsilon v + \nabla u_\varepsilon \cdot \nabla v) dx \qquad (3.14) \\[2mm]
\quad = \int_\omega fv dx + \int_\gamma gv d\gamma.
\end{cases}
$$

It can be shown that

$$
\lim_{\epsilon \to 0} \| u_\epsilon - u \|_{H^1(\Omega)} = 0, \qquad (3.15)
$$

where u is the solution of (3.12). The wavelet implementation of the above regularization method is under investigation.

3.3.3 A Lagrange multiplier method

We associate to (3.11) a *Lagrange multiplier* λ defined over Ω. Indeed, solving problem (3.12) (and therefore problem (3.11)) is equivalent to find a pair $\{\tilde{u}, \lambda\} \in V \times V$ such that

$$
\begin{cases}
\int_\Omega (\alpha \tilde{u} v + \nabla \tilde{u} \cdot \nabla v) dx = \int_\omega (\alpha \lambda v + \nabla \lambda \cdot \nabla v) dx, \ \forall v \in V, \\[2mm]
\int_\omega (\alpha \tilde{u} \mu + \nabla \tilde{u} \cdot \nabla \mu) dx = \int_\omega f\mu dx + \int_\gamma g\mu d\gamma, \ \forall \mu \in V.
\end{cases}
\qquad (3.16)
$$

The *conjugate gradient* solution of problem (3.16), together with the description of finite element and wavelet implementations will be described elsewhere. Remark 3.1 still holds here, i.e., we can combine the regularization and Lagrange multiplier methods.

4. Application to the Solution of the Incompressible Navier-Stokes Equation

4.1 Generalities

The Navier-Stokes equations that we consider are

$$\frac{\partial u}{\partial t} - \nu \nabla^2 u + (u \cdot \nabla)u + \nabla p = 0 \ in \ \Omega \qquad (4.1)$$

$$\nabla \cdot u = 0 \ in \ \Omega \qquad (4.2)$$

completed by initial and boundary conditions.

Concerning the wavelet solution of the above Navier-Stokes equations, we see immediately three sources of potential difficulties, namely:

(i) The treatment of the *incompressibility condition* $\nabla \cdot u = 0$.

(ii) The treatment of the *boundary conditions*.

(iii) The simulation of flow at large Reynolds numbers.

In this section we shall focus on (i); however, the two other issues deserve some comments:

Concerning *boundary conditions*, the *periodic case* is quite easy to implement; on the other hand, other boundary conditions such as Dirichlet and Neumann yield serious difficulties, the main reasons being that in a wavelet expansion the coefficients are not *pointwise* values of the function or of its derivatives, as it is the case with finite elements or finite differences. Among the possible cures let us mention *boundary fitted wavelets* like the ones developed by S. Jaffard and Y. Meyer in [12], or *fictitious domain methods*, in the spirit of Section 3; we are currently investigating the second approach. Another possibility is to couple wavelet approximations (used away from the boundary) with finite elements (used in the neighborhood of the boundary), but the matching problems (at least for nonoverlapping couplings) are essentially as difficult to implement as are boundary conditions.

Concerning now the simulation of flow at large Reynold's numbers we can predict, on the basis of preliminary numerical experiments done with the Daubechies wavelets, that for an equivalent amount of computational work, wavelet based methods *are more stable and accurate than finite element, finite difference and spectral methods.* The above experiments involved the solution of the *Burgers equation* $u_t + u_{xx} = \nu u_{xx}$ (cf. [6], [10]) and of the *Navier-Stokes equations* with *periodic boundary conditions* (cf. [11]). A key property of wavelet based solution methods is that they seem to require much less (if not at all) artificial viscosity for highly advective flow; a possible explanation of this behavior is that it is a consequence of the orthogonality of the basis functions and of their localization properties in the spatial *and* spectral domains.

The treatment of the incompressibility seems to be eventually fairly simple and will be addressed in the next paragraph.

4.2 Wavelet Treatment of the Incompressibility Condition

Operator splitting techniques applied to the solution of the Navier-Stokes equations (4.1), (4.2) lead to the following Stokes equations

$$\alpha u - \nu \Delta u + \nabla p = f \; in \; \Omega, \tag{4.3}$$

$$\nabla \cdot u = 0 \; in \; \Omega. \tag{4.4}$$

We suppose that the boundary conditions are defined by

$$u = g \; on \; \Gamma (with \; \int_\Gamma g \cdot n d\Gamma = 0), \tag{4.5}$$

i.e. are of the *Dirichlet* type.

A *variational formulation* of problem (4.3) - (4.5) is given by

$$\begin{cases} u \in V_g; \; \forall v \in V_o \; we \; have \\[2ex] \int_\Omega (\alpha u \cdot v + \nu \nabla u \cdot \nabla v) dx - \int_\Omega p \nabla \cdot v dx = \int_\Omega f \cdot v dx, \end{cases} \tag{4.6}$$

$$\int_\Omega q \nabla \cdot u dx = 0, \; \forall q \in L^2(\Omega); p \in L^2(\Omega). \tag{4.7}$$

In (4.6), (4.7), we have $v \cdot w = \sum_{i=1}^d v_i w_i, \forall v = \{v_i\}_{i=1}^d, w = \{w_i\}_{i=1}^d; \nabla v \cdot \nabla w = \sum_{i=1}^d \sum_{j=1}^d \frac{\partial v_i}{\partial x_j} \frac{\partial w_i}{\partial x_j}; V_o = (H_o^1(\Omega))^d$ and $V_g = \{v | v \in (H^1(\Omega))^d, v = g \; on \; \Gamma\}$.

It follows from (4.6), (4.7) that the two fundamental spaces in the variational formulation of (4.3), (4.4) are $L^2(\Omega)$ (for the pressure) and $(H^1(\Omega))^d$ (for the velocity). We discuss now the wavelet approximation of the *variational problem* (4.6), (4.7):

From now on we shall denote by φ^N the *scaling function* associated to the positive integer N (the precise definition of the scaling function has been given in Section 1.2); the parameter N plays clearly the role of a polynomial degree. We define next φ_{jl}^N and $\Phi_n^N(\mathbb{R})$ by

$$\varphi_{jl}^N(x) = 2^{j/2}\varphi^N(2^j x - l), \forall x \in \mathbb{R},$$

$$\Phi_n^N(\mathbb{R}) = \text{ closure of the linear space span by } \{\varphi_{nl}^N\}_{l\in\mathbb{Z}},$$

respectively.

In order to apply wavelets to the solution of *multidimensional problems* an obvious approach is to use *tensor products* of one variable function spaces to define the multidimensional ones. We define therefore the spaces $\mathcal{V}_n^N(\mathbb{R}^d)$ and $V_n^N(\mathbb{R}^d)$ by

$$\mathcal{V}_n^N(\mathbb{R}^d) = \bigotimes_{i=1}^{d} \Phi_n^N(\mathbb{R}_{x_i}),$$

$$V_n^N(\mathbb{R}^d) = (\mathcal{V}_n^N(\mathbb{R}^d))^d,$$

respectively.

By restricting to Ω the elements of the two above spaces, we obtain $\mathcal{V}_n^N(\Omega)$ and $V_n^N(\Omega)$; if Ω is bounded, these two spaces are *finite dimensional*. On the basis of the analysis done in [21], concerning the finite difference and finite element approximations of the Stokes/Dirichlet problem (4.3) - (4.5) we shall approximate the *velocity spaces* V_o and V_g by appropriate subspaces of $V_n^N(\Omega)$ (taking into account, in some way or another, the boundary conditions $v = 0$ and $v = g$, respectively), and then the *pressure* space by $V_{n-1}^N(\Omega)$; in order to have $V_n^N(\Omega) \subset H^1(\Omega)$, we have to take $N \geq 3$ (cf., e.g., [3], [6] for this result). We then substitute to V_o, V_g and $L^2(\Omega)$, their wavelet analogues in (4.6), (4.7) to obtain a wavelet/Galerkin approximation of the Stokes problem (4.3) - (4.5).

Conclusion

On the basis of few experiments we can expect compactly supported wavelets to have a most interesting potential for the numerical solution of fluid flow problems. However the practical implementation of this new kind of approximation gives rise to highly nontrivial difficulties which have to be successfully addressed if one wishes to see wavelet-based approximations successfully competing with finite difference and finite element approximations.

Acknowledgment

We would like to acknowledge the helpful comments and suggestions of the following individuals: L. C. Cowsar, C. De la Foye, G. H. Golub, Y. Kuznetsov, P. Lallemand, A. Latto, W. Lawton, P. Le Tallec, J. L. Lions, P. L. Lions, G. Meurant, J. Pasciak, H. Resnikoff, J. Weiss, M. F. Wheeler, O. B. Widlund.

The support of the following corporations or institutions is also acknowledged: AWARE, Dassault Aviation, INRIA, University of Houston, Université Pierre et Marie Curie. We also benefited from the support of DARPA (Contracts AFOSR F49620-89-C-0125 and AFOSR-90-0334), DRET and NSF (Grants INT 8612680 and DMS 8822522). Finally, we would like to thank J. A. Wilson for the processing of this article.

References

[1] J. Morlet, G. Arehs, I. Fourgeau and D. Giard, Geophysics, 47, 203 (1982).

[2] L. F. Bliven and B. Chapron, Naval Research Reviews, 41, 2, 11 (1989).

[3] I. Daubechies, Comm. Pure Appl. Math., 41, 909 (1988).

[4] G. Strang, SIAM Review, 31, 614 (1989).

[5] A. Haar, Math. Ann., 69, 336 (1910).

[6] R. Glowinski, W. Lawton, M. Ravachol and E. Tenenbaum, in Computing Methods in Applied Sciences and Engineering, edited by R. Glowinski and A. Lichnewsky (SIAM, Philadelphia, 1990), p. 55.

[7] M. Marion and R. Temam, Siam J. Num. Anal., 26, 1139 (1989).

[8] R. Banks and H. Yserentant, in Domain Decomposition Methods, edited by T. F. Chan, R. Glowinski, J. Periaux and O. B. Widlund (SIAM, Philadelphia, 1989), p. 40.

[9] W. Lawton, H. Resnikoff, Multidimensional Wavelet Bases (to appear).

[10] A. Latto and E. Tenenbaum, C. R. Acad. Sci. Paris, 311, Serie I, 903 (1990).

[11] J. Weiss, Wavelets and the study of two-dimensional turbulence, AWARE Report, 1991.

[12] S. Jaffard and Y. Meyer, J. Math. Pures et Appl. , 68, 95 (1989).

[13] B. L. Buzbee, F. W. Dorr, J. A. George and G. H. Golub, SIAM J. Num. Anal., 8, 722-736 (1971).

[14] S. A. Finogenov and Y. A. Kuznetsov, Sov. J. Num. Math. Modelling, 3, 301 (1988).

[15] D. P. Young, R. G. Melvin, M. B. Bieterman, F. T. Johnson, S. S. Samant and J. E. Bussoletti, J. Comp. Physics, 92, 1 (1991).

[16] R. Glowinski, Numerical Methods for Nonlinear Variational Problems (Springer, New York, 1984).

[17] R. Glowinski, T. W. Pan, J. Periaux and M. Ravachol, in The Finite Element Method in the 1990's, edited by E. Oñate, J. Periaux, A. Samuelsson (Springer-Verlag, Berlin, 1991), p. 410.

[18] Q. V. Dinh, R. Glowinski, J. He, V. Kwak, T. W. Pan and J. Periaux, Lagrange Multiplier Approach to Fictitious Domain Methods: Application to Fluid Dynamics and Electro-Magnetics (to appear).

[19] M. Fortin and R. Glowinski, Augmented Lagrangian Methods (North-Holland, Amsterdam, 1983).

[20] R. Glowinski and P. Le Tallec, Augmented Lagrangian and Operator-Splitting Methods in Nonlinear Mechanics (SIAM, Philadelphia, 1989).

[21] R. Glowinski, in Vortex Dynamics and Vortex Methods, edited by C. R. Anderson and C. Greengard (AMS, Providence, R. I., 1991), p. 219.

SUBCELL RESOLUTION METHODS FOR VISCOUS SYSTEMS OF CONSERVATION LAWS

Eduard Harabetian

University of Michigan, Ann Arbor, MI 48109

Abstract

A subcell resolution scheme for viscous systems of conservation laws is presented. The scalar model is extended to systems via weakly nonlinear geometrical optics approximations for parabolic perturbations of hyperbolic conservation laws and the Roe field by field decomposition.

1 Introduction

In this paper we consider numerical approximations to viscous perturbations of hyperbolic systems of the form

$$u_t + f(u)_x = \epsilon(B(u)u_x)_x, \tag{1}$$

where $f_u = A(u)$ has real eigenvalues $\lambda_i(u), i = 1, ..., n$ and $B(u)$ is positive. The aplications we consider are the compressible Navier-Stokes equations.

When ϵ is small, solutions of (1) typically develop viscous shock layers where the solution changes rapidly over very narrow zones. Sophisticated methods have been developed for the purely hyperbolic case ($\epsilon = 0$) [6, 10, 9, 11]; these schemes capture shock discontinuities over very narrow zones without overshoots or undershoots. So far, numerical shemes which approximate (1) have relied on splitting the hyperbolic and parabolic parts and approximating the viscous term separately by the centered difference approximation. To avoid this splitting error, we are considering a numerical scheme that uses a subcell model for which one can obtain a time accurate evolution. This scheme is designed to work on a coarse grid, i.e, where the viscous layers are not resolved.

We start with the scalar version of (1) with a convex flux function $f(u)$ and for simplicity we take $B(u) = 1$. Let us consider a spatial grid with uniform spacing h and cells $I_j = [x_{j-1/2}, x_{j+1/2}]$. The solution is approximated at cell centers $u(x_j) \approx u_j$. A subcell function $R_{j+1/2}(x)$ is a function defined on the interval $[x_j, x_{j+1}]$ and such that $R_{j+1/2}(x_j) = u_j$, and $R_{j+1/2}(x_{j+1}) = u_{j+1}$. Let us also denote by $S(t)$ an approximate, or possibly exact, time evolution solver. Any choice of S should be motivated by the fact that one would like to be able to easily compute $S(t)R_{j+1/2}(x)$. With these ingredients one can design a conservation form scheme with numerical flux $h_{j+1/2}$ by

$$\frac{u_j^{n+1} - u_j^n}{\Delta t} + \frac{h_{j+1/2} - h_{j-1/2}}{\Delta x} = 0$$

$$h_{j+1/2} = f(S(\frac{\Delta t}{2})R_{j+1/2}(x_{j+1/2}) - \epsilon(S(\frac{\Delta t}{2})R_{j+1/2})_x(x_{j+1/2})$$

Let us now consider a few choices of $R_{j+1/2}$ and S.

2 Subcell Models

a) *Linear subcell functions*. The subcell function $R_{j+1/2}$ in this case is a linear interpolant. This is a reasonable choice when

$$\frac{h}{\epsilon} \ll 1$$

i.e. the grid has enough points so that the solution "looks" smooth. The reasonable choice of S, given the linear nature of R, is

$$S(t)R_{j+1/2}(x) = R_{j+1/2}(x - a_{j+1/2}t)$$

where $a_{j+1/2}$ is the approximate convective speed:

$$a_{j+1/2} = \frac{f(u_{j+1}) - f(u_j)}{u_{j+1} - u_j}$$

It is easy to see that this scheme consists of a Lax-Wendroff type approximation for the hyperbolic part and a centered difference approximation for the viscous part. In fact, this is a scheme that treats the two parts separately as the viscosity is not taken into account in the evolution operator. This error should be small for "smooth" solutions, in fact the scheme is second order accurate there; however, one could anticipate large errors in nonsmooth regions.

b) *Step subcell functions.* The subcell function $R_{j+1/2}$ in this case is the step function with a jump at $x_{j+1/2}$. This is a reasonable choice if

$$\frac{h}{\epsilon} \gg 1$$

i.e. the grid is coarse and the solution "looks" like it has jump discontinuities. For S we choose the exact solution operator for the hyperbolic part $\epsilon = 0$, i.e the Riemann solver; this is again reasonable since $R_{j+1/2}$ is piecewise constant, hence, assuming we evaluate the numerical flux away from the jump, the viscous contribution is zero.

This scheme is the Godunov scheme for the hyperbolic equation with zero viscosity. It is only first order accurate, but it is stable in the total variation norm.

c) *Travelling wave subcell functions.* This is a nontrivial choice of a subcell function that leads to the new scheme which is the topic of this paper. This is supposed to be the reasonable choice when

$$\frac{h}{\epsilon} \sim 1$$

We consider only the case $u_j > u_{j+1}$ (shock). One looks for travelling wave solutions to the viscous problem (a.k.a. viscous profiles) that interpolate the data between the 2 grid points x_j, x_{j+1}. Specifically, travelling wave solutions are solutions of the form: $u(t,x) = \phi(x - st)$ where the function ϕ must satisfy the following second order ordinary differential equation

$$\epsilon\phi'' = -s\phi' + f(\phi)'$$

The following boundary conditions are imposed:

$$\phi(x_j) = u_j, \quad \phi(x_{j+1}) = u_{j+1}$$

To determine the unknown speed s, one additional condition must be imposed; if f is an even function, (e.g. Burgers equation, $f(u) = u^2$), it is reasonable to impose the condition that the profile is symmetric:

$$\phi(x_{j+1/2} + x) - \hat{u} = -(\phi(x_{j+1/2} - x) - \hat{u})$$

where $\hat{u}$ is the average of u_j, u_{j+1}. In the next section we will consider weakly nonlinear geometric optics approximations which reduce the problem to the case of a Burgers equation.

The viscous profile ϕ is determined uniquely; in the case of the Burgers equation, an explicit formula is available. We let the subcell function be a viscous profile: $R_{j+1/2}(x) = \phi(x)$, and define the solver S as follows

$$S(t)R_{j+1/2} = R_{j+1/2}(x - st)$$

We make two remarks:

1. The subcell function depends on the ratio h/ϵ and tends to the linear subcell function when this ratio goes to zero, and it tends to the step function when this ratio goes to infinity.

2. The solver $S(t)$ is an exact solver since ϕ is a travelling wave solution.

We can show that this scheme is second order accurate (formally) in the following sense:

$$|h_{j+1/2} - (f(u) - \epsilon u_x)(x_{j+1/2}, \frac{\Delta t}{2})| \leq C(\frac{h}{\epsilon})h^2$$

where u is the exact smooth solution which interpolates u_j, u_{j+1} at time zero; the constant $C(\frac{h}{\epsilon})$ depends only on derivatives up to second order of the solution at time zero.

We can show this scheme is TVD when parameters $\frac{h}{\epsilon}$ and $\frac{\Delta t}{h}$ belong to a certain region in the parameter plane.

The proof of these are in [2]. In that paper, we also showed that one cannot hope for both second order accuracy and TVD stability unless the parameters $\frac{h}{\epsilon}$ and $\frac{\Delta t}{h}$ are constrained to lie in a certain region of the plane.

3 Systems

We start with the Roe decomposition:

$$u_{j+1} - u_j = \sum_k b_k r_k(u_0)$$

where u_0 is the Roe avearage state and $r'_k s$ are the eigenvectors of $f'(u_0)$. In the interval $[x_j, x_{j+1}]$, we consider the following ansatz

$$u(t, x) = u_0 - \frac{1}{2}\sum_k b_k \phi_k r_k + ...$$

$$\phi_k = \phi_k(\frac{x - \lambda_k t}{\epsilon/b_k})$$

where $\lambda'_k s$ are the eigenvalues of $f'(u_0)$ and $\phi'_k s$ will be determined. We consider only genuinely nonlinear fields. Upon substituting the ansatz into the equation (1), one obtains the following equation for ϕ_k (cf. [4]):

$$(\phi_k)_t + (\phi_k^2)_x = \epsilon C_k(\phi_k)_{xx}$$

By choosing r_k appropriately one can make $C_k = 1$. Finally, one can solve these Burgers equations with the boundary conditions $\phi_k(x_j) = 1$ and $\phi_k(x_{j+1}) = -1$

$$\phi_k = \frac{1}{tanh(s_k)} tanh(-2s_k \frac{x - x_{j+1/2} - \lambda_k t}{h})$$

$$s_k = \frac{h|b_k|}{\epsilon}$$

(Note that $\phi_k's$ are symmetric.)

The numerical flux is now given by

$$h_{j+1/2} = f(u_0) - \frac{1}{2}\sum_k \phi_k^{s_k}\lambda_k b_k r_k - \frac{\epsilon}{h}\sum_k B(u_0)(\phi_k^{s_k})' b_k r_k$$

$$\phi_k^{s_k}(\xi_k) = \frac{tanh(s_k\xi_k)}{tanh(s_k)} = \phi_k(x_{j+1/2}, \Delta t/2), \ \xi_k = \frac{\lambda_k \Delta t}{h}$$

The first sum in the numerical flux is the artificial viscosity, the second sum is the real viscosity. The balance between the two is achieved as follows: The artificial viscosity is minimized in smooth regions and maximized inside shock layers. The opposite effect occurs with the real viscosity. The parameter s_k serves as a detector of "non-smooth" regions.

The subcell approximation is supposed to work for small $|b_k|$, i.e. for weak viscous shocks. However, in practice strong viscous shocks are also well approximated.

These asymptotic expansions have been studied by Hunter-Keller [7], Majda-Rosales [8] and others for purely hyperbolic problems.

4   Numerical Results

Our first numerical results concern the scalar Burgers equation. Comparisons with the Lax Wendroff - Centered and the second order upwind (centered difference for viscosity) methods were performed. The maximum errors show a marked improvement when the travelling wave subcell method is used.

Our second set of numerical results concern the Navier-Stokes system in 1D with reactive terms. We have computed a CJ detonation on a coarse mesh (200 points) and repeated the computation on a much finer mesh to demonstrate convergenge.

5 Conclusions

In this paper we developed a subcell resolution method for viscous perturbations of hyperbolic conservation laws. Our goal is to obtain an accurate approximation when the grid size is of the same order as the viscosity. The solution cannot be resolved on such coarse grids.

Other subcell resolution methods have been developed for related problems. For example Harten has developed a method for subcell shock capturing [5], and Engquist and Sjogreen have developed a method for inviscid combustion [1].

References

[1] B. Engquist and B. Sjogreen, *Robust Difference Approximations of Stiff Inviscid Detonation Waves*, UCLA CAM Report 91-03

[2] E. Harabetian, *A Numerical Method for Viscous Perturbations of Hyperbolic Conservation Laws*, SIAM J. of Num. Anal., 27 (1990)

[3] E. Harabetian, *A Numerical Method for for Computing Viscous Shock Layers*, Notes on Num. Fluid Mechanics, Josef Ballman, Rolf Jeltsch (Eds.), 24, pp. 220-229

[4] E. Harabetian, A Subcell Resolution Method for Viscous Systems and Computations of Combustion Waves, *Journal of Computational Physics*, *(submitted for publication)*, (1990)

[5] Ami Harten, *ENO Schemes with Subcell Resolution*, UCLA Comp. and Appl. Math Report 87-13, (1987)

[6] Ami Harten, *High Resolution Schemes for Hyperbolic Conservation Laws*, J. Comp. Phys., 49, (1983)

[7] J. Hunter and J. Keller, *Weakly Nonlinear High Frequency Waves*, Comm. Pure and Appl. Math., 36 (1983)

[8] A. Majda and R. Rosales, *Resonantly Interacting weakly nonlinear hyperbolic waves I*, Stud. Appl. Math. 71:149-179 (1984)

[9] S. Osher, *Riemann Solvers, The Entropy Condition, and Difference Aproximations*, SIAM J. Num. Anal. 21 (1984)

[10] P. Roe, *Approximate Riemann Solvers, Parameter Vectors, and Difference Schemes*, J. Comp. Phys. 43 (1981)

[11] P.K. Sweby, *High Resolution Schemes Using Flux Limiters for Hyperbolic Conservation Laws*, SIAM J. Num. Anal. 21 (1984)

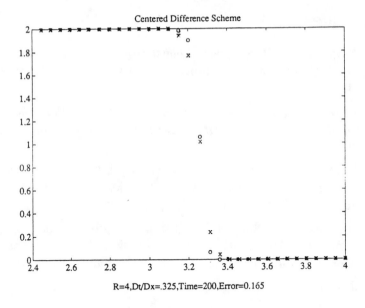

Centered Difference Scheme

R=4,Dt/Dx=.325,Time=200,Error=0.165

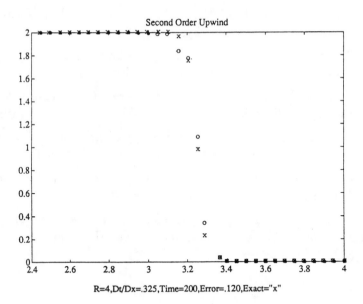

Second Order Upwind

R=4,Dt/Dx=.325,Time=200,Error=.120,Exact="x"

Fig. 1 The Lax-Wendroff - Centered difference (top) and the second order upwind (centered difference for viscosity) (bottom). The parameter R is the numerical Reynolds number $2h/\epsilon$.

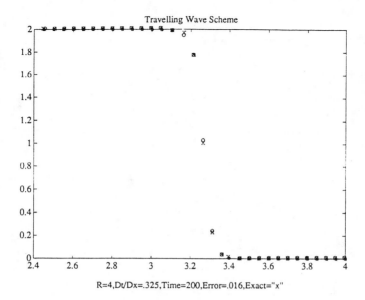

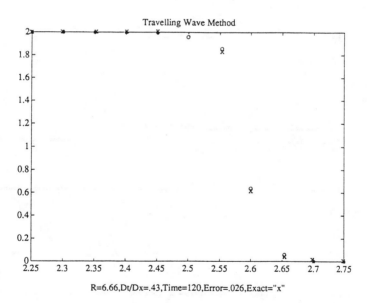

Fig. 2 The Travelling wave subcell method with two different values of $R = 2h/\epsilon$.

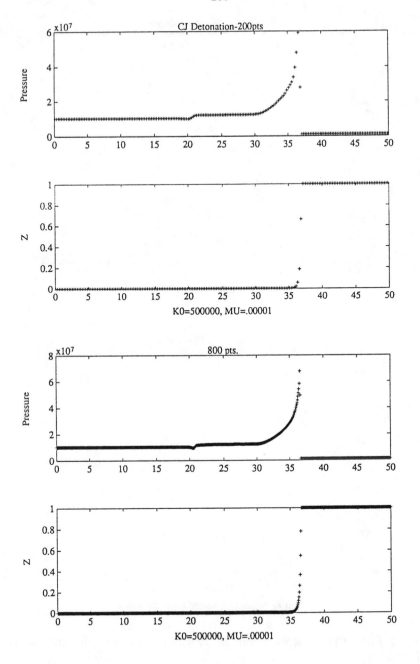

Fig. 3 The pressure and concentration of unburnt gas (Z) in a CJ detonation with a coarse mesh (top) and a fine mesh (bottom).

MULTI-RESOLUTION ANALYSIS FOR ENO SCHEMES

Ami Harten[1]

School of Mathematical Sciences, Tel-Aviv University

and

Department of Mathematics, UCLA

ABSTRACT

Given a function $u(x)$ which is represented by its cell-averages in cells which are formed by some unstructured grid, we show how to decompose the function into various scales of variation. This is done by considering a set of nested grids in which the given grid is the finest, and identifying in each locality the coarsest grid in the set from which $u(x)$ can be recovered to a prescribed accuracy.

We apply this multi-resolution analysis to ENO schemes in order to reduce the number of numerical flux computations which is needed in order to advance the solution by one time-step. This is accomplished by decomposing the numerical solution at the beginning of each time-step into levels of resolution, and performing the computation in each locality at the appropriate coarser grid. We present an efficient algorithm for implementing this program in the one-dimensional case; this algorithm can be extended to the multi-dimensional case with cartesian grids.

1. Introduction

In this paper we consider the Initial-Boundary Value Problem (IBVP) for hyperbolic systems of conservation laws in s-space dimensions:

$$u_t + \operatorname{div} f(u) = 0, \quad x \epsilon \mathcal{D} \subset I\!\!R^s, \quad t > 0 \qquad (1.1a)$$

$$u(x,0) = u_0(x), \quad x \epsilon \mathcal{D} \qquad (1.1b)$$

---

[1]This research was partially supported by the National Aeronautics and Space Administration under NASA Contract No. NAS1-18605 while the author was in residence at the Institute for Computer Applications in Science and Engineering (ICASE), NASA Langley Research Center, Hampton, VA 23665. Additional support was provided by ONR Grant N00014-86-K-0691, DARPA Grant in the ACMP Program and NSF Grant DMS 88-11863.

with given boundary condition on ∂D, the boundary of D. We assume that the problem is well-posed and denote its evolution operator by $E(t)$; note that it includes the influence of the boundary conditions.

The computational domain D is divided into cells $\{C_j\}$

$$D = \cup_j C_j, \quad C_j \cap C_k = \phi \quad \text{for} \quad j \neq k \qquad (1.2a)$$

and we assume that there is a refinement parameter h such that the largest sphere contained in each of the cells is of radius $O(h)$, and that the ratio between the largest cell to the smallest one in the computational domain remains bounded under refinement.

Let $|C_j|$ denote

$$|C_j| = \int_{C_j} dx \qquad (1.2b)$$

and let $\bar{u}_j$ denote the cell-average of $u(x)$ over C_j

$$\bar{u}_j = \frac{1}{|C_j|} \int_{C_j} u(x) dx \equiv A(C_j) \cdot u(x); \qquad (1.3)$$

here $A(C_j)$ denotes the cell-averaging operator.

Given cell-averages $\bar{u} = \{\bar{u}_j\}$ of $u(x)$ in D, we denote by $R(x; \bar{u})$ an r-th order piecewise-polynomial reconstruction of u from $\bar{u}$, i.e.,

$$R(x; \bar{u}) = R_i(x; \bar{u}) \quad \text{for} \quad x \epsilon C_i \qquad (1.4a)$$

where $R_i(x; \bar{u})$ is a polynomial of degree $r - 1$. Expressing $R_i(x; \bar{u})$ as a finite Taylor series around the centroid $c_i = A(C_i) \cdot x$

$$R_i(x; \bar{u}) = D_0 + \sum_{k=1}^{r-1} \frac{1}{k!} \sum_{|\ell|=k} (x - c_i)^\ell D_\ell \qquad (1.4b)$$

where

$$h^{|\ell|} D_\ell = h^{|\ell|} \frac{\partial^\ell u}{\partial x^\ell}(c_i) + O(h^r), \quad 1 \leq |\ell| \leq r - 1, \qquad (1.4c)$$

(accuracy)

$$D_0 = \bar{u}_i - \sum_{k=1}^{r-1} \frac{1}{k!} \sum_{|\ell|=k} \left[A(C_i) \cdot (x - c_i)^\ell \right] D_\ell \qquad (1.4d)$$

(conservation).

Note that (1.4d) implies that $A(C_i) \cdot R_i(x; \overline{u}) = \overline{u}_i$. In (1.4) we have used a multi-index notation

$$\ell = (\ell_1, \cdots, \ell_s), \quad |\ell| = \ell_1 + \cdots + \ell_s \quad (\ell_i \geq 0)$$

with the standard convention

$$y^\ell = (y_1)^{\ell_1} \cdots (y_s)^{\ell_s}, \quad \frac{\partial^\ell}{\partial x^\ell} = \frac{\partial^{|\ell|}}{\partial x_1^{\ell_1} \cdots \partial x_s^{\ell_s}}.$$

We consider the numerical solution of (1.1) by the class of schemes

$$v_j^{n+1} = A(C_j) E(\tau) R(\cdot; v^n), \quad v_j^0 = A(C_j) u_0(x) \qquad (1.5a)$$

where v_j^n is an r-th order approximation to the cell-average of the solution u at time t_n

$$v_j^n \approx A(C_j) \cdot u(x, t_n). \qquad (1.5b)$$

Due to the divergence-free form of the PDE (1.1), the scheme (1.5a) takes the conservation form

$$v_j^{n+1} = v_j^n - \frac{1}{|C_j|} \int_0^\tau \oint_{\partial C_j} f(E(t) \cdot R(\cdot; v^n)) \cdot N \, ds \, dt \qquad (1.5c)$$

where ∂C_j is the boundary of the cell C_j and N is its outward normal. We refer the reader to [3] for details.

The purpose of this paper is to present some preliminary results regarding the application of multi-resolution analysis to the numerical solution of hyperbolic systems of conservation laws. Typically these solutions contain discontinuities (shocks, sliplines, material boundaries, combustion fronts) which may move around and also some localized high-frequency smooth behavior which is associated with shedding of vortices. In such situations we have to use a fine grid in order to resolve the details of discontinuities and vortices, while in other parts of the solution the use of a coarser grid is adequate.

The traditional solution to this computational problem is to use a nonuniform adaptive grid where at each time-step the discretization points are redistributed with the goal of minimizing the truncation error of the scheme. In the context of the schemes (1.5) this means that the cells $C_j = C_j^n$ are redefined at each time and that their size is highly nonuniform: they are very small around discontinuities and rapid smooth variation of the solution, and a lot larger elsewhere.

This is certainly a reasonable solution to this computational problem and there are several such computer codes that accomplish this goal successfully. However, the adaptive grid approach is complicated and the programming effort required for its implementation is formidable.

In the following we describe our concept of multi-resolution analysis which is borrowed from the realm of image-compression techniques. There the functions are defined on a uniform fine grid and are assumed to be over-resolved in some parts of it. The purpose of the multi-resolution analysis in image compression is to determine appropriate levels of resolution for the various parts of the image in order to eliminate superfulous information. In the context of the numerical solution (1.5) this means that our computational grid (1.2) is rather uniform with respect to the parameter h and is assumed to be fine enough to capture all the details that we are interested in, and that the solution is in fact over-resolved on this fine grid in large parts of the computational domain. In order to apply multi-resolution analysis we construct an hierarchy of nested grids as follows:

Given cells $\{C_j^k\}$ of size h_k and cell-averages (1.3) $\{\bar{u}_j^k\}$ of $u(x)$ on them, we define a coarser grid $\{C_j^{k+1}\}$ of size $h_{k+1} > h_k$ by joining some of the cells in the k-th grid into a single larger cell

$$C_i^{k+1} = \cup_j C_j^k \qquad (1.6a)$$

and define

$$\bar{u}_i^{k+1} = \frac{1}{|C_i^{k+1}|} \sum_j |C_j^k| \bar{u}_j^k. \qquad (1.6b)$$

The original grid is thus the finest in the hierarchy, and we denote its cells by $\{C_j^0\}$ and their size by $h_0 = h$. To each of these grids we can apply the reconstruction (1.4) which we denote by $R^k(x; \bar{u})$ and also apply the numerical scheme (1.5) in order to advance its solution in time.

What we mean by multi-resolution analysis is the assignment $k(i)$ such that

$$|R^{k(i)}(x; \bar{u}) - R^0(x; \bar{u})| < \varepsilon \quad \text{for} \quad x \epsilon C_i^0 \qquad (1.7)$$

where $k(i)$ is the largest integer for which (1.7) holds. In Section 2 we shall describe a method to determine $k(i)$.

The general idea is to use a multi-resolution reconstruction $\tilde{R}(x; \bar{u})$ which is defined on the finest grid by

$$\tilde{R}(x; \bar{u}) = R^{k(i)}(x; \bar{u}) \quad \text{for} \quad x \epsilon C_i^0 \qquad (1.8a)$$

and to consider the numerical scheme

$$v_i^{n+1} = A(C_i^0)E(\tau)\tilde{R}(\cdot; v^n). \tag{1.8b}$$

We observe that the scheme (1.8) can be thought of as an adaptive-grid calculation in which the cells are a combination of cells of various scales. We would like to avoid this interpretation because of its inherent complexity, and we look for a simple implementation.

In Section 3 we develop such an implementation by studying a related problem and in Section 4 we describe an efficient multi-resolution ENO scheme for the one-dimensional case. In Section 5 we summarize the present state of affairs and point out directions for future development.

2. Determination of Resolution Levels

In this section we describe how to decompose a given set of cell-averages $\{\overline{u}_j\}$ of a function $u(x)$ over cells $\{C_j\}$ of size h into a set of scales

$$h = h_0 < h_1 < \cdots < h_K \tag{2.1a}$$

given by the hierarchy (1.6). This is done by determining the largest $k = k(i)$ so that

$$|R^k(x; \overline{u}) - R_i(x; \overline{u})| < \varepsilon \quad \text{for} \quad x \epsilon C_i \tag{2.1b}$$

for a prescribed level of accuracy ε. Here $R_i(x; \overline{u})$ is the polynomial of degree $r - 1$ in (1.4) and $R^k(x; \overline{u})$ is the reconstruction which is associated with the scale h_k.

First let us describe the way in which the reconstruction $R(x; \overline{u})$ is defined for the given grid. Expressing $R_i(x; \overline{u})$ as a finite Taylor series around the centroid c_i (1.4b) we have to find coefficients $\{D_\ell\}, 0 \leq |\ell| \leq r - 1$ which satisfy the requirements (1.4c) - (1.4d). Let us denote by d the vector of unknowns

$$\left\{\frac{1}{|\ell|!}h^{|\ell|}D_\ell\right\}, \quad 0 \leq |\ell| \leq r - 1 \tag{2.2}$$

which are ordered in groups of equal $|\ell|$ with $|\ell| = 0, 1, \cdots, r-1$, and denote by $\kappa = \kappa(r)$ the number of unknowns in (2.2). In [3] we show how to select a stencil of κ cells with indices $J(i)$ (including i) so that the relations

$$A(C_j) \cdot R_i(x; \overline{u}) = \overline{u}_j, \quad j\epsilon J(i) \tag{2.3a}$$

result in an invertible system of linear equations

$$Qd = \overline{u}. \tag{2.3b}$$

Here Q is the matrix with entries

$$A(C_j) \cdot (x - c_i)^\ell / h^{|\ell|} \tag{2.4a}$$

and $\overline{u}$ is an appropriate ordering of the RHS of (2.3a). Note that due to the scaling in (2.2), the entries of the matrix Q (2.4a) are 0(1) under refinement and consequently

$$\|Q^{-1}\| \leq \text{const.} \quad \text{as} \quad h \to 0. \tag{2.4b}$$

In [3] we consider two cases: (1) a fixed choice of a centered stencil $J(i)$ which results in an upwind biased "linear" scheme (1.5); (2) an adaptive choice of stencil $J(i) = J(i; v^n)$ where the stencil selected is the one in which $\tilde{u}(x)$ is the smoothest among several candidate stencils. This adaptive selection results in an ENO scheme (1.5).

We turn now to describe the decomposition into scales (2.1). For $k = K, K - 1, \cdots$, we check whether

$$\max_{j \in J(i)} |A(C_j) \cdot R^k(x; \overline{u}) - \overline{u}_j| < \delta(\varepsilon); \tag{2.5}$$

$k(i)$ is the largest k for which (2.5) holds. Next we show that (2.1b) holds for an appropriate choice of $\delta(\varepsilon)$. Let us rewrite the polynomial of degree $(r-1)$ $R^k(x; \overline{u})$ as a finite Taylor series around the centroid c_i as in (1.4b) and denote the vector ordering of its coefficients (2.2) by $\tilde{d}$. We denote by $\tilde{u}^k$ the vector ordering of

$$\tilde{u}_j^k = A(C_j) \cdot R^k(x; \overline{u}), \quad j \in J(i) \tag{2.6a}$$

as in (2.3a). Clearly

$$Q\tilde{d} = \tilde{u}^k \tag{2.6b}$$

where Q is the matrix in (2.3b). It follows therefore that

$$Q(d - \tilde{d}) = \overline{u} - \tilde{u}^k, \tag{2.6a}$$

$$\|d - \tilde{d}\| \leq \|Q^{-1}\| \, \|\tilde{u} - \tilde{u}^k\| \leq \|Q^{-1}\| \delta(\varepsilon) \tag{2.7a}$$

and consequently

$$\max_{x \in C_i} |R^k(x; \overline{u}) - R_i(x; \overline{u})| \leq C\|Q^{-1}\|\delta(\varepsilon) \tag{2.7b}$$

where

$$C = \max_{1 \leq |\ell| \leq r-1} \max_{x \epsilon C_i} \frac{|x - c_i|^\ell}{h^{|\ell|}} \tag{2.7c}$$

and $\| \; \|$ is the ℓ_1-norm. We note that the terms C and $\|Q^{-1}\|$ in (2.7) depend on the geometry of the cell and of the stencil $J(i)$, respectively.

Up to this point we have dealt with the concept of multi-resolution analysis in the context of approximation of functions. We turn now to consider its application to the numerical solution of the IBVP (1.1) by the scheme (1.5). Given $\{v_j^n\}$, the cell-averages of the numerical solution at time t_n over the cells $\{C_j\}$, we apply the multi-resolution analysis (2.5) to find $k(i)$ and define the multi-resolution reconstruction $\tilde{R}(x; v^n)$ (1.8a) by

$$\tilde{R}(x; v^n) = R^{k(i)}(x; v^n) \quad \text{for} \quad x \epsilon C_i \tag{2.8a}$$

and consider the numerical scheme

$$v_j^{n+1} = A(C_j)E(\tau)\tilde{R}(x; v^n). \tag{2.8b}$$

We observe that due to the well-posedness of the problem

$$\|E(\tau)R(x; v^n) - E(\tau)\tilde{R}(x; v^n)\| \leq \text{const.} \|R(x; v^n) - \tilde{R}(x; v^n)\| \tag{2.8c}$$

and therefore the deviation of the values computed by the multi-resolution scheme (2.8) from those of the original scheme (1.5) is of order ε. Roughly speaking, all the values v_j^{n+1} in (2.8b) for which $k(j) = k_0$ can be obtained from the calculation on the coarse grid k_0 by

$$v_j^{n+1} = A(C_j)E(\tau)R^{k_0}(x; v^n), \quad k(j) = k_0. \tag{2.9}$$

What we need now is an efficient algorithm that will enable us to perform the computation in (2.8) using the appropriate coarse grid calculations (2.9) at the computational cost of a corresponding adaptive grid implementation without its inherent complexity. To get ideas for such an algorithm we consider a related computational problem in the next section.

3. A Related Problem

In this section we consider the problem of computing discrete values of a composite function $\phi(v(x))$, where $v(x)$ is a function of the type that we encounter in the solution of hyperbolic conservation laws (1.1), i.e., it is piecewise smooth with different scales of smooth variation. The function $\phi(v)$ is a model for the numerical flux in (1.5c). We assume that $\phi(v)$ is a smooth function of v which is expensive to compute, e.g.,

$$\phi(v) = \int_{-\infty}^{\infty} g(v, x)dx \qquad (3.1)$$

with $g(v, x)$ that depends smoothly on v.

In this model problem we assume that $v(x)$ is defined in $[0,1]$ and its discrete values are given on the uniform grid

$$G^0 = \{x_j\}_{j=0}^{2^J}, \quad x_j = 2^{-J} \cdot j, \quad v_j = v(x_j). \qquad (3.2a)$$

We assume that the grid G^0 is fine enough to resolve $v(x)$ to our satisfaction, and that in fact $v(x)$ is over-resolved in some parts of the grid. Our task is to find an efficient algorithm to obtain values ϕ_j^0, $0 \le j \le 2^J$ which approximate $\phi(v(x_j))$ within a specified tolerance of error ε, i.e.,

$$|\phi_j^0 - \phi(v_j)| < \varepsilon. \qquad (3.2b)$$

We consider the nested set of grids $\{G^k\}$, $0 \le k \le K$, which is defined by

$$G^k = \{x_j\}_{j \in J^k}, \quad J^k = \{2^k \ell\}_{\ell=0}^{2^{J-k}} \qquad (3.3a)$$

in which G^0 is the finest (given) grid and G^K is the coarsest; let h_k denote the spacing of points in the grid G^k

$$h_k = 2^{k-J}. \qquad (3.3b)$$

Clearly

$$G^k \cap G^{k-1} = G^k = \{x_{2\ell}\}_{2\ell \in J^{k-1}}, \qquad (3.3c)$$

$$G^{k-1} - G^k = \{x_{2\ell-1}\}_{(2\ell-1) \in J^{k-1}}. \qquad (3.3d)$$

The restriction $v^k = \{v_j\}_{j \in J^k}$ of the values $\{v_j\}$ to the grid G^k constitutes a level of resolution of $v(x)$ which corresponds to the scale h_k.

We begin the computation on the coarsest grid G^K by calculating the values $\{\phi_j^K\}_{j \in J^K}$ from the given "expensive" expression, say (3.1)

$$\phi_j^K = \phi(v_j), \quad j \in J^K. \qquad (3.4)$$

Then for $k = K, \ldots, 1$ we use the already computed values ϕ^k on the grid G^k to obtain acceptable approximate values ϕ^{k-1} of $\phi(v(x))$ on the finer grid G^{k-1}, as follows:

At all points (3.3c) which are common to both grids, we retain the already computed values

$$\phi_{2\ell}^{k-1} = \phi_{2\ell}^{k}, \quad 2\ell \epsilon J^{k-1}. \tag{3.5a}$$

The points (3.3d) are the ones added to G^k by splitting each of its intervals into two. At each such middle point we make a decision whether to compute ϕ directly, i.e.,

$$\phi_{2\ell-1}^{k-1} = \phi(v_{2\ell-1}), \quad (2\ell-1)\epsilon J^{k-1} \tag{3.5b}$$

or to approximate ϕ there by interpolation from the grid G^k, i.e.,

$$\phi_{2\ell-1}^{k-1} = I(x_{2\ell-1}; \phi^k), \quad (2\ell-1)\epsilon J^{k-1}. \tag{3.5c}$$

We elect to interpolate (3.5c) wherever it can be ensured that the interpolated value of ϕ is accurate to a prescribed tolerance ε_{k-1}

$$|\phi_{2\ell-1}^{k-1} - \phi(v_{2\ell-1})| < \varepsilon_{k-1}, \quad (2\ell-1)\epsilon J^{k-1}. \tag{3.5d}$$

The basic idea behind this algorithm is that $\phi(v_j)$ is to be computed directly until the prescribed level of resolution implied by (3.2b) is obtained. Once the required level of resolution is achieved in a certain locality we cease to use the expensive direct computation and the values of ϕ are transferred to finer grids by interpolation (which is assumed to be considerably less expensive). The basic assumption is that $\phi(v)$ is smooth and does not vary much in $[\min v(x), \max v(x)]$. Thus the quality of approximating $\phi(v(x))$ on G^{k-1} by interpolation from the coarser G^k is determined up to a scaling factor by the error in doing so for $v(x)$. Since $v(x)$ is given on all grids, our success in interpolating v^{k-1} from v^k can be easily measured. The criterion for acceptable interpolation is formulated by

$$\max_{\ell} |I(x_{j+\ell}; v^k) - v_{j+\ell}| < \delta_k \Rightarrow |I(x_j; \phi(v^k)) - \phi(v_j)| < \varepsilon_k \tag{3.6a}$$

Here the "$\max_{\ell}$" denotes checking the quality of the interpolation at the point $j \epsilon J^{k-1}$ ($\ell = 0$) and possibly at neighboring points of the finest grid ($\ell = \pm 1, \ldots$). δ_k is the corresponding tolerance for the v-interpolation, and it is to be determined by analytical methods.

We recall that the values of ϕ^k that are actually used in the interpolation (3.5c) are an ε_k approximation to the exact values $\phi(v^k)$, thus

$$|I(x; \phi^k) - I(x; \phi(v^k))| < C_k \cdot \varepsilon_k, \qquad (3.6b)$$

and therefore in (3.5d)

$$|I(x_j; \phi^k) - \phi(v_j)| \le |I(x_j; \phi^k) - I(x_j;$$

$$\phi(v^k))| + |I(x_j; \phi(v^k)) - \phi(v_j)| \le (1 + C_k) \cdot \varepsilon_k. \qquad (3.6c)$$

Consequently the tolerance levels $\{\varepsilon_k\}$ satisfy

$$\varepsilon_0 = \varepsilon > \varepsilon_1 > \cdots > \varepsilon_K = 0, \qquad (3.7a)$$

$$(1 + C_k) \cdot \varepsilon_k \le \varepsilon_{k-1}. \qquad (3.7b)$$

We turn now to consider the practical implementation of the multi-resolution algorithm (3.4) - (3.6). The computational effort of executing it is composed of $N(1 - 2^{-K-1})$ checks, $M \le N$ direct calculations of ϕ and $(N - M)$ interpolations of ϕ; $N = 2^J$. The number of checks is fixed and is almost N, so we want to make the test in (3.6a) as simple as possible. On the other hand we wish to reduce M, the number of direct evaluations $\phi(v)$; this calls for an elaborate test so as to not miss points in which interpolation is acceptable. These considerations lead to a compromise in the level of sophistication to be used in the interpolation $I(x, v^k)$ and $I(x; \phi^k)$. The simplest and least expensive interpolation is based on the use of a fixed central stencil. However, $v(x)$ is a discontinuous function and using a fixed central stencil means that we shall miss all the points near the discontinuity which otherwise could be well-approximated by the more expensive ENO interpolation that uses an adoptive stencil [1], [2]. Investing even more in an ENO interpolation by using subcell resolution [4] will enable us to get a good approximation even in a cell which contains a discontinuity.

A reasonable strategy is to use a hierarchy of checks: first to try the simplest interpolation with fixed central stencil which is a good enough test for most of the domain. If the test in (3.6a) fails for the simple interpolation we may try to get an acceptable approximation by an ENO interpolation. In cases where the computation of $\phi(v)$ is so expensive that it justifies an additional investment in checks, we may try to use also a subcell resolution approach.

4. Multi-Resolution for One-Dimensional Conservation Laws

In this section we present a multi-resolution algorithm for the solution of one-dimensional hyperbolic conservation laws which is motivated by the algorithm (3.4) - (3.6) for the model problem.

We consider the IBVP for the one-dimensional conservation law

$$u_t + f(u)_x = 0, \quad 0 \le x \le 1, \quad t > 0 \qquad (4.1a)$$

$$u(x,0) = u_0(x), \quad 0 \le x \le 1 \qquad (4.1b)$$

with appropriate boundary conditions at $x = 0, x = 1$.

We discretize functions in [0,1] by taking their cell-averages on the grid G^0 (3.2a)

$$\bar{u}_j = A([x_{j-1}, x_j]) \cdot u(t), \quad 1 \le j \le 2^J.$$

Given cell-averages $\bar{u} = \{\bar{u}_j\}_{j \in J^0}$ we consider the reconstruction $R(x; \bar{u})$ which is defined in a piecewise manner in [0,1] and satisfies

$$R(x; \bar{u}) = u(x) + 0(h^r), \quad \text{wherever } u(x) \text{ is smooth}, \qquad (4.2a)$$

$$A([x_{j-1}, x_j]) \cdot R(x; \bar{u}) = \bar{u}_j, \quad 1 \le j \le 2^J. \qquad (4.2b)$$

We use the numerical scheme (1.5) which can be written in this case as

$$v_j^{n+1} = A([x_{j-1}, x_j]) \cdot E(\tau) \cdot R(x; v^n),$$

$$v_j^0 = A([x_{j-1}, x_j]) \cdot u_0(x), \ 1 \le j \le 2^J, \qquad (4.3a)$$

and has the conservation form

$$v_j^{n+1} = v_j^n - \lambda_0(\bar{f}_j - \bar{f}_{j-1}), \quad \lambda_0 = \tau/h_0 \qquad (4.3b)$$

with the numerical flux

$$\bar{f}_j = \frac{1}{\tau} \int_0^\tau f(E(t) \cdot R|_{x_j}) dt. \qquad (4.3c)$$

Note that the computation in (4.3) is to be performed with λ_0 which is restricted by the CFL condition, and that $\bar{f}_0$ and $\bar{f}_{2^J}$ include the influence of the corresponding boundary conditions at $x = 0$ and $x = 1$.

In solving one-dimensional problems, as well as in multi-dimensional problems on cartesian grids, it is convenient to work with the primitive function

$$U(x,t) = \int^x u(\xi,t)d\xi. \qquad (4.4a)$$

Since

$$\frac{1}{h}[U(x_j,t_n) - U(x_{j-1},t_n)] = v_j^n, \qquad (4.4b)$$

$$\frac{1}{\tau}[U(x_j,t_n+\tau) - U(x_j,t_n)] = \overline{f}_j, \qquad (4.4c)$$

it is convenient to define the reconstruction $R(x,v^n)$ by

$$R(x;v^n) = \frac{d}{dx}I(x;U^n) \qquad (4.5)$$

where $I(x;U^n)$ is a piecewise interpolation of $\{U(x_j,t_n)\}$ (see [2]) and to compute the numerical fluxes from (4.4c).

We turn now to describe a multi-resolution algorithm for carrying out the computation of the numerical solution (4.3). We are given $v^n = \{v_j^n\}_{j\epsilon J^0}$ on the grid G^0 and our task is to compute all the numerical fluxes $\{\overline{f}_j\}$ on this grid within a prescribed tolerance for error. Our basic assumption is that the solution is well-resolved on the grid G^0, and in fact it may be over-resolved in some parts of it. The computational task is thus similar to the one entailed in the model problem, except that here $\overline{f}_j$ is really a functional of $u(x,t_n)$ and not a function like $\phi(v(x))$ in Section 3. We use the same set of nested grids (3.3) $G^k, 0 \le k \le K$, and initialize the algorithm by evaluating $\overline{f}_j$ in (4.3c) on the coarsest grid G^K. These fluxes, which we denote by

$$\overline{f}_j = \overline{f}(v^n)_j, \quad j\epsilon J^K, \qquad (4.6)$$

are actually an r-th order approximation to (4.3c) which uses the reconstruction $R(x;v^n)$ of the finest grid G^0. Thus we use the same numerical fluxes that we would use in a straight forward fine grid calculation, except that these fluxes are computed only in the few locations which correspond to the coarsest grid G^K.

The algorithm for the numerical flux computation proceeds in analogous way to (3.5). For $k = K,\ldots,1$ we use the already computed values of the numerical flux $\overline{f}^k$ on the grid G^k to obtain acceptable values $\overline{f}^{k-1}$ on the finer grid G^{k-1} as follows:

At all points (3.3c) which are common to both grids we retain the already computed values

$$\overline{f}_{2\ell}^{k-1} = \overline{f}_{2\ell}^{k}, \quad 2\ell\epsilon J^{k-1}. \tag{4.7a}$$

At the new middle points (3.3d) we make a decision whether to compute the numerical flux directly

$$\overline{f}_{2\ell-1}^{k-1} = \overline{f}(v^n)_{2\ell-1}, \quad (2\ell-1)\epsilon J^{k-1} \tag{4.7b}$$

or to approximate it by interpolation of $\overline{f}^k$, the already computed values of the numerical flux on the grid G^k

$$\overline{f}_{2\ell-1}^{k-1} = I(x_{2\ell-1}, \overline{f}^k), \quad (2\ell-1)\epsilon J^{k-1}. \tag{4.7c}$$

The decision whether to interpolate or compute at a certain locality depends on whether we can obtain the prescribed accuracy for the numerical flux from the level of resolution which is offered by the grid G^k. As is indicated by (2.8c), the error

$$\overline{f}_j^k - \overline{f}_j = \frac{1}{\tau}\int_0^\tau [f(E(t)\cdot R^k|_{x_j}) - f(E(t)\cdot R|_{x_j})]dt \equiv \varepsilon_j^f \tag{4.8a}$$

which is committed by replacing the exact numerical flux (4.3c) by $\overline{f}_j^k$, the numerical flux corresponding to the level of resolution which is available in G^k, can be bounded by the corresponding error in reconstruction

$$\max_{x_{j-1}\leq x\leq x_{j+1}} |R^k(x;v^n) - R(x;v^n)| \equiv \varepsilon_j^R. \tag{4.8b}$$

In the constant coefficient case $f(u) = au$, we get

$$|\varepsilon_j^f| \leq \frac{|a|}{\tau}\int_0^\tau |R^k(x_j - at; v^n) - R(x_j - at; v^n)|dt \leq |a|\varepsilon_j^R, \tag{4.9}$$

and similar estimates can be obtained in the general nonlinear case. As is indicated by (2.5) and (2.7), ε_j^R (4.8b) can be estimated from ε_j^u

$$\varepsilon_j^u = \max_{i\epsilon J(j)\cup J(j-1)} |A([x_{i-1}, x_i])\cdot R^k(x; v^n) - v_i^n|. \tag{4.10a}$$

This check is foolproof but expensive. In our numerical computations we have experimented with a simplified version of (4.10a)

$$\varepsilon_j^u = \max\{|A([x_{j-1}, x_j])\cdot R^k(x; v^n) - v_j^n|, \\ |A([x_j, x_{j+1}])\cdot R^k(x; v^n) - v_{j+1}^n|\} \tag{4.10b}$$

which takes into account only the two adjacent fine grid intervals that determine the numerical flux at x_j. These numerical experiments indicate that the test (4.10b) may be adequate enough.

¿From our numerical experiments we have also learned that it is better not to interpolate the numerical fluxes directly (4.7c), but rather use the following indirect way: Given $\bar{f}^k$ we can compute $v^{n+1,k}$, the cell-averages at time t_{n+1} on G^k by

$$v_j^{n+1,k} = v_j^{n,k} - \frac{\tau}{h_k}(\bar{f}_j^k - \bar{f}_{j-2^k}^k), \quad j \epsilon J^k. \qquad (4.11a)$$

¿From $v^{n+1,k}$ we get the values of the primitive function $U^{n+1,k}$ by using the relation (4.4b). Using interpolation $I(x; U^{n+1,k})$ for the primitive function at time t_{n+1} and possibly a different interpolation $\tilde{I}(x; U^{n,k})$ for the primitive function at time t_n we can now obtain the value of the numerical flux at the middle points (3.3d) from the relation (4.4c). Thus we replace (4.7c) by

$$\bar{f}_{2\ell-1}^{k-1} = \frac{1}{\tau}[I(x_{2\ell-1}; U^{n+1,k}) - \tilde{I}(x_{2\ell-1}; U^{n,k})], \quad (2\ell - 1)\epsilon J^{k-1}.$$
$$(4.11b)$$

We remark that unlike the model problem of Section 3, the computation of the numerical flux involves propagation of discontinuities. Consequently a stencil of cells which did not contain discontinuities of $u(x, t_n)$ may contain a discontinuity of $u(x, t_{n+1})$ at its two extreme cells. To overcome this problem we use a central stencil for R^k in the test (4.10b) and for the interpolation $\tilde{I}$ for the primitive function at time t_n; in order to account for the movement of discontinuities we take I in (4.11b) to be an ENO interpolation for the primitive function at time t_{n+1}.

5. Summary

This work on multi-resolution analysis has been motivated by the current interest in wavelets. The concept of wavelets has developed into a beautiful theory which is rich in structure. The endowment of wavelets with many properties is also their main handicap when it comes to practical computations: Indeed wavelets are more local than Fourier analysis, but they are not really local. The desire to have an orthonormal basis with wavelet functions of compact support results in wavelet functions which are of fractal nature. The attempt to make these wavelet functions smoother results in considerably enlarging their support.

In the present work we have attempted to take from wavelet only its most important idea – the multi-resolution analysis – and to strip it of any other structure. In doing so we have developed a multi-resolution analysis which is devoid of beautiful theory, but we have gained flexibility and the ability to adapt locally. Given a function $u(x)$ which is represented by cell-averages on an arbitrary partition into cells, we have shown how to decompose it into various scales. This general method provides a very useful tool for removing superfulous information, especially when the data is only piecewise smooth, e.g., application of this methodology to image-compression has proven to be very successful.

We remark that this multi-resolution analysis can be easily extended to functions $u(x)$ which are represented by point-values on an unstructured grid. The reason for presenting only the cell-average version is that we wish to combine multi-resolution analysis with high order of accuracy, and this forces us to choose the scheme (1.5) for accurate cell-averages. We cannot use point-value schemes which are both high-order accurate and in conservation form. The need to fictitiously write f_x in (4.1a) as a difference of a numerical flux results in a strong dependence of the numerical flux of the point-value scheme on the spacing of the grid. Consequently we cannot retain the already computed numerical fluxes on the grid G^k when refining to G^{k-1} (4.7a) for point-value schemes with order of accuracy larger than 2.

The multi-resolution analysis presented in this paper has points in common with both multi-grid and adaptive-grid methods. It resembles multi-grid in its use of nested grids. However multi-grid is primarily an iteration scheme which is designed to accelerate convergence, while multi-resolution is primarily a data-compression scheme which is designed to avoid superfulous computations by discarding insignificant information. Its goal is the same as that of adaptive-grid methods but it differs in its concept: Adaptive-grid methods tell you where to refine the mesh, while multi-resolution analysis tells you where not. The difference in the practical implementation of the two concepts reminds one of the difference between shock-fitting and shock-capturing. In adaptive-grid methods you have to worry about missing spontaneously generated features. On the other hand you can do an optimal job in resolving features that you know of, although at the cost of complex programming. In multi-resolution methods you cannot miss new features because conceptually we always compute on a fine enough grid. The main issue is efficiency:

You want to make sure that you do not compute on too fine a grid, when you do not have to. An additional advantage is the automatic way in which the data is handled in multi-resolution analysis which translates into simplicity of programming.

The one-dimensional algorithm of Section 4 can be extended in a straightforward way to the multi-dimensional case with cartesian grids. Based on our one-dimensional results we are led to believe that realistic savings by factors between 10 to 100 are quite possible in the three-dimensional case.

References

Harten, A., "On high-order accurate interpolation for non-oscillatory shock capturing schemes," IMA Program on Continuum Physics and its Applications, 1984-85, pp. 72-105; also MRC Technical Summary Report No. 2829, June 1985.

Harten, A., Engquist, B., Osher, S. and Chakravarthy, S. R., "Uniform high-order accurate ENO schemes, III," J. Comput. Phys., Vol. 71, pp. 231-303, 1987; also ICASE Report No. 86-22, April 1986.

Harten, A. and Chakravarthy, S. R., "Multi-dimensional ENO schemes for general geometries," ICASE Report No. 91-76, September 1991.

Harten, A., "ENO schemes with subcell resolution," J. Comput. Phys., Vol. 83, pp. 148-184, 1989; also ICASE Report No. 87-56, August 1987.

ADAPTIVE-MESH ALGORITHMS FOR COMPUTATIONAL FLUID DYNAMICS

Kenneth G. Powell and Philip L. Roe

The University of Michigan
Ann Arbor, MI 48109-2140

James Quirk

Institute for Computer Applications in Science and Engineering
NASA Langley Research Center
Hampton, VA 23665-5225

ABSTRACT

The basic goal of adaptive-mesh algorithms is to distribute computational resources wisely, by increasing the resolution of "important" regions of the flow, and decreasing the resolution of regions that are less important. While this goal is one that is, more often than not, worthwhile, implementing schemes that have this degree of sophistication remains more of an art than a science. In this paper, the basic pieces of adaptive-mesh algorithms are described, and some of the possible ways to implement them are discussed and compared. These basic pieces are: the data structure to be used; the generation of an initial mesh; the criterion to be used to adapt the mesh to the solution; and the flow-solver algorithm on the resulting mesh. Each of these is discussed, with particular emphasis on methods suitable for the computation of compressible flows.

1 Introduction

In order to compute complex flows, researchers in computational fluid dynamics have found it necessary to build an increasing amount of sophistication into their codes. While rapid advances in the capacity of computers have drastically improved the quality of solutions that can be obtained, the increased sophistication of algorithms has had at least as big an effect.

One particular approach to making wise use of computational resources is that of allowing the mesh to adapt to the solution in some way, so that high-gradient regions are not under-resolved, and low-gradient regions are not over-resolved. A rather unfortunate side-effect of implementing this approach is that the mesh-generation algorithm and the flow-solution algorithm become inextricably coupled. This side-effect is one of the primary reasons that adaptive-mesh algorithms are inherently more difficult to implement than fixed-mesh schemes. The potential savings, both in memory and compute-time, that can be gained from adaptive-mesh schemes is large enough to warrant the challenging programming task for many types of flows, however. Virtually any flow with multiple scales, which tend to lead to localized regions of high gradients, is a candidate for an adaptive scheme. The larger the difference in scales, the greater the potential savings. These differing scales can be temporal or spatial in nature, or both. In extreme cases, adaptive refinement can yield solutions that would be prohibitively expensive on fixed meshes. An example of this is shown in Figures 1 and 2, which show grid and Mach contours for a Mach stem formed by the $M_\infty = 2.0$ flow past a 15° compression followed by a 15° expansion [DP91]. The resulting Mach stem is too small to be captured sufficiently without substantial refinement.

Because of their wide range of applicability, adaptive-mesh schemes have been implemented for finite-difference, finite-volume and finite-element flow solvers, on a variety of mesh types. For a researcher setting out to implement an adaptive-mesh scheme for a computational fluids problem, there are many choices to made, including:

- the data structure to be used;

- the initial mesh generation method to be used;

- how to adapt the mesh;

- the determination of where to adapt the mesh;

305

15 Degree Wedge Mach Number Line Contours.

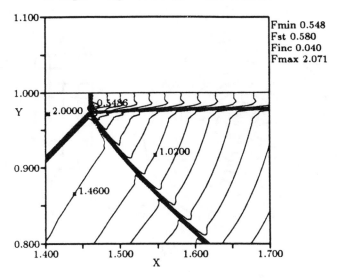

Figure 1: Mach Contour Detail — $M_\infty = 2.0$ Mach Stem

15 Degree Wedge Grid Plot.

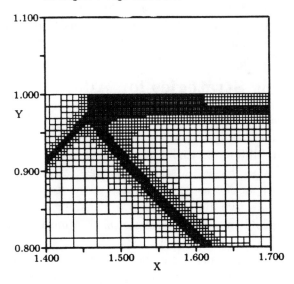

Figure 2: Grid Detail — $M_\infty = 2.0$ Mach Stem

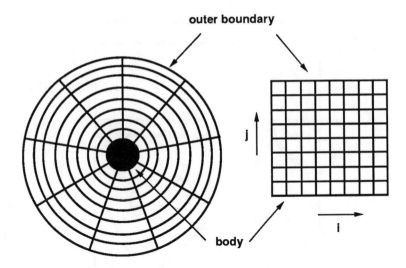

Figure 3: Example of Mapping of a Structured Mesh from Physical Space to Computational Space

- the flow-solver algorithm on the adapted mesh.

The goal of this paper is to list and describe some of the possible choices, and to explain some of the trade-offs that can help guide one to suitable choices for the class of problems being solved.

2 Various Strategies for Adaptation

There is a very strong tie between the strategy chosen for mesh adaptation and the data structure used in the calculation. The majority of finite-difference and finite-volume methods developed for fluid dynamics are based on structured meshes that can be mapped from physical space to a computational space in which they appear as a square, for two-dimensional problems, or a cube, for three-dimensional problems. An example of such a mesh is shown in Figure 3.

The fact that the mesh can be mapped in this way allows the mesh to be thought of as "structured," with the physical coordinates of the mesh points, and the flow variables, stored as arrays. The point in physical space corresponding to the mesh point (i, j) in

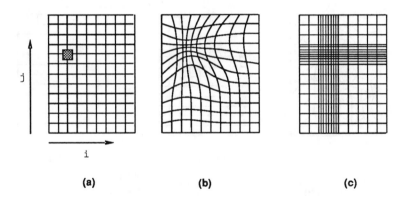

Figure 4: Refinement of a Structured Mesh

computational space is then $\mathbf{x}_{i,j}$, and its four immediate neighbors are $\mathbf{x}_{i+1,j}$, $\mathbf{x}_{i-1,j}$, $\mathbf{x}_{i,j-1}$, and $\mathbf{x}_{i,j+1}$. Thus, due to the array-based data structure, the neighbors of a mesh point are easily determined. Typically, in structured-mesh calculations, a convention is taken in which the cell made up of the four vertices (i, j), $(i+1, j)$, $(i, j+1)$, $(i+1, j+1)$ is labeled as cell (i, j), so that cells' neighbors are also easily determined.

Structured meshes lend themselves to a specific type of mesh adaptation; that of mesh-point movement. An example of this is shown in Figure 4. On the structured mesh of part (a) of the figure, a region of interest is flagged (the shaded square). Mesh points can then be redistributed as in (b), so that there is a higher density of points in the vicinity of the region of interest. This type of refinement, called *r-refinement*, is natural on structured meshes. An alternate type of refinement, *h-refinement*, consists of adding extra mesh points in the region of interest. On a structured mesh, adding a new *point* necessitates adding the entire *line* along which the point lies. This has the detrimental result of introducing new mesh points where they are not needed, as shown in part (c) of Figure 4. For this reason, mesh point addition/deletion is not practical on structured meshes.

To allow the addition (or deletion) of points in a region, without the problem outlined above, more sophisticated data structures must be used than the simple array-type addressing of structured meshes. Three examples of h-refinement on unstructured meshes are shown

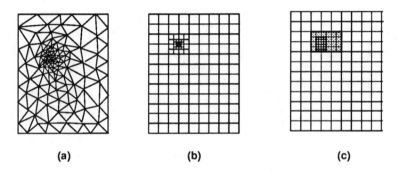

Figure 5: Refinement of Unstructured Meshes

in Figure 5. In part (a) of the figure, a mesh of triangles is shown. This mesh cannot be mapped to a computational space with (i, j) indexing. Instead, *connectivity information* must be stored in some way, so that a cell "knows" which cells are its neighbors, and which vertices define the cell. Meshes of this type are typically addressed by "flat" data structures, in which mesh cells and/or faces and/or vertices are stored in linked lists.

In part (b) of Figure 5, a mesh of quadrilaterals is shown. This mesh could be addressed similarly to that in part (a); i.e. with a "flat" data structure of linked lists. An alternate approach is a "hierarchical" data structure, i.e. a tree structure. In a cell-based tree structure, for example, depicted in Figure 6, a cell on a given level of the tree is refined, spawning four children cells at the next level of the tree. Some or all of these children cells may then spawn their own children, and so on until the final mesh is built. In part (b) of Figure 5, the mesh resulting from spawning three levels of children cells in the region of interest is shown. Cells are the obvious entity to store in a tree; vertices or faces require a "forest of trees," that is, they cannot be traced back to a single root. Cell-based trees have the added advantage of being well-suited to multigrid [HvdME90].

In part (c) of Figure 5, a tree of structured meshes is shown. That is, a region of the initial structured mesh is marked for refinement, and a single child (in this example) structured mesh, of finer mesh spacing, is embedded in the parent mesh. This process may be repeated indefinitely, leading to the amount of refinement desired. This approach, implemented by Berger and Colella [BC87] and Quirk [Qui91], can result in very efficient adaptive algorithms

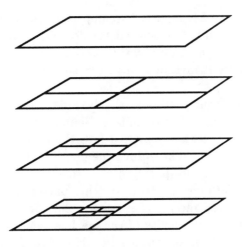

Figure 6: A Cell-Based Tree Data Structure

for unsteady flows.

Each of the data structures listed above has advantages and disadvantages; there is no one "correct" data structure for a given problem. The choice of the appropriate data structure for a depends upon, among other things:

- the complexity of the problem geometry;

- whether or not h-refinement will be done;

- whether compute-time or available memory is the primary computational limitation;

- the frequency with which the grid will be adapted.

Basically, simpler data structures result in a loss of generality, but can greatly simplify the code.

Once a data structure has been chosen, the first step to be taken is the generation of an initial, yet-to-be adapted, mesh. For structured meshes, these techniques are well known. Whether based on algebraic techniques or the solution of partial differential equations (elliptic, hyperbolic or parabolic), the primary consideration is that the mesh fit the boundaries.

Techniques for the generation of unstructured meshes are less developed, but are currently being heavily researched. For fully unstructured meshes of triangles, two steps are necessary: the generation of a cloud of points to discretize the flow-field, and the generation of a connectivity of these points. The two can be done hand-in-hand, as in advancing-front methods, or as two separate steps, as in Delaunay methods.

In advancing-front methods, the boundaries of the domain are discretized, and each collection of faces defining a boundary is treated as a "front," which is then marched into the interior of the domain. The marching procedure introduces new points, whose locations are determined based on the spacing of a coarse, non-body-fitted "background mesh" that gives control over the spacing and smoothness of the final, body-fitted mesh. These new points are incorporated into the front, and the connectivity of the mesh is generated as the fronts march towards each other, ending in triangulation of the entire domain. Some of the issues in implementing advancing-front grid generation techniques have been discussed by Löhner and Parikh [LP88], Löhner [Löh88], and Peraire et al [PVMZ87].

The primary advantages of advancing front methods are the ability to control mesh spacing by means of the background mesh, and the fact that the point cloud in the interior is generated in the advancing-front procedure. Both of these advantages are double-edged, however; specification of the background mesh requires user input, and the point cloud generated by the advancing-front procedure, and the connectivity of the cloud, might be far from optimal.

One particular connectivity of a given set of points has a number of mathematical properties that make it optimal for some problems. This is the Delaunay triangulation. Some of its properties are:

- the Delaunay triangulation of a given point cloud is unique, unless there are degeneracies (four points lying on a circle) within the cloud;

- no triangle in a Delaunay triangulation has a circumcircle containing any point other than those that define the triangle;

- the Delaunay triangulation maximizes the minimum interior angle of the triangles in the mesh;

- a piecewise linear approximation to a function defined at the points is smoothest for the Delaunay triangulation;

- solutions to Laplace's equation on a two-dimensional Delaunay mesh satisfy a numerical maximum principle.

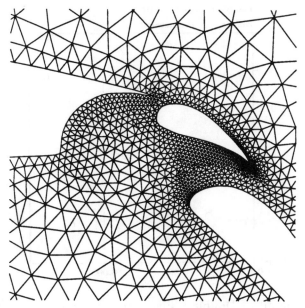

Figure 7: Detail of Advancing-Front/Delaunay Mesh

These properties are discussed by Rippa [Rip89], Barth [Bar90] and Baker [Bak91]. Although this list of special properties is impressive, and although there exist a number of methods for efficiently constructing the Delaunay triangulation of a given set of points, Delaunay meshes have their drawbacks. First, these methods require a cloud of points, that must be generated by another procedure; Delaunay is just a way of connecting these points. Second, Delaunay methods are, by their nature, best suited for isotropic problems. If highly skewed meshes are desired, other triangulations could be more suitable. Finally, some of the properties for two-dimensional Delaunay meshes do not carry over to three dimensions. In particular, the numerical maximum principle for solutions to Laplace's equation does not, in general, hold on three-dimensional Delaunay meshes.

Some very nice initial meshes can be obtained by combining advancing front and Delaunay techniques. In a method developed by Müller et al [MRD91], a point cloud is generated by an advancing-front technique, and the points are introduced into the triangulation so as to give the Delaunay connectivity. An example mesh is shown in Figure 7.

Meshes of triangles or quadrilaterals can also be generated by tree-based procedures [She88, Bae89, DP91]. A quadtree-based generation of a body-cut Cartesian mesh, for example, is carried out in three steps. These steps are depicted in Figures 8–10. First,

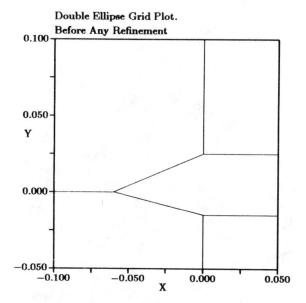

Figure 8: Detail of Double-Ellipse Mesh after Refinement to Meet Background Δs Criterion

a quadrilateral that encloses the entire flow domain is generated, and enough levels of children cells are spawned that a user-specified cell size is reached. The intersections of this mesh of quadrilaterals with the body (or bodies) in the flow are computed. The mesh produced in this way for a certain geometry is shown in Figure 8. At this point, all of the cells which are cut by the body spawn children cells, until a user-specified maximum cell size on the body is reached. The result is shown in Figure 9. Finally, slopes of neighboring faces on the body are compared, with children cells spawned until a maximum body-curvature threshold is reached. The result is shown in Figure 10. The resulting mesh, obtained with only three simple criteria set by the user, is an acceptable initial mesh, with the geometry of the problem captured well. Similar quadtree [ML91] and tritree [Ken88] (three children triangles spawned from each parent) techniques for two-dimensional triangulations have been developed, and octree-based methods for three dimensions have recently been carried out [She88, YMB+91].

Once the initial mesh has been generated, by whatever technique, there remains the issue of how to improve the mesh adaptively. If an h-refinement approach is taken, two choices need to be made. First, a decision must be made as to whether a re-meshing procedure must be done at each adaptation step, or if h-refinement should be done

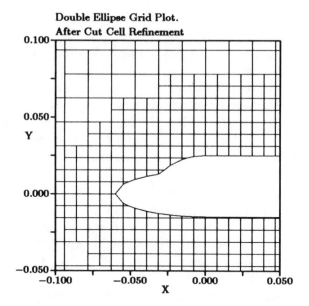

Figure 9: Detail of Double-Ellipse Mesh after Refinement to Meet Body Δs Criterion

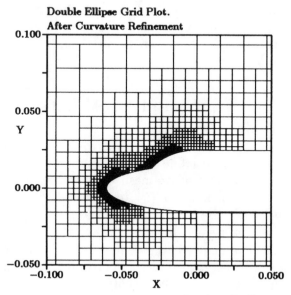

Figure 10: Detail of Double-Ellipse Mesh after Refinement to Meet Body Curvature Criterion

locally at each adaptation step. While the re-meshing procedure, generating a new connectivity of the points in the domain each time new points are added, guarantees that certain properties of the mesh (e.g. Delaunay connectivity) are maintained, it can be prohibitively expensive. It is also difficult to do the re-meshing in such a way as to guarantee a conservative transfer of the discrete solution from the old mesh to the new [GP91]. Unless the number of adaptation steps to be made is very small, it is usually more convenient simply to carry out local h-refinement. In local h-refinement, points are introduced, but the mesh is only reconnected in the immediate vicinity of the new points. If, after several steps, the mesh gets too skewed or stretched, a re-meshing can be done at that point.

The second decision to be made is whether to carry out the refinement in an isotropic or anisotropic manner. Isotropic and anisotropic refinements of triangles, quadrilaterals, and body-cut Cartesian cells are shown in Figure 11. Isotropic refinement of a triangle can be done by connecting the midpoints of each face of the parent triangle, resulting in four children; anisotropic refinement can be done by connecting the midpoint of one face of the parent to the opposite vertex of the parent, resulting in two children. Isotropic refinement (four children) and anisotropic refinement (two children) of a quadrilateral, or a body-cut Cartesian cell, follows similarly. Detail of an isotropically-refined body-cut Cartesian mesh is shown in Figure 12; detail of an isotropically refined mesh of triangles is shown in Figure 13.

3 The Determination of Where to Adapt the Mesh

The decision of where to refine the mesh, and where to coarsen it, is at the very heart of adaptive-mesh algorithms. The fact that a consensus has not formed as to how to make this decision points to the conclusion that there is no uniquely "correct" approach. Some of the approaches in use are described below.

In r-refinement schemes and h-refinement schemes, the adaptation is basically a two-step process. The first step is the same in both methods; each location in the flow-field must be assigned an analog value related to its "importance" in terms of the solution fidelity. For the r-refinement schemes, this analog measure must be used in some mesh-movement scheme to push the mesh points away from

315

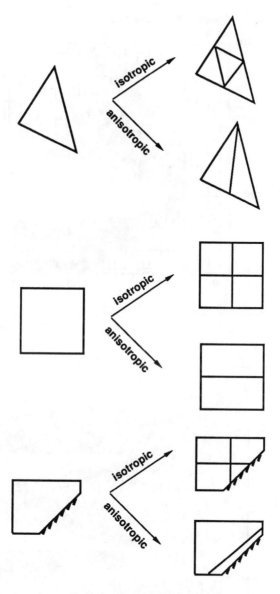

Figure 11: Isotropic and Anisotropic Refinements of Various Types of Cells

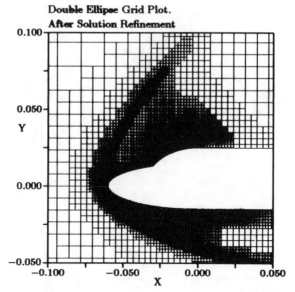

Figure 12: Detail of Double-Ellipse Mesh after Solution-Adaptive Refinement

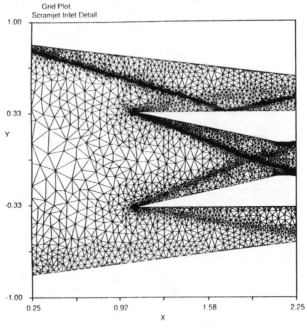

Figure 13: Isotropically-Refined Mesh of Triangles

each other in relatively unimportant regions, and pull them together in relatively important regions. In the h-refinement schemes, the analog measure must be converted into a digital decision at each location, as to whether to add a point, delete a point, or leave things as they are. A recent review of r-refinement methods is that of Hawken et al [HGH91], and a comparison of r-refinement and h-refinement methods for steady, transonic flows is given by Dannenhoffer [Dan91]. In the following, h-refinement schemes will primarily be discussed.

The measures of the importance of various regions of the flow are typically based on one of two approaches. In one approach, some method of estimating the local truncation error is used. The idea behind this approach is to equidistribute, at least approximately, the numerical error. While this approach is theoretically very attractive, there are some rather daunting obstacles in implementing it on a relatively complex problem. The fundamental problem is that the truncation error is (obviously) not known, so that it must be approximated in some way. Typically, a Richardson-type extrapolation is used. This can be very inaccurate, however, due to:

- large variations in mesh spacing on stretched and/or unstructured meshes;

- variations in order of the numerical scheme to facilitate shock capturing.

The other approach is more heuristic. Foreknowledge of the flow features of interest that might occur, and the way that they might manifest themselves, is used to define an adaptation criterion. These criteria are usually combinations of first and second derivatives of flow quantities, normalized in various ways. Examples of some of these are tabulated in Table 1. They have been used/advocated by Dannenhoffer [Dan87], Peraire et al [PVMZ87], and Paillere et al [PPZ86], among others.

One of the difficulties in using these heuristic criteria is clear from studying the table. Various flow features (e.g. shocks, shears, contacts) show up in the solution in various ways. So, a criterion that is well suited for shocks (e.g. pressure gradient) may be poorly suited for shears. One possible solution to this dilemma is to devise a "catch-all" criterion, that detects as many features as possible. This, however, does not address another potential problem — that of flow features of disparate strengths. If a single catch-all criterion is used for a flow with, for example, a strong shear and a weak shock, the

Quantity	Comments
$\epsilon_i \propto |\Delta q|_i$	Se, So, Dd
$\epsilon_i \propto (|\rho_{xx}| + |\rho_{xy}| + |\rho_{yy}|)$	Co, So, Di
$\epsilon_i \propto |\nabla p_0|$	Co, So, Di
$\epsilon_i \propto |\nabla \cdot \mathbf{u}|$	So, Di
$\epsilon_i \propto |\nabla \times \mathbf{u}|$	Se, Di

Dd   Directionally dependent

Di   Directionally independent

Co   Detects contacts

Se   Detects shears

So   Detects shocks

Table 1: Various Adaptation Criteria and Their Attributes

shock might not be detected. So, for complex flows, separate criteria, each of which detects one type of feature well, may be better. An example of this is shown in Figures 14 –18. The Mach contours for an under-expanded axisymmetric jet are shown in Figure 14. The jet boundary is a strong shear; relatively weak shocks are present in the flow as well. The result of refining 15% of the cells in the mesh, based on a criterion that captures both shocks and shears, is shown in Figure 15. The shear is detected, but portions of the shocks are missed. The mesh resulting from refining 10% of the cells, based on a criterion that is tailored to capturing shears is shown in Figure 16; only the shear is detected. The mesh resulting from refining 10% of the cells, based on a criterion that is tailored to capturing shocks and expansions is shown in Figure 17; the shear is entirely ignored. The combination of these two criteria, such that the two together yield a refinement of 15% of the cells, the same number of cells as with the "catch-all" criterion, gives a very good mesh; the shocks, the shear, and the expansion are all detected, as shown in Figure 18.

Another problem that can occur is that the detection criterion may cause undue refinement of a feature that is already resolved, at the expense of a region of the flow that needs refinement. An excellent (and sobering) depiction of this problem has been given by Warren et al [WATK91], along with a remedy. The remedy is a

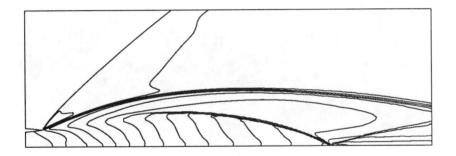

Figure 14: Under-Expanded Jet Mach Contours

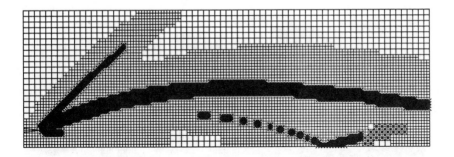

Figure 15: Under-Expanded Jet Grid — Catch-all Criterion

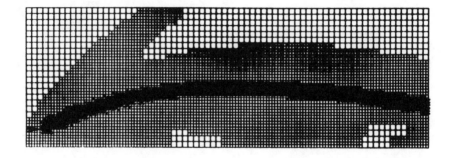

Figure 16: Under-Expanded Jet Grid — Velocity-Curl Criterion

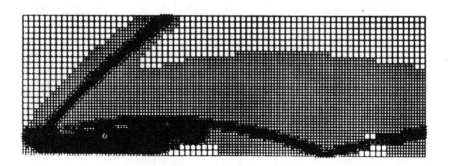

Figure 17: Under-Expanded Jet Grid — Velocity-Divergence Criterion

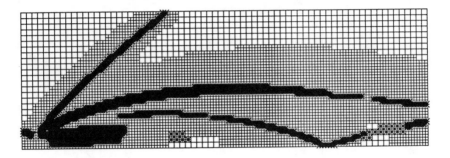

Figure 18: Under-Expanded Jet Grid — Combined Curl/Divergence Criterion

careful choice of the constants of proportionality in the criteria listed in Table 1. First-derivative quantities must be multiplied by a factor h_i^p, where h_i is a characteristic length for cell i, and the power p is chosen so that, as the mesh is refined to the point that the feature is resolved, the value of ϵ_i vanishes. For refinement criteria based on first derivatives, the requirement is that $p > 1$; for criteria based on second derivatives, it is $p > 2$.

If h-refinement is to be done, then, once a criterion is chosen, thresholds must be set so as to determine which cells are to be refined, which coarsened, and which left alone. There are basically two approaches to setting the thresholds:

1. choosing thresholds "by experience;"

2. choosing thresholds based on an analysis of the distribution of the criterion ϵ on the mesh.

In the first method, a numerical value ϵ_r is set so that all cells with $|\epsilon| > \epsilon_r$ are refined. Similarly, a numerical values ϵ_c is set so that all cells with $|\epsilon| < \epsilon_c$ are coarsened. All remaining cells are left alone (neither refined nor coarsened). This method has the advantage of simplicity, but requires tuning of the values ϵ_r and ϵ_c; values that work well for one flow may not work well for another flow.

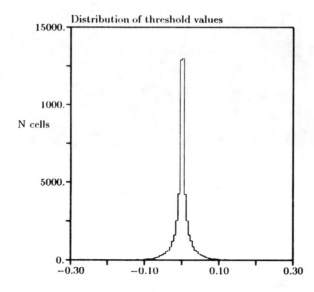

Figure 19: Distribution of the Refinement/Coarsening Criterion

The second method is most easily understood by looking at the distribution function for the criterion ϵ on the mesh [Dan87]. In Figure 19, the distribution of ϵ for a typical case is shown. Adaptation may be carried out by:

- refining all cells with a value of ϵ falling outside the bounds of x_1 standard deviations above or below the mean;

- coarsening all cells with a value of ϵ falling inside the bounds of x_2 standard deviations above or below the mean;

- leaving all remaining cells alone.

The values of x_1 and x_2 may be chosen so that a certain percentage of the cells will be flagged for refinement, and a certain percentage flagged for coarsening. This method is more general than the first, but requires the added work of calculating the distribution, its mean and its standard deviation.

For approaches based on a tree of structured grids (as in part (c) of Figure 5), one additional step is required. The cells flagged by the methods outlined above must be "grouped" into rectangular blocks. A depiction of a depth-first recursive procedure, taken from [Qui91], is given in Figure 20. In the top left corner, the originally flagged cells are denoted by crosses. The extent of these flagged cells is found

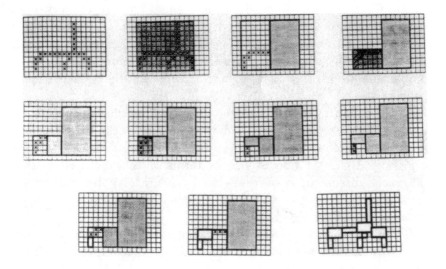

Figure 20: Grouping of Flagged Cells

(top row, second from left), and divided in half (top row, third from left). This process of determining the extent of the flagged cells, and halving the bounding rectangle, is repeated until the final grouping of the flagged cells (bottom right corner) is determined.

4 Techniques for Flow Solvers on Adapted Meshes

The advantage of adapted meshes — the ability to resolve multiple scales — also implies the disadvantage of adapted meshes — that the flow solver must be carefully constructed to account for large variations in the sizes of neighboring cells. This difficulty, however, arises in *all* unstructured mesh calculations, and so any algorithm suited for solving the equations on an unstructured mesh may be used on an adapted mesh. The desired attributes of a numerical scheme for the Euler or Navier-Stokes equations are conservation, consistency, accuracy and positivity. If care is taken, all of these can be obtained on an adapted mesh. There is a price to be paid, however; the resulting algorithms are somewhat more complex than comparable algorithms that *do not* account for the variable mesh spacing.

In the following, algorithms for the two-dimensional Euler equations will be discussed. The methods carry over to three dimensions, with the triangles and quadrilaterals becoming tetrahedra, hexahedra or prisms. Discretization of viscous terms on unstructured meshes will not be discussed; however, positive, accurate approximations to the viscous terms in the Navier-Stokes equations can be constructed by the methods described below [Bar91].

The Euler equations in two dimensions represent the conservation of mass, momentum and energy for an inviscid, compressible fluid. They may be written as

$$\frac{\partial \mathbf{U}}{\partial t} + \frac{\partial \mathbf{F}}{\partial x} + \frac{\partial \mathbf{G}}{\partial y} = 0 \tag{1a}$$

where

$$\mathbf{U} = (\rho, \rho u, \rho v, \rho E)^{\mathrm{T}} \tag{1b}$$

$$\mathbf{F} = \left(\rho u, \rho u^2 + p, \rho u v, \rho u \left(E + \frac{p}{\rho} \right) \right)^{\mathrm{T}} \tag{1c}$$

$$\mathbf{G} = \left(\rho v, \rho u v, \rho v^2 + p, \rho v \left(E + \frac{p}{\rho} \right) \right)^{\mathrm{T}} \tag{1d}$$

An equation of state of the form

$$E = E(p, \rho) \tag{1e}$$

is necessary to close the equations; typically the ideal gas relation

$$\frac{p}{\rho} = (\gamma - 1) \left[E - \frac{u^2 + v^2}{2} \right] \tag{1f}$$

is used.

In discretizing the partial differential equations, the first decision that has to be made is how the discrete data will be represented on the mesh. In Figure 21, three possible ways of representing data on a mesh are shown. In part (a), the components of the state vector

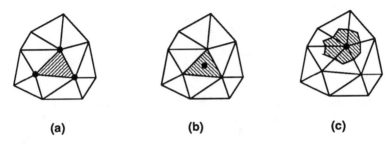

(a)   (b)   (c)

Figure 21: Three Ways to Represent Data on a Mesh of Triangles

U are stored at nodes, and are assumed to vary linearly over the triangles. This is an example of a finite-element viewpoint; other distributions may be assumed, leading to higher-order elements. In part (b), the cell-averages of the components of U are stored for each cell; in essence the mass, momentum and energy for a control-volume coinciding with a cell of the mesh are stored. This is one way of implementing a finite-volume scheme. In part (c), a combination of the two approaches is taken. Average values of the components of U are stored, as in part (b), but the averages are done over cells of the *median dual* of the mesh, resulting in the unknowns being associated with the vertices of the mesh, as in part (a). This is another way of implementing a finite-volume scheme. Choosing between the representations shown in parts (b) and (c) for a finite-volume scheme depends on the types of cells in the mesh: for a mesh of quadrilaterals (2D) or hexahedra (3D), the number of cells in the mesh is approximately equal to the number of nodes in the mesh; for a mesh of triangles (2D) or tetrahedra (3D), the number of cells in the mesh is typically several times the number of nodes in the mesh.

Finite-element methods are well-suited for unstructured and adapted meshes. Shape functions (N) and test functions ($\tilde{N}$) are chosen, and the Euler equations are integrated over the solution domain, yielding the semi-discrete system [Sha91]

$$\mathbf{M}\frac{d\mathbf{U}_i}{dt} = -\left(\mathbf{R_x}\mathbf{F}_i + \mathbf{R_y}\mathbf{G}_i\right) \tag{2a}$$

where

$$\mathbf{M} = \int_i \tilde{\mathbf{N}}^T \mathbf{N}\, dx\, dy \tag{2b}$$

$$\mathbf{R_x} = \int_i \tilde{\mathbf{N}}^T \frac{\partial \mathbf{N}}{\partial x}\, dx\, dy \qquad (2c)$$

$$\mathbf{R_y} = \int_i \tilde{\mathbf{N}}^T \frac{\partial \mathbf{N}}{\partial y}\, dx\, dy \qquad (2d)$$

are integrals that can be carried out analytically, and evaluated specifically for element i, once the form of the shape and test functions is defined. This is a set of coupled differential equations in time, with the coupling produced by the mass matrix $\mathbf{M}$. If steady-state solutions are the goal, the mass matrix may be "lumped" (replaced by a diagonal matrix in which each diagonal entry is the sum of the elements in the corresponding row of $\mathbf{M}$) to decouple the equations. If time-accurate solutions are required, the coupled system must be solved iteratively. This can be done with a few iterations of a pre-conditioned conjugate-gradient method [LMZ84, Wat87].

Finite-element methods for smooth problems, e.g. the Galerkin method in which $\tilde{\mathbf{N}} = \mathbf{N}$, do not ensure positivity, rendering them less useful in solving for inviscid or high Reynolds number flows. Large overshoots can occur in the vicinity of high-gradient regions of the flow, leading, in the extreme, to non-physical negative values of pressure and density. To remedy this, various methods may be employed. One method is to choose "smart" test functions that avoid oscillations. This nonlinearity, introduced into the schemes through the test functions, is at the heart of the Petrov-Galerkin and Streamwise Upwind Petrov-Galerkin (SUPG) methods [Joh90]. Another method to avoid oscillations in finite-element schemes is the FCT (flux-corrected transport) method [Zal79]. In the FCT method, two approximations to the change $\Delta \mathbf{U}_{i,j}$ at node i due to triangle j are calculated. The first is a high-order approximation to the change, calculated by, for instance, a Galerkin procedure. The second is a low-order approximation to the change, calculated by a procedure that includes an artificial viscosity. The update at a node is then carried out in the form

$$\Delta \mathbf{U}_{i,j} = \Delta \mathbf{U}_{i,j}^L + \phi \left(\Delta \mathbf{U}_{i,j}^H - \Delta \mathbf{U}_{i,j}^L \right) \qquad (3)$$

where a superscript L denotes the low-order scheme, a superscript H denotes the high-order scheme, and $\phi \in [0,1]$ is a "limiter" calculated so as to avoid spurious oscillations.

Finite-volume techniques can also be implemented fairly naturally on unstructured meshes, if care is taken. The most successful approaches have been extensions of Van Leer's MUSCL algorithm [vL79]. The basic building blocks of a MUSCL-type scheme are:

1. calculation of the distribution of the flow variables within a "control-volume" by a reconstruction procedure;

2. calculation of the fluxes arising from the interaction of neighboring control-volumes, by means of an approximate solution of the Riemann problem;

3. summation of the fluxes into each control-volume, and integration in time of the changes to the flow variables.

The second step is not affected by the fact that the grid is unstructured; the third step is mildly affected; the first step is profoundly affected.

To illustrate the reconstruction procedure, a scheme in which the primitive variables are taken to vary linearly over the cells will be described. There are many variations on this theme: the conserved variables, rather than the primitive variables, can be interpolated, and/or higher-order reconstructions can be done [BF90]. The methods described, though a special case, are illustrative of the entire class of reconstruction methods.

If the vector of primitive variables is

$$\mathbf{W} = (\rho, u, v, p)^{\mathrm{T}} \tag{4}$$

and is assumed to vary linearly over a cell, then the value of $\mathbf{W}$ at any point in the cell is given by

$$\mathbf{W}(\mathbf{x}) = \bar{\mathbf{W}} + (\mathbf{x} - \bar{\mathbf{x}}) \cdot \nabla \mathbf{W} \tag{5}$$

where $\bar{\mathbf{W}}$ is the mean value of the primitive variables in the cell, and $\bar{\mathbf{x}}$ is the cell centroid. It is clear, then, that an approximation to the gradient of $\mathbf{W}$ is necessary to carry out the reconstruction. Since only volume averages of $\mathbf{W}$ are given, this gradient must be calculated from a combinations of these averages. One way to do this is by use of Gauss' theorem

$$\nabla \mathbf{W} = \frac{1}{A} \oint \mathbf{W} \mathbf{n} \, ds \tag{6}$$

where A is the area enclosed by the path of integration, and $\mathbf{n}$ is the unit outward-pointing normal on the path. Some choices for paths on a mesh of triangles are given in Figure 22.

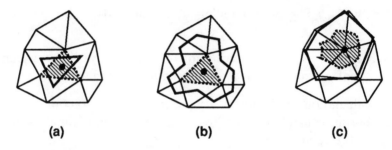

(a)   **(b)**   **(c)**

Figure 22: Some Possible Paths for Gradient Calculations. Control volumes for the flux balance are shown as dashed lines; paths for the gradient calculation are shown as heavy lines.

In part (a), a minimal path is shown, employing only the three face-neighbors of the cell. This approach leads to an inexpensive gradient calculation. If vertex-neighbors of the cell are included as well, as shown in part (b), a more accurate approximation is obtained. Particularly on non-smooth meshes, this more expensive gradient calculation appears to be necessary. Part (c) of the figure shows a gradient calculation for a case in which the variables are associated with cells of the median-dual mesh. Here, the order-one neighbors typically give a good approximation to the gradient. Some gradient-integral paths on a body-cut Cartesian mesh are shown in Figure 23.

An alternate approach to the path-integral approach outlined above is to calculate a least-squares approximation to the gradient in a region [Bar90]. Thus, to get the gradient of $W^{(k)}$, the kth component of the primitive vector $\mathbf{W}$, the system to be solved is

$$\mathcal{L}\nabla W^{(k)} \;=\; f \tag{7a}$$

$$\mathcal{L} = \begin{pmatrix} w_1\Delta x_1 & w_1\Delta y_1 \\ \vdots & \vdots \\ w_N\Delta x_N & w_N\Delta y_N \end{pmatrix} \qquad f = \begin{pmatrix} w_1\Delta W_1^{(k)} \\ \vdots \\ w_N\Delta W_N^{(k)} \end{pmatrix} \tag{7b}$$

where

$$\Delta x_i \;=\; x_i - x_0 \tag{7c}$$
$$\Delta y_i \;=\; y_i - y_0 \tag{7d}$$
$$\Delta u_i \;=\; u_i - u_0 \tag{7e}$$

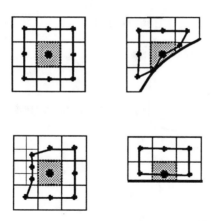

Figure 23: Some Possible Paths for Gradient Calculations on a Body-Cut Cartesian Mesh.

and the points are numbered so that 0 is the origin, as shown in Figure 24. The w_i's are weight factors, and can be taken to be one, although other values of the weight factors can give gradient approximations that meet certain desirable properties. This approach, while slightly more expensive computationally than the path-integral approach, is very general, and does not require any particular ordering of the data used in the gradient calculation.

Once the gradient has been calculated, values of the primitive variables at the faces may be obtained. However, for the constraint of positivity to be met, the gradient must be limited. That is, the reconstruction formula used is

$$\mathbf{W}(\mathbf{x}) = \bar{\mathbf{W}} + \phi(\mathbf{x} - \bar{\mathbf{x}}) \cdot \nabla\mathbf{W} \tag{8}$$

where $\phi \in [0,1]$ is a limiter. This limiter is calculated so that the values of $\mathbf{W}$ reconstructed within the cell are bounded by the data used in the gradient calculation. Typical values of the limiter are shown in Figure 25, for a transonic airfoil case. As can be seen, the scheme is higher order ($\phi \approx 1$) virtually everywhere, reverting to first order ($\phi \approx 0$) only in the vicinity of the shocks and the wake.

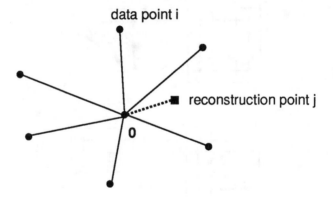

Figure 24: Numbering of data points

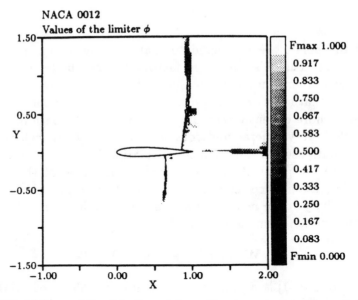

Figure 25: Values of the Limiter ϕ for a Representative Flow Calculation

The finite-volume form of the Euler equations can be written as

$$\frac{d\mathbf{U}}{dt} = -\frac{1}{A} \sum_{faces} (\mathbf{F}\Delta y - \mathbf{G}\Delta x) \tag{9}$$

where A is the area of that cell, Δx and Δy are the changes of x and y along a face (defined so that the integral is carried out in a counter-clockwise sense). Defining the face length and the normal and tangential velocities as

$$\Delta s = \sqrt{(\Delta x)^2 + (\Delta y)^2} \tag{10a}$$

$$u_n = \frac{(u\Delta y - v\Delta x)}{\Delta s} \tag{10b}$$

$$u_t = \frac{(u\Delta x + v\Delta y)}{\Delta s}, \tag{10c}$$

the flux through a face may be written as

$$(\mathbf{F}\Delta y - \mathbf{G}\Delta x) = \begin{pmatrix} \rho u_n \\ \rho u_n u + p\frac{\Delta y}{\Delta s} \\ \rho u_n v - p\frac{\Delta x}{\Delta s} \\ \rho u_n H \end{pmatrix} \Delta s \equiv \boldsymbol{\Phi}\Delta s . \tag{11}$$

This flux is a function of the values of the state variables to the "left" and "right" of the face, as calculated from the reconstruction step. Using Roe's approximate Riemann solver to calculate the flux gives

$$\boldsymbol{\Phi}(\mathbf{U}_L, \mathbf{U}_R) = \frac{1}{2}[\boldsymbol{\Phi}(\mathbf{U}_L) + \boldsymbol{\Phi}(\mathbf{U}_L)]$$

$$- \frac{1}{2} \sum_{k=1}^{4} |\hat{a}_k|^* \Delta V_k \hat{\mathbf{R}}_k \tag{12}$$

where the a_k's, ΔV_k's and $\hat{\mathbf{R}}_k$'s are the speeds, strengths and eigenvectors, respectively, associated with the waves of the solution to the Riemann problem [Roe81].

With this or another flux function, the sum of the fluxes for a cell may be calculated from Equation 9, or from a higher-order quadrature in which, for instance, fluxes are calculated at Gauss points of each face [BF90]. This sum yields the time rate of change of the solution within the cell. These temporal changes may be integrated by explicit or implicit time-marching procedures. Explicit time-marching algorithms for adaptive meshes work exactly as do those on structured meshes. Multi-stage schemes are particularly easy to implement, and can be very efficient, if they are chosen so as to be well-suited for the spatial-discretization scheme [vLTP89]. Implicit schemes that are popular for structured meshes (e.g. SLOR and ADI) cannot be implemented in obvious ways on on unstructured or adaptive meshes, although some fairly intricate and very clever ways of defining sweep directions have been developed [HMP91]. Even direct solvers can be implemented; the resulting matrices are not banded, as in structured-mesh codes, but are sparse [Ven88].

If time-accuracy is not necessary, various convergence-acceleration techniques (local time-stepping, preconditioning [vLLR+91], multigrid, etc.) may be employed. If unsteady calculations are the goal, the options for making the time integration more efficient are fewer. One useful technique for unsteady flows on adaptive meshes is that of *adaptive time-stepping*. In this technique, several small time-steps are taken on the finer cells of the mesh for each (larger) time-step on the coarser cells of the mesh. One implementation of this has been presented by Quirk [Qui91], who used it to carry out calculations that would otherwise have been prohibitively expensive. An alternate method has been developed by Chiang [Chi91]. A representative calculation using an adaptive time-stepping scheme is shown in Figure 26, which depicts the axisymmetric Mach number distribution due a strong shock wave exiting a cylinder. The resolution is such that the shear layer shows an instability.

5   Concluding Remarks

In this paper, several choices have been given for each piece of an adaptive-mesh scheme. Some of these are:

- The data can be stored in a linked-list structure or a tree structure;

- Delaunay, advancing-front or Cartesian-based grid generation methods may be used;

- finite-volume or finite-element formalisms may be used to construct a flow algorithm;

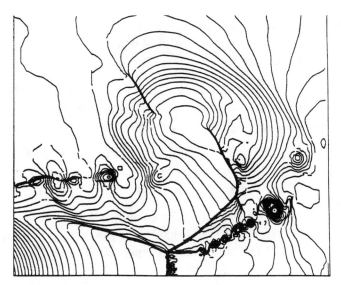

Figure 26: Axisymmetric Calculation of a Strong Shock Exiting a Cylindrical Tube

- any one of a number of adaptation criteria can be used to determine where to adapt the mesh.

The reason for this dizzying array of choices is that *no one approach is best*. The types of problems to be solved should guide the choice of what data structure to use, what adaptation criterion to use, and so on.

The first decision to be made, even before the ones listed above, is whether an adaptive-mesh scheme will pay off for the problem being solved. In general, problems with multiple length scales and/or multiple time scales are the ones for which adaptive schemes will yield the largest gains. For problems that will not have localized regions that are more "interesting" than other regions of the flow, the gains will be modest, and probably not worth the added complexity introduced into the code. For a great number of problems, however, the gains can be substantial.

References

[Bae89]   P. L. Baehman. *Automated Finite-Element Modeling and Simulation*. PhD thesis, Rensselaer Polytechnic Institute, 1989.

[Bak91]   T. Baker. Unstructured meshes and surface fidelity for
           complex shapes. In *AIAA 10th Computational Fluid
           Dynamics Conference*, 1991.

[Bar90]   T. J. Barth. On unstructured grids and solvers. In
           Computational Fluid Dynamics. Von Kármán Institute
           for Fluid Dynamics, Lecture Series 1990-04, 1990.

[Bar91]   T. J. Barth. Numerical aspects of computing viscous
           high Reynolds number flows on unstructured meshes.
           AIAA Paper 91-0721, 1991.

[BC87]   M. J. Berger and P. Colella. Local adaptive mesh
           refinement for shock hydrodynamics. Technical Re-
           port UCRL-97196, Lawrence Livermore National Lab-
           oratory, 1987.

[BF90]   T. J. Barth and P. O. Frederickson. Higher order solu-
           tion of the Euler equations on unstructured grids using
           quadratic reconstruction. AIAA Paper 90-0013, 1990.

[Chi91]   Y.-L. Chiang. *Simulation of Unsteady Inviscid Flow on
           an Adaptively Refined Cartesian Grid*. PhD thesis, Uni-
           versity of Michigan, 1991.

[Dan87]   J. F. Dannenhoffer III. *Grid Adaptation for Complex
           Two-Dimensional Transonic Flows*. ScD thesis, Mas-
           sachusetts Institute of Technology, 1987.

[Dan91]   J. F. Dannenhoffer III. A comparison of adaptive-grid
           redistribution and embedding for steady transonic flows.
           *International Journal for Numerical Methods in Engi-
           neering*, 32:653–663, 1991.

[DP91]   D. De Zeeuw and K. G. Powell. An adaptively-refined
           carterian mesh solver for the Euler equations. To appear
           in *Journal of Computational Physics*, 1991.

[GP91]   A. Goswami and I. Parpia. Grid restructuring for mov-
           ing boundaries. In *AIAA 10th Computational Fluid Dy-
           namics Conference*, 1991.

[HGH91]   D. F. Hawken, J. J. Gottlieb, and J. S. Hansen. Review
           of some adaptive node-movement techniques in finite-

element and finite-difference solutions of partial differential equations. *Journal of Computational Physics*, 95:254–302, 1991.

[HMP91]   O. Hassan, K. Morgan, and J. Peraire. An implicit finite-element method for high-speed flows. *International Journal for Numerical Methods in Engineering*, 32:183–205, 1991.

[HvdME90] P. W. Hemker, H. T. M. van der Maarel, and C. T. H. Everaars. BASIS: a data structure for adaptive multigrid computations. Center for Mathematics and Computer Science Report NM-R9014, Amsterdam, 1990.

[Joh90]   C. Johnson. The streamline diffusion finite element method for compressible and incompressible fluid flow. In *Computational Fluid Dynamics*. Von Kármán Institute for Fluid Dynamics, Lecture Series 1990-04, 1990.

[Ken88]   S. R. Kennon. A vectorized Delaunay triangulation scheme for non-convex domains with automatic nodal point generation. AIAA Paper 88-0314, 1988.

[LMZ84]   R. Löhner, K. Morgan, and O. Zienkiewicz. Adaptive grid refinement for the compressible Euler equations. In *Accuracy Estimates and Adaptivity for Finite Elements*. Wiley, 1984.

[Löh88]   R. Löhner. Some useful data structures for the generation of unstructured grids. *Communications in Applied and Numerical Mathematics*, 4:123–135, 1988.

[LP88]   R. Löhner and P. Parikh. Generation of three-dimensional unstructured grids by the advancing-front method. AIAA Paper 88-0515, 1988.

[ML91]   N. Maman and B. Larrouturou. On the use of dynamical finite-element mesh adaptation for 2d simulation of unsteady flame propagation. In K. W. Morton, editor, *Proceedings of the 12th International Conference on Numerical Methods in Fluid Dynamics*. Springer, 1991.

[MRD91]   J.-D. Müller, P. L. Roe, and H. Deconinck. A frontal approach for internal point cloud creation in Delaunay triangulations. Submitted to *Journal of Computational Physics*, 1991.

[PPZ86]   H. Paillère, K. G. Powell, and D. L. De Zeeuw. A wave-model-based refinement criterion for adaptive-grid computation of compressible flows. AIAA Paper 92-0322, 1986.

[PVMZ87]   J. Peraire, M. Vahdati, K. Morgan, and O. C. Zienkiewicz. Adaptive remeshing for compressible flow calculations. *Journal of Computational Physics*, 72:449–466, 1987.

[Qui91]   J. Quirk. *An Adaptive Grid Algorithm for Computational Shock Hydrodynamics*. PhD thesis, Cranfield Institute of Technology, 1991.

[Rip89]   S. Rippa. *Minimal Roughness Property of the Delaunay Triangulation*. PhD thesis, Tel-Aviv University, 1989.

[Roe81]   P. L. Roe. Approximate Riemann solvers, parameter vectors and difference schemes. *Journal of Computational Physics*, 43, 1981.

[Sha91]   R. A. Shapiro. *Adaptive Finite-Element Solution Algorithm for the Euler Equations*. Vieweg, 1991.

[She88]   M. S. Shepard. Approaches to the automatic generation and control of finite-element meshes. *Applied Mechanics Review*, 41(4):169–185, 1988.

[Ven88]   V. Venkatakrishnan. Newton solution of inviscid and viscous problems. AIAA Paper 88-0413, 1988.

[vL79]   B. van Leer. Towards the ultimate conservative difference scheme. V. A second-order sequel to Godunov's method. *Journal of Computational Physics*, 32, 1979.

[vLLR+91] B. van Leer, W. T. Lee, P. L. Roe, K. G. Powell, and C. H. Tai. Design of optimally-smoothing schemes for the Euler equations. *Journal of Applied and Numerical Mathematics*, 1991.

[vLTP89]   B. van Leer, C. H. Tai, and K. G. Powell. Design of optimally-smoothing multi-stage schemes for the Euler equations. In *AIAA 9th Computational Fluid Dynamics Conference*, 1989.

[Wat87]   A. J. Wathen. Realistic eigenvalue bounds for the Galerkin mass matrix. *IMA Journal of Numerical Analysis*, 7:449–457, 1987.

[WATK91]   G. Warren, W. K. Anderson, J. Thomas, and S. Krist. Grid convergence for adaptive methods. In *AIAA 10th Computational Fluid Dynamics Conference*, 1991.

[YMB+91]   D. P. Young, R. G. Melvin, M. B. Bieterman, F. T. Johnson, and S. S. Samant. A locally refined rectangular grid finite element method: Application to computational fluid dynamics and computational physics. *Journal of Computational Physics*, 62:1–66, 1991.

[Zal79]   S. Zalesak. Fully multidimensional flux-corrected transport algorithm for fluids. *Journal of Computational Physics*, 31, 1979.

INHERENTLY MULTIDIMENSIONAL
SCHEMES

BEYOND THE RIEMANN PROBLEM, PART I

Philip L. Roe

Department of Aerospace Engineering
The University of Michigan
Ann Arbor, MI 48109-2140

1   Introduction

Over the past four decades, the speed and capacity of computing machinery, as well as the effectiveness of CFD algorithms, have all increased by about one order of magnitude per decade. We have become accustomed to, and learned how to live with, this giddy progress. The parallel revolution promises (perhaps, more appropriately, threatens) to raise the speed of computers by three orders of magnitude or even more during the decade to come. In consequence, three-dimensional calculations will become feasible in the time now required for two-dimensional ones. In addition, there will be available the power needed to resolve truly unsteady flows, and very complex configurations. More detail will be available, and adaptive meshes will be used to pursue it. Flow solvers will be coupled with optimizers to attempt design problems. Do we believe that our existing codes are capable of all this?

At present it seems that people in the research establishments or in industry who spend their days using CFD codes to solve the problems that lie at the leading edge of todays technology, devote much of that time to 'babysitting' the code. They make constant small adjustments to the mesh, or to the parameters of the scheme, in order to get the best results; sometimes just to get any results at all. In the more ambitious contexts described above, this will be very much harder, and perhaps impossible. The chief requirement for the schemes will become robustness, and the 90's may very well be the decade in which algorithms do not contribute their own order of magnitude to the overall progress, but merely hold their ground within an unfamiliar and swiftly changing environment.

There will certainly be a tendency to obtain accuracy from refined meshes, rather than tuning the algorithm for maximum performance. New methods will gain acceptance only after their reliability has been very convincingly demonstrated, but an increase in the operation count per mesh point may then be quite acceptable. Certainly, if massively parallel architectures win the day, then operation count per point will in fact signify rather little provided requirements for message passing can be held down.

It is chiefly due to their reputation for reliability that upwind schemes have enjoyed such popularity during the past decade. In view of the remarks above one might expect this popularity to continue. However, there are some important classes of problem looming ahead for which the present upwind schemes, based on one-dimensional physical modelling, are not entirely satisfactory. These are largely to do with massively separated flow, turbulence, and mixing. The reasons for this will be reviewed below. I will then describe some of the attempts that have recently been made to devise improvements whilst still employing merely one-dimensional physical models. All of these methods have enjoyed enough success to suggest strongly that there is something 'out there' well worth exploring.

Very recently, though, we have come within sight of methods that preserve an even more intimate relationship between the numerical model and the physics of multidimensional flows. I will give a somewhat impressionistic overview of these, leaving the technical details to be covered in the companion paper presented here by Deconinck. These methods employ very compact stencils and will therefore incur very low overheads for communication. They follow a philosophy of extracting the maximum amount of physics from a small amount of information. They resolve discontinuities over very small distances and without overshoots, despite containing no limiter functions or artificial viscosity, and they seem to have good convergence properties. I feel very excited and optimistic about these methods. If bad physics has achieved so much in the past, what might not good physics accomplish in the future?

2   The Case For Upwinding

There is a sense in which upwinding is inevitable. Suppose we want to (surely everybody wants to!) solve the linear advection equation,

$$u_t + au_x = 0. \tag{1}$$

A scheme for doing this at Courant number ν is

$$u_j^{n+1} = u_j^n - \nu \sum_{k=0}^{k=K} d_k^-(u_{j-k} - u_{j-k-1}) + d_k^+(u_{j+k+1} - u_{j+k}) \quad (2)$$

which is conservative and at least first-order accurate provided

$$\sum_{k=0}^{k=K} (d_k^- + d_k^+) = 1.$$

In fact, any scheme, explicit, multilevel or implicit, can be manipulated into this form, although the number of weights $\{d_k\}$ may become infinite for implicit schemes.

It is easy to show that a necessary condition for stability at small wavenumbers is that

$$\nu \sum_{k=0}^{k=K} (k + 1/2)(d_k^- - d_k^+) \geq \frac{\nu^2}{2} \quad (3)$$

Thus, for flow from left to right (ν positive) the gradients on the left must be weighted more than the gradients on the right, and vice versa. In other words, anyone who devises a consistent, stable method for solving eqn(1) by any method whatsoever has devised an upwind scheme. Of course, they may have done so implicitly by introducing dissipation! In any case, for equation(1) there is no question about what should be done, merely about what to call it.

The names 'upwinding' and 'dissipation' are nevertheless significant because they suggest completely different mechanisms by which the scheme should be generalised to a system of equations, or to multidimensional problems, bearing in mind that we really want to make *both* extensions.

If we were content to make only the extension to one-dimensional systems there is not much doubt that we should use some form of the characteristic equations to decouple our problem as much as possible into a set of independent scalar problems. The extent of the upwind bias, or the amount of numerical dissipation, or whatever language is used to describe the neccessary stabilising mechanism, can then be chosen independently and optimally for each of the finite number of waves comprising the flow. To me, it this independent treatment of the various waves that that is the real hallmark of an upwind scheme. I would therefore classify the recent "matrix viscosity" schemes of Turkel [27] as upwinded, despite any protests by the author!

Strict independence is not usually possible for nonlinear problems, but wave interactions are usually handled extraordinarily well, just by ensuring conservation. Conservative schemes that compute fluxes by considering wave propagation are firmly grounded on the two most important mathematical properties of hyperbolic conservation laws. Having done such a good job of treating one-dimensional systems, however, it has to be admitted that upwind schemes lose much of their theoretical support in higher dimensions. I have argued this for some time, and will return to the theme below.

By contrast, the schemes which rely on added dissipation suffer no substantial extra loss when extended to higher dimension. Fig. 1 is a graphical way to express the fact that each philosophy loses its cogency at a different stage of the process. Nevertheless, there is

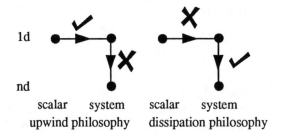

Figure 1: A diagrammatic representation of algorithm development for fluid flow equations. The extension from scalar to system is expressed as a horizontal motion; from one to more dimensions as a vertical motion.

one benefit of upwinding still felt for multidimensional flows. This relates to nonlinear stability. One can easily show that if the gasdynamic equations are solved by a first-order Godunov method using the exact Riemann solution, then no nonphysical states (negative pressures or densities) ever appear in the solution [7]. The proof is valid for any fluid with a convex equation of state, defined here to mean that the region of phase space, occupied by the conserved variables, is convex. For higher-order methods which compute fluxes by solving Riemannn problems for reconstructed data (MUSCL [11], PPM [3], or ENO/UNO [8]), it is required that the reconstruction is also physically meaningful. And even if the Riemann problem is not solved exactly, incorporating some of the nonlinear physics into the code seems to add valuable properties of *robustness*.

3   The Case Against Upwinding

Computational cost and slow convergence are the two most frequent complaints at the present time. Cost, of course, needs to be judged in relation to reliability, and it is argued above that the emphasis is about to shift. Slow convergence is frequently associated with an algorithm that contains internal conflicts, and this is what seems to me to be the trouble with current upwind schemes.

Without significant loss of generality we can consider the case of a two-dimensional Cartesian mesh. All upwind schemes in current practice employ updates of the cell-centered finite-volume form (see Fig. 2)

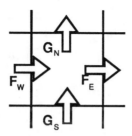

Figure 2: Fluxes through the boundaries of a generic computational cell.

$$\mathbf{u}^{n+1} = \mathbf{u}^n - \frac{\Delta t}{\Delta x}\left[\mathbf{F}_E - \mathbf{F}_W\right] - \frac{\Delta t}{\Delta y}\left[\mathbf{G}_N - \mathbf{G}_S\right] \tag{4}$$

where the fluxes $\mathbf{F}_{E,W}$ are derived by considering wave motion in the x-direction, and $\mathbf{G}_{N,S}$ by considering wave motion in the y-direction. In the case of a more general mesh, either quadrilateral or triangular, the fluxes are at present usually found by considering wave motion normal to each interface.

Consider, in this light, a region of flow occupied by a wave whose level lines are not parallel to the grid. For example when solving the Euler equations we may encounter a shear wave inclined at some angle θ. In such a wave, variations of the conserved variables are proportional to

$$\begin{vmatrix} 0 \\ -a\sin\theta \\ a\cos\theta \\ av\cos\theta - au\sin\theta \end{vmatrix} \tag{5}$$

which is an eigenvector of $\mathbf{F_u}\cos\theta + \mathbf{G_u}\sin\theta$. The fluxes, however, will be calculated, in any upwind scheme of the present type, essentially by projecting the cell differences onto the eigenvectors of $\mathbf{F_u}$ for $\mathbf{F}_{E,W}$, and onto those of $\mathbf{G_u}$ for $\mathbf{G}_{N,S}$. Neither set of eigenvectors has the oblique wave (5) as a member. In general, the 'oblique eigenvectors' only belong to the 'orthogonal eigenvectors' if the x- and y- operators happen to commute, which is not the case for the Euler equations, or most other physically interesting systems. Such a projection destroys what I take to be the foundation of upwinding, namely accurate *recognition* of the physical processes present. For example, in calculating $\mathbf{F}$ the identity

$$
\begin{vmatrix} 0 \\ -a\sin\theta \\ a\cos\theta \\ av\cos\theta - \\ au\sin\theta \end{vmatrix} \equiv \frac{\sin\theta}{2} \begin{vmatrix} 1 \\ u-a \\ 0 \\ h-ua \end{vmatrix} - \frac{\sin\theta}{2} \begin{vmatrix} 1 \\ u+a \\ 0 \\ h+ua \end{vmatrix} + \cos\theta \begin{vmatrix} 0 \\ 0 \\ a \\ av \end{vmatrix} \quad (6)
$$

will cause the shear to be misrepresented as the sum of two acoustic waves that bring about the change of velocity u and a shear wave that brings about the change of velocity v. A more physical way to

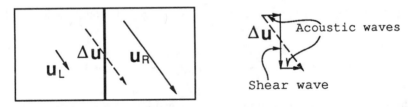

Figure 3: Misrepresentation of an oblique shear wave by an interface-oriented Riemann solver.

explain the effect is shown in Fig. 3. Resolving the observed velocities normal to the cell interface creates an appearance of rarefaction waves, which causes the pressure calculated on the interface to be too low. Anomalous pressure disturbances due to this cause can be clearly seen in a separated flow first computed by Venkatakrishnan [28], and shown in Fig. 4, taken from [24], where the mesh has been chosen to exaggerate the effect. For this aspect of this particular flow, one must admit that central differencing does better.

The theme has been elaborated in [21], where a linear problem is proposed that can either be solved exactly, or by operator splitting with each component step solved exactly. The discrepancy, due solely

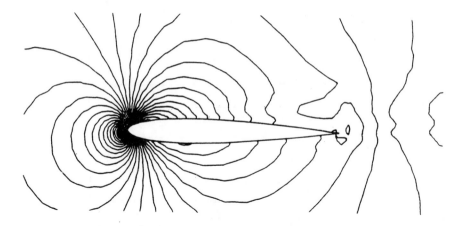

Figure 4: Computation of the viscous separated flow over a NACA 0012 airfoil at Reynolds number 5000

to the splitting, is dispersive and non-dissipative (see Fig. 5). We may deduce that as the components of an operator split method are improved, the results will get worse!

This discussion has concentrated on shear waves, since this appears to be the most acute case. Shock waves, at least of the kind encountered in gasdynamics, have a tendency to pull numerical error along converging characteristics into their interior structure, where it often gets lost. They are, in Gino Moretti's descriptive phrase 'garbage collectors', and capable of handling their own pollution. Even so, it is possible that some of the anomalies reported for blunt-body calculations [16, 13] can be traced to poor physical modelling of shocks almost aligned with the mesh. As for contact discontinuities, either between different fluids, or betwen the same fluid at different entropies, these are eigenvectors of both operators, and to treat them by splitting is perfectly legitimate. (They might be called the commuting part of the problem: all the complexities occur in the subspace that excludes them.)

Oblique shear waves, however, are treated very badly by existing upwind schemes, and it is for this reason that separated flows, turbulence, and mixing were particularly mentioned in the Introduction as requiring improved methods for their treatment.

The widespread adoption of unstructured meshes will, if it comes

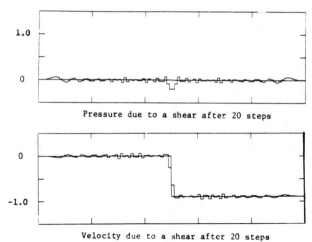

Pressure due to a shear after 20 steps

Velocity due to a shear after 20 steps

Figure 5: Dispersive errors in the solution to a shear flow solved by exact split operators

about, exacerbate the situation. On a square mesh, if a discontinuity is aligned with the grid, then the Riemann problem in one direction will treat the situation correctly, and the other will introduce no error because it will see no gradients. On a triangular mesh, however, even a perfectly regular one, although one set of Riemann solvers may recognize the problem correctly, at least one of the other two directions will produce wrong interpretations of finite gradients.

4 $1\frac{1}{2}$-dimensional Physics

In this section I want to review some recent attempts to improve upwind schemes whilst still solving merely one-dimensional Riemann problems. Two of the schemes also remain within the well-known context of equation(4). This would have the advantage that most parts of an existing code could be left untouched. Only the subroutine that computes the interface fluxes need be replaced.

4.1 More relevant Riemann problems

We require the normal component $(\mathbf{F}, \mathbf{G}) \cdot (n_x, n_y)$ of the flux through a typical interface. Like any other quantity with a vectorial character, the flux $(\mathbf{F}, \mathbf{G})$ itself can be represented by its components in arbitrary coordinate directions, such as (see Fig. 6)

$$\begin{pmatrix} \mathbf{F} \\ \mathbf{G} \end{pmatrix} \Leftrightarrow \begin{pmatrix} \mathbf{F}\cos\theta + \mathbf{G}\sin\theta \\ \mathbf{G}\cos\theta - \mathbf{F}\sin\theta \end{pmatrix} = \begin{pmatrix} \mathbf{F}_\perp \\ \mathbf{F}_\parallel \end{pmatrix} \qquad (7)$$

Then we can express (n_x, n_y) in the same coordinate system and there will be no change in the computed inner product.

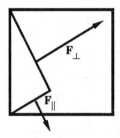

Figure 6: Illustrating the resolution of an interface flux into two orthogonal directions.

Suppose, however, that *numerically* the flux components $(\mathbf{F}_\perp \mathbf{F}_{\parallel})$ are found from different formulae, or using different data? Then the rotation will make a real difference, and we can try to exploit this. The earliest attempt along these lines was made by Davis [5], who concentrated on the capturing of oblique shock waves. He noted that since the velocity component parallel to a shockwave does not change across it, then any isolated shockwave must be moving in the direction of the vector Δq, defined to be the velocity change across the interface. He therefore chose that direction to define a "principal direction" θ, for use in (6). (For the precise formulae used the original paper should be consulted.) Davis subsequently came to prefer a direction based on the local pressure gradient, which must of course be computed from the states in more than two cells.

Davis also found, as have many others who have followed this path, that wherever the flow was nearly uniform, and the privileged directions therefore ill-defined, his calculations tended to become unstable, or at best very slowly convergent. To overcome this he adopted a strategy of smoothing out the predicted directions.

More recently, Levy [12], and, independently, Dadone and Grossman [4] have revived and improved Davis' ideas, creating schemes that are second-order accurate. A general framework that covers both their methods is illustrated in Fig. 7. The objective is to compute the flux across the marked face of the shaded cell. In each method the first step is to compute the gradient of some sensitive quantity, such as pressure [4] or velocity magnitude [12], by applying the Divergence Theorem to the inner (rectangular) control volume, which connects the centers of the six nearest cells. This direction, and the one orthogonal to it, are then selected as the directions along

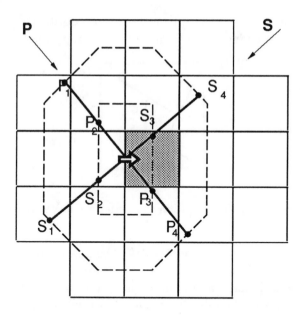

Figure 7: Stencil used for computing second-order rotated fluxes

which the fluxes $\mathbf{F}_\perp, \mathbf{F}_\parallel$ will be evaluated. To save time, and also avoid convergence problems, these directions need only be evaluated once every hundred iterations or so. To first order, each component of the flux is found using data from this same inner control volume. For example, one solves a Riemann problem with data interpolated to the points P_2, P_3, and another Riemann problem with data interpolated to S_2, S_3.

Specialised to linear advection,

$$u_t + \vec{a} \cdot \vec{\nabla} u = u_t + a u_x + b u_y = 0, \tag{8}$$

and with data close to a steady solution, one of these privileged directions would be the characteristic direction $\vec{a}$ and the other would be the solution gradient $\vec{\nabla} u$. The component of flux along $\vec{a}$ is $|\vec{a}|u$, and the component normal to this vanishes. The flux can then be seen to come from interpolating u to the intersection of the characteristic with the inner control volume. Somewhat surprisingly, this scheme is *second*-order accurate in the steady state, and hence non-monotone!

Thus, some form of limiting is needed even at this lowest level. Levy [12] took a hint from the work of Sidilkover [25] by refusing to interpolate along any line making an angle greater than $45°$ with the interface normal. Instead, the data on the $45°$ line is used. Dadone and Grossman, without any supporting analysis, avoid interpolation entirely by taking their data from whichever points on the control volume lie closest to the lines P and S. In fact, the whole issue of limiting and 'monotonicity' is much more complex in higher dimensions, and so far we merely seem to have scratched the surface.

Both methods achieve second-order accuracy by means of MUSCL-type interpolation along the lines P and S, now using information that derives from the outer (octagonal) control volume. Levy again interpolates linearly to P and Q, whereas Dadone and Grossman employ a nearest neighbour strategy. This second choice presumably makes the scheme less sensitive to small variations in the estimated angles, and Dadone and Grossman do not report any need to 'freeze' these in order to stabilize their convergence.

With both methods there is a huge improvement over classical upwinding for the first-order schemes, but in higher dimensions the classical scheme is somewhat of a 'straw man', being in a fairly strict sense the *worst*[1] monotone algorithm. It is more impressive that a classical second-order upwind scheme can also be improved substantially. In a standard oblique-shock computation Dadone and Grossman report that the width of the captured shock was roughly halved, with only two or three internal points. Other things being equal, the same degree of flow resolution is then available on a mesh of twice the spacing, requiring, in three dimensions, only one eighth of the storage.

Work similar to the above can also be found in [14, 10].

A rather different approach has been described by Tamura and Fujii [26]. Their scheme is of the cell-vertex type, known to have generally superior performance on stretched and irregular meshes, but so far not usually associated with upwind techniques. Fig. 8 shows a typical cell, taken here for simplicity to be square. They compute, on this cell, the normalized gradients of density, pressure, and dynamic pressure, and choose the one whose value is greatest. This gradient then defines the privileged direction for that cell. Depending on which of four possible sectors contains this direction, they select

---

[1]Consider solving equation(8) on the square domain $(x, y) \in [0, 1]^2$ with $\vec{a}$ in the first quadrant, subject to boundary conditions on $x = 0, y = 0$. The steady solution of the classical upwind method can be found directly by rearranging the problem as one involving marching along the diagonals $x + y = const$. The marching scheme involved is the Lax-Friedrichs scheme, which has the maximum allowable artificial dissipation. For a detailed comparison of monotone advection schemes, see [22]

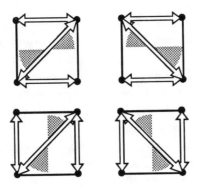

Figure 8: Illustrating Tamara and Fujii's cell-vertex method

three out the six possible pairs of points in the cell, and then solve one-dimensional Riemann problems for each of those pairs. In each Riemann problem, the waves travelling toward either input state are used to update that state. Again, higher-order accuracy is achieved by a MUSCL-type extension, and results are given from a third-order scheme. The cell-vertex formulation results in a slightly more compact scheme. A given vertex is influenced by at most fifteen other vertices, whereas the stencil of Fig. 7 shows one cell influenced by twenty others.

An isolated shear wave is usually a difficult test for the Euler equations, and Tamura and Fujii give some dramatic comparisons for a wave travelling at an angle of 45°. I will try to summarize these by constructing a table from their graphical plots. Since the exact solution is a pure discontinuity we again determine resolution by just counting the points inside the numerical transition.

Displacement	Classical first-order	New first-order	Classical third-order	New third-order
3	10	6	6	3
9	20	8	9	3
13	25	9	12	3

The entry in the first column is the number of mesh points over which the wave has travelled since its formation. Not only are the profiles from the new schemes much crisper after a given displacement, but their growth is much slower, and for the new third-order scheme there appears to be no growth. In addition, it is reported that the new schemes allowed significantly larger time steps.

This result is typical of several of the new approaches, in that it becomes possible to capture very sharply waves that are aligned with the *diagonals* of the mesh, although for classical schemes this is usually the worst case.

4.2   Interface Reconstructions

In existing schemes of the MUSCL or ENO type the usual philosophy is to attempt a reconstruction of the solution *inside* each computational cell, from information about the current average values in that cell and its neighbours. The idea then is to let this reconstructed solution evolve in time as accurately as possible. The evolution will then not be responsible for the numerical errors, almost all of which can be attributed to, and controlled by, the reconstruction process. This approach has led to many outstandingly successful codes. It is also attractive because it leads to a systematic, modular code design technique that can be applied to unstructured meshes, as demonstrated in the contributions by Harten and by Powell *et al* to these proceedings. See also [1].

Nevertheless, I believe that this idea is not well suited to a genuinely multi-dimensional approach. It leads inevitably to reconstructions that are discontinuous along cell boundaries, and accurate evolution then demands the solution of interface-oriented Riemann problems, with all their attendant loss of realism.

One possibility is to attempt the reconstruction, not inside the cells, but around their interfaces. The basic question is, what is the flow that crosses the interface doing? It is not possible, in more than one dimension, to answer this question unambiguously by referring simply to the two states on either side of the interface, but an attempt can be made. We can apply the philosophical principle (Occam's Razor) that simple explanations are more likely than complicated explanations.

Recall that in primitive variables (ρ, u, v, p) the eigenvectors describing two-dimensional wave motion in a direction θ are

$$
\mathbf{r}_a = \begin{vmatrix} 1 \\ a\cos\theta \\ a\sin\theta \\ \rho a^2 \end{vmatrix}, \mathbf{r}_v = \begin{vmatrix} 0 \\ -a\sin\theta \\ a\cos\theta \\ 0 \end{vmatrix}, \mathbf{r}_s = \begin{vmatrix} 1 \\ 0 \\ 0 \\ 0 \end{vmatrix}. \tag{9}
$$

Here r_a is an acoustic wave, r_v is a shear, or vorticity, wave, and r_s is an entropy wave. Since the entropy wave does not couple with any other wave, we will ignore it, and consider a three-dimensional phase space (u, v, p). In this space, changes due to acoustic waves are generators of the cone

$$(\Delta u)^2 + (\Delta v)^2 = \frac{(\Delta p)^2}{\rho^2 a^2}, \qquad (10)$$

whereas changes due to shear waves are generators of the plane

$$\Delta p = 0. \qquad (11)$$

The total change between the states u_L, u_R in two neighbouring cells can be represented in phase space as a path from u_L to u_R that everywhere satisfies either (10) or (11). Classical upwinding imposes the further (artificial) constraint that only those paths corresponding to waves travelling normal to the interface are allowable. This makes the path unique, except for the ordering of its segments. Such a path is shown in Fig. 9.

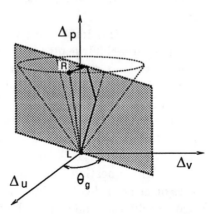

Figure 9: The $\Delta p, \Delta u, \Delta v$ plane. The surface of the cone defines all states that can be reached by acoustic waves from the state at the origin. The half of the cone shown represents compression waves. The shaded plane lies in the direction normal to a particular grid interface. A one-dimensional Riemann solver will only use acoustic waves whose eigenvectors lie in this plane, and shear waves whose eigenvectors are normal to it.

When all directions are allowable, uniqueness comes from some minimization principle. In accordance with Occams Razor, we take the path with the fewest number of segments, and if the choice is still ambiguous, mimimize some measure of the path length. It is easily shown that all segments should then lie in the plane that contains the Δp-axis and the vector $(\Delta u, \Delta v, 0)$, with a subsequent division of the choice into two cases, as shown in Fig. 10. I discussed such diagrams

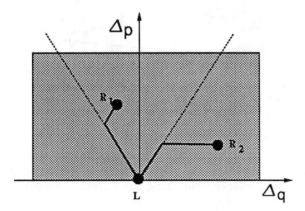

Figure 10: Minimum length paths connecting L with a state R_1 inside the acoustic cone, or a point R_2 outside the acoustic cone

with Dr.M.J.Baines at the University of Reading in the early eighties, but neither of us could see how to create a viable code out of the idea. Later I mentioned the concept to Dr. B.van Leer during the summer of 1988, and he added enough fresh thoughts to fuel a Ph.D. project, whose results are summarized in [23, 24]. Around the same time, Dr I.Parpia at the University of Texas at Arlington independently conceived the idea of a reconstruction using the fewest number of two-dimensional waves [15].

Such a reconstruction can be represented as

$$\mathbf{u}(\mathbf{x}) = \mathbf{u}_0 + \sum \alpha_k (\mathbf{n}_k \cdot \mathbf{x}) \mathbf{r}_k \qquad (12)$$

where $\mathbf{n}_k$ is the unit normal in the direction of the kth wave. Thus, the solution in the vicinity of the interface would be expected to evolve according to

$$\mathbf{u}(\mathbf{x}, t) = \mathbf{u}_0 + \sum \alpha_k (\mathbf{n}_k \cdot \mathbf{x} - \lambda_k t) \mathbf{r}_k \qquad (13)$$

and the problem is how to use this information (perhaps better described as a conjecture!) to compute the flux. Most work to date has been based on a simple modification of the usual upwind flux formula,

$$\Phi = \frac{1}{2}[\Phi_{\mathbf{L}} + \Phi_{\mathbf{R}}] - \frac{1}{2}\sum \Omega_k |\lambda|_k \mathbf{r}_k. \qquad (14)$$

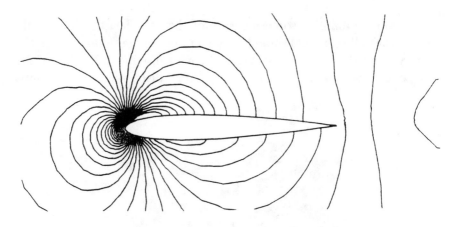

Figure 11: Computation of the viscous separated flow over a NACA 0012 airfoil at Reynolds number 5000, using fluxes from an interface reconstruction.

Here $\boldsymbol{\Phi}$ is the flux in the direction normal to the interface, *i.e.* $\mathbf{F}\cos\theta_g + \mathbf{G}\sin\theta_g$, $\{\mathbf{r}_k\}$ are the eigenvectors of $A\cos\theta_g + B\sin\theta_g$, $\{\lambda_k\}$ are the corresponding eigenvalues, and $\{\Omega_k\}$ are the coefficients of the eigenvector expansion of $\boldsymbol{\Phi}_{\mathbf{L}} - \boldsymbol{\Phi}_{\mathbf{R}}$. In the modified version, the $\{\lambda_k\}$ are the projections of the wavespeeds normal to the interface.

Experience with this approach has been mixed. For simple shear flow the results from first- and second-order codes were comparable to those already described, due to Tamura and Fujii, although not quite so clean, and third-order results were not obtained. The edges of the discontinuity were marked by oscillations. To remove these, an elaborate monotonicity analysis was carried out [24], which led to a prescription for limiting the angles allowed between the wave directions and the cell normals. The Euler code that grew out of this analysis was remarkable for showing very low entropy production around the leading edges of airfoils. The Navier-Stokes version, applied to the separated flow shown in Fig.4 produced much cleaner pressures (see Fig. 11) and a less grid-dependent drag. Unfortunately, the *lift* predicted by the new method was *more* strongly dependent on the mesh. Possibly this is because the limiting procedures developed for the Euler calculations were not employed in the Navier-Stokes calculations, since they were not needed to remove any noise. This may have resulted in an *under*dissipative code.

I feel that any further work on these methods should concentrate on the flux formulae and the limiting procedures. These issues cannot be resolved by appealing to equation(8), because that equation only involves scalar differences that cannot be decomposed. I suspect that the right questions have not yet been asked. One rather simple possibility is to recognize (see Fig. 12) that a wave carrying a jump

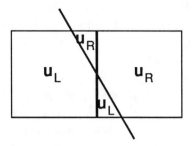

Figure 12: The jump in cell averages produced by an oblique wave.

from $\mathbf{u}_L$ to $\mathbf{u}_R$ and crossing an interface between square cells at an angle θ will contribute to the difference of cell averages by an amount

$$\left[\mathbf{u}_L \left(1 - \frac{\tan\theta}{8}\right) + \mathbf{u}_R \frac{\tan\theta}{8}\right] - \left[\mathbf{u}_R \left(1 - \frac{\tan\theta}{8}\right) + \mathbf{u}_L \frac{\tan\theta}{8}\right],$$

or, more simply,

$$(\mathbf{u}_L - \mathbf{u}_R) \left(1 - \frac{\tan\theta}{4}\right)$$

Thus the actual wavestrength is greater by a factor $(1 - 0.25\tan\theta)^{-1}$ than indicated by the difference of cell averages, and if such a factor is included in the expression (14) it adds dissipation and removes oscillations. Some unpublished calculations by Nelson Carter at Ann Arbor show substantial improvement without resorting to the complex limiting used in [24], but the idea needs much more testing, and the correction will not be so simple on a non-square mesh.

5 Fully Multidimensional Methods

Here there are just two ideas currently being developed. Each is based on a local representation of the flow as a sum of simple waves not aligned with any grid directions.

5.1 Local Diagonalization

This proposal is due to Hirsch [6], and is based on an attempt to diagonalize, in some modified sense, the multi-dimensional Euler equations. In general, there is no transformation of variables that reduces a multidimensional system of hyperbolic partial diferential equations to an uncoupled set of scalar equations, unless, as discussed in Section 3, the individual operators in each space dimension commute. What this means is that there is no decomposition into scalar subproblems that works for all data, and this difficulty is present even at the level of *linear* systems. In one dimension, the step from linear scalar to linear system is simple; we just put the equations into characteristic form. In two dimensions, the corresponding step comprises the essence of the difficulty.

The key is to realize that we do not need a trick that works for all data. We just need a trick that works for the particular data that we have. To illustrate this remark, consider the equation that holds along bicharacteristics in two dimensions. The linear combination of governing equations

$$\rho_0 a_0 \cos\theta(u_t + u_0 u_x + v_0 v_x + (1/\rho_0)p_x) +$$
$$\rho_0 a_0 \sin\theta(v_t + u_0 v_x + v_0 v_x + (1/\rho_0)p_y) +$$
$$p_t + u_0 p_x + v_0 p_y + \rho_0 a_0^2(u_x + v_y) \quad = \quad 0 \qquad (15)$$

can be rearranged as

$$\left(\frac{D}{Dt}\right)_b (p + \rho_0 a_0(u\cos\theta + v\sin\theta)) =$$

$$- \rho_0 a_0^2 \left[\sin\theta\frac{\partial}{\partial x} - \cos\theta\frac{\partial}{\partial y}\right](u\sin\theta - v\sin\theta) \qquad (16)$$

where the operator

$$\left(\frac{D}{Dt}\right)_b = \left[\frac{\partial}{\partial t} + (u_0 + a_0\cos\theta)\frac{\partial}{\partial x} + (v_0 + a_0\sin\theta)\frac{\partial}{\partial y}\right]$$

acts along the bicharacteristic. The terms on the right are often thought of as a kind of source term, defining the extent of departure from one-dimensionality. Hirsch seeks those special values of θ for which these terms vanish, or are minimized. Such values of θ will, of course, vary from place to place, and that is why I describe the process as a local diagonalization. They are the roots of

$$(u_x - v_y)\cos 2\theta + (v_x + u_y)\sin 2\theta = (u_x + v_y)$$

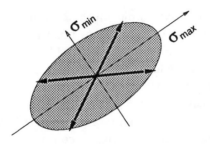

Figure 13: Relationship of the Hirsch bicharacteristic directions to the principal strain axes. These are the axes of an ellipse that shows the distortion, for small times, of an initially circular fluid parcel.

They are symmetrically disposed (se Fig. 13) about the principal strain axes of the velocity field, which are given by

$$(u_x - v_y)\cos 2\phi + (v_x + u_y)\sin 2\phi = 0.$$

It is easy to show that within a simple wave in steady supersonic flow the Hirsch prescription recovers as one root the correct (characteristic) orientation of the wave, which then coincides with one of the strain axes.

The recognition of such isolated waves seems to be about the best that we can hope for from a multidimensional scheme. In a compound wave, neither Hirsch's nor any of the other methods, will give reliable indications of all the components that might be present. This is because there are infinitely many possible wave directions, and thus infinitely many possible explanations for any finite set of observations. The explanation cannot therefore be perfect everywhere, but we can hope that almost everywhere it will be a great deal better than dimensional upwinding.

Implementations of these ideas have been reported in [9]. For shock reflection problems there is an increase of resolution comparable to that observed in the other methods discussed, but as far as I am aware there has not yet been a successful application to subsonic flow.

5.2 Wave Modelling

This is an idea that I personally have been trying to get to work for around ten years. Although it may ultimately prove to be a very robust and conceptually simple method, it has seemed that every one of several independent ingredients must be harmonized with the others before success can be achieved. Thus, it is only within the last year or two that I have felt confident enough to describe the ideas in anything but very tentative terms. Much of this progress is due to fruitful collaboration with people at the von Karman Institute, and many details of the recent progress, together with some representative results, are given in the companion paper by Herman Deconinck.

The idea of the "wave model" was introduced in [18], and is similar in some respects to the ideas discussed in Section 4.2. The data in some small region of space is again assumed to vary linearly,

$$\mathbf{u} = \mathbf{u}_0 + \vec{\nabla}\mathbf{u} \cdot (\mathbf{x} - \mathbf{x}_0), \tag{17}$$

but in the version that is coming to seem most natural, this variation is within a computational element that is triangular or tetrahedral, with solution values stored at its vertices. So, whereas (11) was a conjecture based on insufficient data, (17) is firmly established. And whereas the linear gradients that are used in MUSCL-type schemes give rise to reconstructions that are discontinuous along cell boundaries, these gradients are just those of the natural linear interpolants; no reconstruction is involved.

Again, the gradient of $\mathbf{u}$ is decomposed into parts, each of which represents a plane wave solution of the equations of interest. Thus

$$\vec{\nabla}\mathbf{u} = \sum_k \alpha_k \mathbf{r}_k \vec{n}_k$$

where $\mathbf{r}_k$ is an eigenvector for a wave propagating in the direction of the unit vector $\vec{n}_k$ with speed λ_k. Then the solution evolves in time according to

$$\partial_t \mathbf{u} = \sum_k \alpha_k \lambda_k \mathbf{r}_k$$

In this way, the evolution of the data is represented as the sum of independent scalar problems, each of which represents a wave moving in a direction that does not follow any feature of the grid. A way to visualize this process is provided in Fig. 14.

To create a viable numerical algorithm out of these ideas, three ingredients are needed.

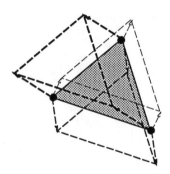

Figure 14: Each triangular element can be thought of as containing several independent waves that will travel independently of each other at different speeds and in different directions.

- A strategy for making the above decomposition, i.e. a "wave model",

- A strategy for updating the solution in response to each scalar component, i.e. an "advection scheme", and

- A strategy for ensuring conservation.

These are independent, in the sense that each can be studied separately, and we seem to have reached the stage where each component, even if not yet perfect, is sound enough to support the others.

Note that in MUSCL- or ENO-type schemes the responsibility for "monotonicity" rests with the reconstruction step. Here there is no such step, and responsibility rests with the update step. In one dimension these two versions can merely be restatements of the same algorithm [17], but in higher dimensions they may be radically different. This is because there is more than one way to regard multi-dimensional wave motion.

Consider, as in Fig.(15), an acoustic wave that passes through the origin at $t = 0$, and at $t = T$ is tangential somewhere to the Mach circle with radius aT and center (uT, vT). The wave speed can be plausibly defined as either $(OP)/T$, the "ray speed", or as $(OQ)/T$, the "frontal speed". Even for the linear advection equation

$$u_t + \vec{a} \cdot \vec{\nabla} u$$

this ambiguity exists; the wavespeed is either $\vec{a}$ or

$$\left(\frac{\vec{a} \cdot \vec{\nabla} u}{\vec{\nabla} u \cdot \vec{\nabla} u} \right) \vec{\nabla} u.$$

By using *both* definitions, monotone advection codes can be created

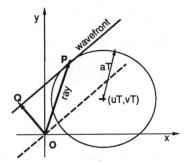

Figure 15: Geometry of an acoustic wave

that do not involve the use of limiter functions, and employ very compact stencils [20]. In fact, the flow may be processed triangle-by-triangle with no reference to other triangles at any stage, except that each node will accumulate update contributions from all the triangles to which it belongs.

These properties are responsible for two rather nice tricks that usefully speed up the later stages of convergence. If a triangle is found to have a sufficiently small residual, it requires no further processing, at least on this time step, so that in regions where the flow is almost uniform or nearly converged all the wave decompositions can be avoided. Also, the Courant restriction for stable explicit time stepping becomes based on the frontal speed rather than the ray speed, and this allows larger time steps as the waves slow down.

Two recent items of progress are worth reporting. The accuracy of the advection schemes is hard to analyse, but a recent numerical experiment by Jens Müller (University of Michigan, private communication) is striking. The test is convection in a circle around the origin, with a Gaussian concentration as input along the negative x-axis, which should reappear as ouput on the positive x-axis. Using a class of schemes first suggested in [19], and subsequently refined at the VKI where they were christened "low diffusion" schemes, Müller has demonstrated third-order accuracy, provided the grid is adapted to the flow in a very simple way. The grid is basically square, but triangulated by added "uprunning" diagonals for negative x, and "downrunning" diagonals for positive x. With all of the other triangulation strategies tried, the accuracy appears to be second-order. A sample of the results is shown in Fig. 16.

The message that I draw from this execise is that we have a fur-

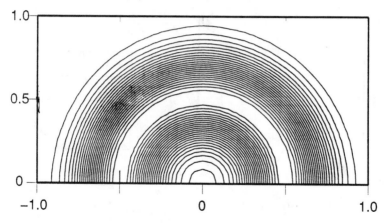

Figure 16: Solution contours for circular advection of a Gaussian distribution on a 40 × 20 mesh

ther stage to go through in adaptive mesh design. Heresy though it may be, I do not believe that the Delaunay meshes are truly optimal. None of their elegant properties depend on the equations we may wish to solve, or the data we wish to represent. For specific equations, or specific data, both the placing of the mesh points and their connections should reflect as much knowledge of that individuality as we can generate. and it may not turn out to be that hard to do this.

I also want to mention a new development in the wave model itself. The basic requirement for a wave model is that it should have enough degrees of freedom to match an arbitrary linear variation of the initial data, and that it should return the correct solution when presented with data corresponding to any isolated plane wave. To deal with the Euler equations in n space dimensions the model must account for n scalar gradients of $n + 2$ scalar quantities, and so must provide for $n(n + 2)$ degrees of freedom. This can be done by generalising the wave model in [18] to provide $2n$ mutually orthogonal acoustic waves that require $n(n - 1)/2$ parameters to specify their orientation, together with one entropy wave needing $n - 1$ parameters to orient it. This leaves $n(n-1)/2$ degrees of freedom to specify rotational effects, and this is in fact precisely the number of independent vorticity components that can be found in n dimensions.

Although this book-keeping works out very neatly, it has not been evident just how the rotational effects should be fitted in to the update strategy, which is built around the propagation of plane waves. Previously the neccessary shear waves had been thought of

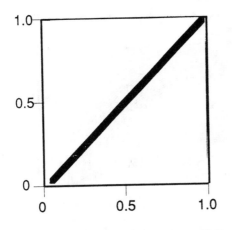

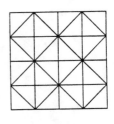

Figure 17: Mach number contours for a 45° shear wave. An anlarged portion of the mesh is shown at right.

as moving orthogonal to the pressure gradient, which is ill-defined if there is no pressure gradient, or else the vorticity was regarded as having an isotropic effect.

The new insight is that all of the required waves (except for the entropy wave, which does not couple to the others) can be anchored to the strain axes of the local velocity gradients (as in Fig. 13). An isolated acoustic wave gives rise to strain axes aligned with itself, and an isolated shear wave gives rise to strain axes at 45° to itself [[2], p.83]. Thus, given the orientation of the strain axes, and supposing the disturbance to be due to a single plane wave, that wave must be either an acoustic wave propagating along one of the $2n$ axis directions, or else a shear wave propagating in one of the $n(n-1)$ directions that bisect some pair of axes. This gives twice as many shear waves as needed, but it turns out that half of them can be eliminated by choosing the proper roots to certain quadratic equations. Incidentally, the number of waves associated with a given n-dimensional simplex is then $(n+1)(n+2)/2$, which (by coincidence?) is also the number of waves to be computed in a cell-centered method where the cells are the simplices.

An encouraging feature of the new wave model, which has been implemented by Lisa Beard at the University of Michigan, is that it seems to eliminate a convergence problem experienced with the older models. Convergence down to machine zero is now routine. It is also encouraging that the new scheme is able to resolve very well the isolated shear waves that have largely motivated the effort. Fig. 17 shows Mach number contours for a shear wave propagating through

through the mesh shown at right, and dividing a Mach 2.0 stream from a Mach 2.5 stream. The captured shear spans just two triangles and does not grow at all with distance from the origin. This is also the solution to which the code converges, given any initial data. If all squares of the mesh are triangulated parallel to the wave, then the captured shear spans only one triangle. For isolated discontinuities at least, the new wave model permits solutions to the Euler equations that are as well resolved as those to the scalar model.

The practical potential implied by this result for the sharp capturing of shear and contact discontinuities and boundary layers is enormous. The present generation of upwind schemes can, as mentioned in Section 3, acomplish something similar to Fig. 17 on square meshes if the discontinuity aligns with one grid direction, but results deteriorate substantially if the alignment is not perfect. They cannot produce such results on any triangular (unstructured) grid, since there is always at least one set of cell boundaries that "sees" the wrong physics. With the present method, perfect capturing of the discontinuity requires only that one set of boundaries be aligned with the wave, which is a fairly modest target for a solution-adaptive mesh to achieve. There is of course some deterioration as the alignment fails, but this occurs 'gracefully'.

Although all of this discussion has turned on the Euler equations, the extension to Navier-Stokes is of course crucial. Only very preliminary results are available at present, but a scheme which uses the same data structure as the wave modelling algorithm just described has been worked out. Very promising results for Laplace's equation and for a two-dimensional advection diffusion problem have been found (Tim Tomaich, private communication, University of Michigan). There seems to be a fairly secure knowledge base now available on which some very ambitious codes can be constructed.

6   Concluding Remarks

Robust, accurate flow algorithms will be a continuing requirement, and it is not clear that existing methods will be adequate for future needs. Several quite radical new ideas have been proposed recently, and all of them encourage the prospect of further progress. They differ in so many respects from existing methods that it is hard to say when they will become ready for practical application; the very foundations need to be rebuilt. These possibilities need to be investigated rather urgently, so that they will be sufficiently developed to play a rôle when the time comes to rewrite CFD codes for parallel architectures.

References

[1] T.J. Barth. On unstructured grids and solvers. VKI Lecture Series 1990-03, 1990.

[2] G.K. Batchelor. *An Introduction to Fluid Dynamics*. Cambridge, 1967.

[3] P. Colella and P. Woodward. The piecewise parabolic method (PPM) for gasdynamical simulations. *J. Comput. Phys.*, 54:174–201, 1984.

[4] A. Dadone and B. Grossman. A rotated upwind scheme for the Euler equations. *AIAA Journal*, to appear.

[5] S. Davis. A rotationally biased upwind difference scheme for the Euler equations. *J. Comp. Phys.*, 56:65–92, 1984).

[6] H. Deconinck, Ch. Hirsch, and J. Peutemann. Characteristic decomposition methods for multidimensional Euler equations. In *Lecture Notes in Physics, 264*. Springer, 1986.

[7] B. Einfeldt, C.D. Munz, P.L. Roe, and B. Sjögreen. On Godunov-type methods near low densities. *J.Comput. Phys.*, 92:273–295, 1991.

[8] A. Harten B.E. Engquist, S.J. Osher, and S.R. Chakravarthy. Uniformly high-order accurate essentially non-oscillatory schemes III. *J. Comput. Phys.*, 71:231–303, 1987.

[9] Ch. Hirsch and C. Lacor. Upwind algorithms based on a diagonalization of the multidimensional Euler equations. AIAA paper 89-1958, 1989.

[10] D.A. Kontinos and D.S. McRae. An explicit, rotated upwind algorithm for solution of the Euler/Navier-Stokes equations. AIAA paper 89-1531CP, 1991.

[11] B. van Leer. Towards the ultimate conservative differencing scheme, V. a second-order sequel to godunov's method. *J. Comput. Phys.*, 32:101–136, 1979.

[12] D.W. Levy. *Use of a rotated Riemann solver for the two-dimensional Euler equations*. PhD thesis, University of Michigan, 1990.

[13] H-C. Lin. Dissipation additions to flux-difference splitting. AIAA paper 91-1544, 1991.

[14] S. Obayashi and P.M. Goorjian. Improvements and applications of a streamwise upwind algorithm. AIAA paper 89-1957, 1989.

[15] I. Parpia. A planar oblique wave model for the Euler equations. AIAA paper 91-1545, 1991.

[16] K.M. Peery and S.T. Imlay. Blunt-body flow simulations. AIAA paper 88-2904, 1988.

[17] P.L. Roe. Some contributions to the modelling of discontinuous flow. In B.E. Engquist, S.J. Osher, and R.J.C. Somerville, editors, *Large-Scale Computations in Fluid Mechanics*. Am. Math. Soc, 1985.

[18] P.L. Roe. Discrete models for the numerical analysis of time-dependent multidimensional gas dynamics. *J. Comp. Phys.*, 63:458–476, 1986.

[19] P.L. Roe. Linear advection schemes on triangular meshes. Technical Report 8720, College of Aeronautics, Cranfield, 1987.

[20] P.L. Roe. 'Optimum' upwind advection on a triangular mesh. Technical Report Rept.90-75, ICASE, 1990.

[21] P.L. Roe. Discontinuous solutions to hyperbolic systems under operator splitting. *Num. Meth. PDEs.*, 7:277–297, 1991.

[22] P.L. Roe and D. Sidilkover. Optimum positive schemes for linear advection in two and three dimensions. *SINUM*, to appear.

[23] C. Rumsey, B. van Leer, and P.L. Roe. A grid-independent approximate riemann solver with applications to the Euler and Navier-Stokes equations. AIAA paper 91-0239, 1991.

[24] C.R. Rumsey, B. van Leer, and P.L. Roe. Effect of a multidimensional flux function on the monotonicity of Euler and Navier-Stokes computations. AIAA paper 91-1530, 1991.

[25] D. Sidilkover. *Numerical solution to steady-state problems with discontinuities.* PhD thesis, Weizmann Institute, Israel, 1990.

[26] Y. Tamara and K. Fujii. A multi-dimensional upwind scheme for the Euler equations on structured grid. In M.M. Hafez, editor, *4th ISCFD Conference, U.C. Davis*, 1991.

[27] E. Turkel. Improving the accuracy of central-differencing schemes. In D.L. Dwoyer, M.Y. Hussaini, and R.G. Voigt, editors, *Proc. 11th Int. Conf. Num. Meth. Comput. Fluid. Dyn.* Springer, 1989.

[28] V. Venkatakrishnan. Viscous computations using a direct solver. *Computers and Fluids*, 18:191–204, 1990.

BEYOND THE RIEMANN PROBLEM, PART II

H. Deconinck

von Karman Institute for Fluid Dynamics
Sint-Genesius-Rode, Belgium

1   Introduction

In a first contribution on upwind methods 'beyond the Riemann problem', Phil Roe discussed the basic motivation and principles of upwind methods which aim for more than one-dimensional physics.

A first class of such methods is based an an improved definition of the interface flux function in a standard finite volume context. The improvement consists in not simply considering the two states on both sides of the interface as initial values for a one dimensional Riemann problem in the directioni of the normal. Instead, some multidimensional physics is added allowing a more accurate prediction of the time evolution of the solution at the interface in a multidimensional space. The multidimensional physics may consist in simply using a flow dependent orientation for the definition of the one-dimensional Riemann problem [7], [4], [21] ; alternatively, the time evolution at the interface is reconstructed starting from a fully multidimensional wave decomposition [22], [17], [8]. Common to these methods is that the multidimensionsal physics is added on a per-interface basis.

A second class to be discussed in more detail in this paper is based on totally different concepts. Instead of concentrating on what happens at finite volume interfaces located in between the unknowns, one considers the average time evolution of a complete cell with the unknowns located on its vertices, and the vertex values of each cell are updated by the effect of linear waves solutions evolving from the piecewise linear initial data over the cell. This approach avoids to look at finite volume interfaces which almost inevitably privileges the direction of the normal.

Although the concepts used are radically different from the standard finite volume methods, the resulting schemes are identical when applied in one space dimension. This is due to the fact that in one space dimension cell vertices and cell interfaces are both points, while the upwinding direction is necessarily fixed along the x-axis. In two or three space dimensions, interfaces and vertices have a totally different nature, and the upwinding direction has one or two degrees of freedom. As a consequence, the schemes resulting from the two approaches are totally different, generalizing the one-dimensional schemes from two different conceptual viewpoints. To show the way the generalization works for the second approach, we will start with recalling Roe's one-dimensional Riemann solver following the second methodology. Then we will step by step introduce the generalizations which lead to the two- and three-dimensional schemes.

2 Roe's one-dimensional flux difference splitting scheme

In this section we briefly recall the fluctuation-splitting formulation of Roe's scheme as it was initially proposed in 1981, [10], [9]. Contrary to what has been done in later interpretations, this formulation does not necessarily appeal Finite Volumes nor to the Riemann problem. Given the solution at a certain time level n as a discrete mesh function $\{u_i, i = 1, ..., N\}$ on an irregular grid with meshpoints $\{x_i\}$, one assumes a *continuous* piecewise linear representation in between the meshpoints, precisely as when using linear Finite Elements in space. Once this viewpoint has been taken there is no room left for Riemann problems, since the initial data are piecewise linear everywhere.

An upwind space discretization is obtained as follows. Consider first a nonlinear scalar conservation equation

$$u_t + F_x = 0, \tag{1}$$

with advection speed $\lambda = F_u$. Assuming piecewise linear initial data at time level n allows us to define linear wave solutions for each cell $[x_i, x_{i+1}]$ with length $\Delta x_{i+\frac{1}{2}} = x_{i+1} - x_i$. Such wave solutions are of the form

$$u(x, t) = u_i^n + \frac{u_{i+1}^n - u_i^n}{x_{i+1} - x_i} (x - x_i - \overline{\lambda}(t - t^n)), \tag{2}$$

where $\overline{\lambda}$ is a suitable averaged speed over the interval.

If no information is used from neighbouring cells, solution (2) contains the maximum of information about the time evolution for this cell. The following issues rise :

- How to use this information to update the meshpoint values on both sides ? This is the scalar advection scheme.

- How to define the averaged speed $\overline{\lambda}$ such that discrete conservation and 'shock trapping' is obtained for a nonlinear conservation law.

- How to extend the results to systems

2.1 scalar advection scheme

To explain the general idea, we limit ourselves to a simple updating scheme equivalent to first order upwinding : Suppose that the averaged speed is positive. Then, the linear initial value distribution has shifted to the right over a distance $\Delta x - \overline{\lambda}\Delta t$. Hence, the solution u_i at the left is not affected, while the solution u_{i+1} at the right is changed according to (2). Conversely, for an averaged speed which is negative, the linear initial value distribution has shifted to the left and the solution u_{i+1} at the right is not affected. Taking into account the non-uniformity of the mesh, this leads to the following updating scheme for meshpoint i, by adding the contributions from the two sides :

$$u_i^{n+1} = u_i^n + \frac{\Delta t}{\Delta x_i}(\beta_{i+\frac{1}{2}}^i \Phi_{i+\frac{1}{2}} + \beta_{i-\frac{1}{2}}^i \Phi_{i-\frac{1}{2}}), \tag{3}$$

where $\Delta x_i = \frac{1}{2}(\Delta x_{i+\frac{1}{2}} + \Delta x_{i-\frac{1}{2}})$ is the median dual cell around x_i and $\Phi_{i+\frac{1}{2}}$ is the cell residual or cell-fluctuation for a cell, given by

$$\Phi_{i+\frac{1}{2}} = -\overline{\lambda}(u_{i+1}^n - u_i^n). \tag{4}$$

The downstream distribution coefficients β summing up to one for a given cell, are defined as follows :

$$\begin{aligned}
\beta_{i+\frac{1}{2}}^i &= 0, \quad \beta_{i+\frac{1}{2}}^{i+1} = 1 \text{ for } \overline{\lambda} \geq 0 \\
\beta_{i+\frac{1}{2}}^i &= 1, \quad \beta_{i+\frac{1}{2}}^{i+1} = 0 \text{ for } \overline{\lambda} < 0
\end{aligned} \tag{5}$$

Stability and monotonicity is easily enforced by requiring that the coefficients in the updating are positive. Scheme (3) is identical to first order upwinding. Other schemes based on a three-point stencil such as central, Lax-Wendrov, Fromm etc. can be obtained by simply redefining the distribution coefficients β [10]. For example, a central space discretization is obtained by taking all coefficients equal to one half.

2.2 Discrete conservation and discontinuity capturing

Discrete conservation is easely ensured by imposing a constraint on the linearization. Indeed, summing up equation (3) for all meshpoint one obtains canceling of all fluxes at the interior cell boundaries provided that

$$F_{i+1} - F_i = \overline{\lambda}(u_{i+1} - u_i), \tag{6}$$

This is of course the standard Roe-linearization. A procedure suitable for generalization in two and three space dimensions is the following : Make use of the assumption that u has linear variation in space and redefine the fluctuation as

$$\Phi_{i+\frac{1}{2}} = -\int_{x_i}^{x_{i+1}} F_x(u(x))dx. \tag{7}$$

Using the chain rule and taking u_x out of the integration leads to

$$\overline{\lambda} = \frac{1}{\Delta x_{i+\frac{1}{2}}} \int_{x_i}^{x_{i+1}} F_u(u(x))dx. \tag{8}$$

The fluctuation splitting concept allows an easy way to analyse the shock capturing properties of the scheme [9] : For example, consider a steady shock captured over the mesh, such that $F(u_0) = F(u_N)$ but $u_0 \neq u_N$. The $\overline{\lambda}$ have opposite signs on both sides and will send the information towards the middle of the domain. Capturing of the shock will occur in a layer of one or two cells : The case of one cell occurs if $\overline{\lambda} = 0$ for the cell separating the left and right states, such that at steady state all cells are in equilibrium individually. To show the existence of capturing with one intermediate point, suppose an intermediate state u_m, and assume speeds on both side that are opposite and non zero, pointing towards x_m. Equilibrium at x_m is then obtained for u_m solution of

$$\Phi_{m+\frac{1}{2}} + \Phi_{m-\frac{1}{2}} = -\overline{\lambda}_{m+\frac{1}{2}}(u_N - u_m) - \overline{\lambda}_{m-\frac{1}{2}}(u_m - u_0) = 0. \tag{9}$$

Because of the existence of shocks captured over one cell, the scheme is not entropy-satisfying : The scheme accepts as a steady state also the situation where the two speeds point away from the shock cell which itself is in equilibrium. As discussed in [9] a two-cell capturing like with Oshers Riemann solver is always entropy satisfying. This will prove an important consideration in the multi-dimensional generalization.

2.3 The system of one-dimensional Euler equations

Assuming piecewise linear variation of a set of variables $\mathbf{u}$, the gradient of general linear initial data can be decomposed into parts corresponding to linear wave solutions by projecting along the eigenvectors $\mathbf{r}^k$ of the Jacobian $\mathbf{F_u}$, giving

$$\frac{\mathbf{u}_{i+1}^n - \mathbf{u}_i^n}{x_{i+1} - x_i} = \sum_{k=1}^{3} \alpha^k \bar{\mathbf{r}}^k. \tag{10}$$

Solving for the coefficients α^k, the global solution for the linear initial data is given by

$$\mathbf{u}(x,t) = \mathbf{u}_i^n + \sum_{k=1}^{3} \alpha^k \bar{\mathbf{r}}^k (x - x_i - \bar{\lambda}^k(t - t^n)), \tag{11}$$

where $\bar{\mathbf{r}}^k$ and $\bar{\lambda}^k$ are again suitable averages over the cell. Hence, simply assuming linear variation of a set of variables, e.g. the conservative variables, one can compute the update to the left and right vertex due to a given cell by superposing the effect of each linear wave, allowing to generalize the scheme (3) to :

$$\mathbf{u}_i^{n+1} = \mathbf{u}_i^n + \frac{\Delta t}{\Delta x_i} \sum_{k=1}^{3} ([\beta_{i+\frac{1}{2}}^i]^k \Phi_{i+\frac{1}{2}}^k + [\beta_{i-\frac{1}{2}}^i]^k \Phi_{i-\frac{1}{2}}^k), \tag{12}$$

where the cell residual has been splitted in parts

$$\Phi_{i+\frac{1}{2}} = \sum_{k=1}^{3} \Phi_{i+\frac{1}{2}}^k = -\Delta x_{i+\frac{1}{2}} \sum_{k=1}^{3} \bar{\lambda}^k \alpha^k \bar{\mathbf{r}}^k. \tag{13}$$

The requirement of discrete conservation imposes a constraint on the linearization, which just like before leads to Roe's linearization satisfying property U [11] :

$$\frac{\mathbf{F}_{i+1} - \mathbf{F}_i}{x_{i+1} - x_i} = \sum_{k=1}^{3} \bar{\lambda}^k \alpha^k \bar{\mathbf{r}}^k. \tag{14}$$

An elegant way to construct Roe's linearization which is ready for generalization in 2 and 3D makes use of definitions (7) and (8) [5], assuming that the parameter vector $\mathbf{z} = \sqrt{\rho}(1, u, H)^\tau$ is the variable having linear change in space :

$$\Phi_{i+\frac{1}{2}} = -\int_{x_i}^{x_{i+1}} \mathbf{F}_x(\mathbf{z}(x)) dx = -\int_{x_i}^{x_{i+1}} \mathbf{F}_\mathbf{z}(\mathbf{z}(x)) dx \; \mathbf{z}_x. \tag{15}$$

This can be rewritten in terms of the conservative variable $\mathbf{u}$ using the relation $\mathbf{u}_x(\mathbf{z}) = \mathbf{u}_\mathbf{z}(\mathbf{z})\mathbf{z}_x$. Since the two matrices involved are linear in the components of $\mathbf{z}$, which itself is linear in space, the integrations are just arythmetic averages, and one obtains :

$$\Phi_{i+\frac{1}{2}} = -\int_{x_i}^{x_{i+1}} \mathbf{F}_x(\mathbf{z}(x)) dx = -\mathbf{F}_\mathbf{u}(\bar{\mathbf{z}}) \left(\mathbf{u}_{i+1} - \mathbf{u}_i\right), \tag{16}$$

where $\bar{\mathbf{z}} = \frac{(\mathbf{z}_{i+1} + \mathbf{z}_i)}{2}$. Hence the averaged speeds, eigenvectors and strengths α used in the above expressions are the analytical expressions for the Jacobian matrix $\mathbf{F}_\mathbf{u}(\mathbf{z})$, evaluated at the particular Roe-averaged state $\mathbf{z} = \bar{\mathbf{z}}$.

The discontinuity capturing properties of the scalar scheme carry over to the system because the linearization reduces to the Rankine-Hugoniot relations in the case that the states $\mathbf{u}_{i+1}$ and $\mathbf{u}_i$ can be connected by a single shock or contact. This means that effectively only one of the scalar waves contributes in this case, and its speed is the speed of the discontinuity.

3 Scalar conservation laws in two and three space dimensions

The generalization of the concepts introduced before to two and three space dimensions is straight forward. Consider the scalar conservation equation

$$u_t + \vec{\nabla}\vec{F}(u) = 0, \qquad \vec{F} = F\vec{1}_x + G\vec{1}_y + H\vec{1}_z, \tag{17}$$

with advection speed vector $\vec{\lambda} = F_u \vec{1}_x + G_u \vec{1}_y + H_u \vec{1}_z$. For the 2D case will assume that $H = 0$ and $H_u = 0$. The assumption of continuous, piecewise linear space variation requires triangular cells in 2D and tetrahedral cells in 3D, with the solution stored at the vertices. Considering a triangle or tetrahedron, the initial value problem is defined by the constant gradient $\vec{\nabla} u^n$ at time level n, and the solution evolves in time according to

$$u(\vec{x}, t) = u_0 + \vec{\nabla} u^n (\vec{x} - \vec{\lambda} t),$$ (18)

where $\vec{\lambda}$ is a suitable averaged advection speed vector over the cell, determined such that discrete conservation will be assured. As before, we will use the maximum of information contained in (18) to update the nodes of the cell. This is achieved by splitting the cell residual in parts which will be distributed towards the nodes of that particular cell. Hence, the following scheme is the natural generalization of equation (3) towards two and three space dimensions :

$$u_i^{n+1} = u_i^n + \frac{\Delta t}{V_i} \sum_T \beta_T^i \Phi_T,$$ (19)

where the summation is carried out over all the cells having i as a common vertex. The volume (surface in 2D) V_i generalizes the median dual edge $\Delta x_i = \frac{1}{2}(\Delta x_{i+\frac{1}{2}} + \Delta x_{i-\frac{1}{2}})$ to one fourth of the surrounding tetrahedra or one third of the surrounding triangles. The coefficients β for a given cell T satisfy a generalization of (5) :

$$\sum_{k=1}^{d+1} \beta_T^{i(k)} = 1,$$ (20)

where $\{i(k), k = 1, ..., d+1\}$ are the nodes of element T, with d the number of dimensions, $d = 2$ in 2D and $d = 3$ in 3D. The fluctuation in conservative form Φ_T is a straight forward generalization of (7), obtained by computing the flux balance over a cell (triangle or tetrahedron), supposing linear variation of u :

$$\Phi_T = -\oint_{S_T} \vec{F} \vec{n} dS = -\int_{V_T} \vec{\nabla} \vec{F}(u(\vec{x}) \, dV.$$ (21)

Applying the chain rule $F_x = F_u \, u_x$, $G_y = G_u \, u_y$, $H_z = H_u \, u_z$, one obtains the definition of the averaged advection speed which assures discrete conservation :

$$\vec{\lambda} = \overline{\lambda}_x \, \vec{1}_x + \overline{\lambda}_y \, \vec{1}_y + \overline{\lambda}_z \, \vec{1}_z,$$ (22)

where

$$\begin{aligned}
\overline{\lambda}_x &= \tfrac{1}{V_T} \int_{V_T} F_u(u(\vec{x}))dV, \\
\overline{\lambda}_y &= \tfrac{1}{V_T} \int_{V_T} G_u(u(\vec{x}))dV, \\
\overline{\lambda}_z &= \tfrac{1}{V_T} \int_{V_T} H_u(u(\vec{x}))dV.
\end{aligned} \tag{23}$$

Using this definition, the conservative fluctuation eq. (21), can be rewritten in the linearized form

$$\Phi_T = -V_T \, \vec{\overline{\lambda}} \, \vec{\nabla} u = -\sum_{i=1}^{d+1} k_i u_i, \tag{24}$$

where the summation is over the nodes of the element T. The coefficients k_i are computed (e.g. using linear finite elements) as :

$$k_i = \frac{1}{d} \, \vec{\overline{\lambda}} \, \vec{n}_i \quad \text{satisfying} \quad \sum_{i=1}^{d+1} k_i = 0, \tag{25}$$

where $\vec{n}_i$ is the inward *scaled* normal to the face opposed to vertex i.

The only remaining ingredient needed to define the fluctuation splitting schemes completely is the precise definition of the coefficients β satisfying the constraint (20). Here, we cannot longer rely on any 1D theory, and a new theory of advection schemes in 2D and 3D has been necessary. This theory is strikingly elegant and is now fairly complete. It is not possible to go in all the details in the context of this paper, and the reader is referred to [13] and [18] for the full details.

We define a downstream vertex as a vertex opposite to an inflow face corresponding to $k_i \geq 0$, and an upstream vertex as a vertex for which $k_i \leq 0$. One then observes that there are two different types of triangles and three different types of tetrahedra : triangles with one or two downstream vertices, and tetrahedra with one, two or three downstream vertices. Recalling (18), we will denote a scheme as an *upwind* scheme if in accordance with this solution no contribution is sent to the upstream vertices. Hence, we will consider one and two target distributions for triangles and one, two and three target distributions for tetrahedra.

Investigating the possible choices for linear schemes (schemes in which the update is a linear function of the u_i if $\vec{\lambda}$ is a constant vector), two different classes appear :

i. A first class in which the coefficients β are independent of the u_i. For this class, a triangle or tetrahedron will send no updates to its vertices if it is in equilibrium, corresponding to k-exactness for linear polynomial solutions ($k = 1$). On regular grids, this corresponds to a steady state which is second order in space.

ii. A second class in which the coefficients β do depend on the unknowns in the following way :

$$\beta_T^i = \frac{\gamma_T^i}{\Phi_T} \quad \text{with} \quad \sum_i \gamma_T^i = \Phi_T, \tag{26}$$

where γ_T^i are linear functions of the u_j.

The second class allows a cell to be in equilibrium while sending non zero contributions to the downstream vertices whose sum vanishes. Such a scheme will in general only preserve a *constant* steady state and destroy a linear steady state satisfying $\vec{\lambda}\vec{\nabla}u = 0$ for each triangle. These schemes will be called first order in the present context.

It turns out that *only* in this class one can construct linear positive schemes, thus providing a generalized Godunov type theorem in two and three space dimensions : positivity and k-exactness for linear polynomials are incompatible for linear schemes [18].

Hence, to obtain schemes which are both accurate and positive, it will be necessary to consider nonlinear advection schemes similar as what was done in 1D when TVD properties were introduced. This will be introduced in section 3.2.

First, we present in the next section two linear schemes, one for each of the two classes.

3.1   Linear distribution schemes

The schemes have been developed first for triangles first [13], [18] . Recently, they have been generalized to 3D with the help of Guy Bourgois at the von Karman Institute[1], who discovered the general-

---

[1]research supported by the HERMES RFQ program monitored by Dassault Aviation.

ized formula given below. Indeed, it turns out that all the schemes on triangles have natural extensions on tetrahedra.

A first scheme which is k-exact for linear polynomials has been called the 'Low Diffusion scheme A' in [18]. Here we give a generalized expression of the distribution coefficients, valid both on 2D triangles and 3D tetrahedra :

$$\beta_T^i = \frac{max(0, k_i)}{\sum_{j=1}^{d+1} max(0, k_j)} \qquad i = 1, \ldots, d+1 \qquad (27)$$

First, the scheme will preserve an exact steady state which satisfies $\vec{\lambda} \vec{\nabla} u^n = 0$ because the coefficients are independent of u_j. Further, the scheme is upwind because it sends no contribution to the upstream vertices for which $k_i < 0$. If there is only one downstream vertex, the entire residual is sent to that vertex. If there is more than one downstream vertex, there is a nice geometric interpretation : for example in 2D, the velocity put in the upstream vertex divides the triangle in two subtriangles. The coefficients β sending to the two downstream vertices are then proportional to the surface of each subtriangle. A similar interpretation exists for tetrahedra in 3D. In this case, there are configurations with two and three downstream vertices, and the tetrahedron is split in two respectively three subtetrahedra along the velocity vector. It is easy to check that scheme (27) is not positive, and it produces oscillations around discontinuities.

A second linear scheme belongs to the class ii and satisfies (26). It is studied extensively in [13] and [18] for the 2D case. The generalization valid in 2D and in 3D is obtained if the coefficient γ needed in (26) is defined as follows :

$$\gamma_T^i = \frac{max(0, k_i)}{\sum_{j=1}^{d+1} max(0, k_j)} \sum_{j=1}^{d+1} [min(0, k_j)(u_i^n - u_j^n)] \quad i = 1, \ldots, d+1 \quad (28)$$

It is easy to check that summing up these coeficients reproduces the residual $(\sum_i \gamma_T^i = \sum_i k_i u_i)$. Also as before, the scheme sends only contributions to the downstream vertices $(k_i \geq 0)$. However, it does not preserve a linear steady state because the differences $(u_i^n - u_j^n)$ can be non zero even when the sum of the γ's vanishes for a triangle in equilibrium. On the other hand, because we have complete control over the sign of each coefficient, this scheme is positive for timesteps small enough : substituting in (19), one obtains for the update at a given meshpoint

$$u_i^{n+1} = u_i^n + \frac{\Delta t}{V_i} \sum_T \frac{max(0, k_i^T)}{\sum_{j=1}^{d+1} max(0, k_j^T)} \sum_{j=1}^{d+1} [min(0, k_j^T)(u_i^n - u_j^{n,T})].$$

(29)

The terms contributing from all vertices except u_i have a positive coefficient by construction. Hence, the global scheme is positive if the timestep is restricted such that

$$1 + \frac{\Delta t}{V_i} \sum_T \frac{max(0, k_i^T)}{\sum_{j=1}^{d+1} max(0, k_j^T)} \sum_{j=1}^{d+1} min(0, k_j^T) \geq 0.$$

(30)

This expression can be simplified to a CFL like condition for the local timestep at meshpoint i

$$\Delta t_i \leq \frac{V_i}{\sum_T max(0, k_i^T)}.$$

(31)

As for the previous scheme, the N scheme has a simple geometric interpretation both for triangles and for tetrahedra. This interpretation decomposes the advection speed vector in components parallel with the edges pointing from upstream to downstream vertices. With each component corresponds a part of the residual which is send to the downstream vertex. In ref. [18] it shown that the scheme in 2D is optimum, in the sense that among the linear positive schemes of the form (19) it is the one which allows the largest timestep and has the smallest cross diffusion. For further details we refer to [18] for the 2D and to future publications for the 3D case.

Other distributions have been developed as well, e.g. a Lax-Wendroff derivation corresponding to the finite element SUPG scheme, the distribution corresponding to first order upwinding, the Galerkin Finite Element distribution (corresponding to $\beta_T^i = \frac{1}{d+1}$). A comparison of numerical results obtained with these schemes will be given in section 3.3.

3.2   Nonlinear distribution schemes

None of the above linear schemes can be positive and k-exact at the same time, and as in 1D, nonlinear schemes have to be used for high resolution and non-oscillatory discontinuity capturing. The mechanism by which the nonlinearity is introduced in 1D and in the dimension by dimension extensions relies on nonlinear averaging functions, known as limiters. This usually leads to a widening of the stencil, which is inevitable in 1D.

In multidimensional schemes one can hope to introduce the non-linearity while still maintaining a compact stencil. This was first recognized for structured grids by Sidilkover in his PhD thesis [20] and was achieved by adding to the stencil the nearestby meshpoints along the diagonals of the grid. The limiter functions compare gradients of the solution over this compact stencil.

A remarkable discovery when studying the multidimensional fluc-tuation splitting schemes was the existence of an alternative for lim-iting : By looking at the multidimensional scalar advection equation and the wave solution (18), a new mechanism for introducing nonlin-earity in the scheme became apparent, which is completely different from the limiting approach and which does not exist in 1D because solution gradients and advection speeds are always in the same di-rection. Purely at the level of the linear wave solution (18) we can define a "frontal" advection speed vector in the direction of $\vec{\nabla} u^n$, as is well known. This speed is given by

$$\vec{\lambda}_m = \frac{\vec{\lambda}\, \vec{\nabla} u^n}{|\vec{\nabla} u^n|}\, \vec{m} = \frac{V_i\, \Phi_T}{|\vec{\nabla} u^n|}\, \vec{m} = \lambda_m\, \vec{m}, \tag{32}$$

where

$$\vec{m} = \frac{\vec{\nabla} u^n}{|\vec{\nabla} u^n|} \qquad \text{and} \qquad \lambda_m = \frac{\vec{\lambda}\, \vec{\nabla} u^n}{|\vec{\nabla} u^n|}. \tag{33}$$

It is fully legitimate to use the "frontal" speed $\vec{\lambda}_m$ instead of the advection speed $\vec{\lambda}$ in any of the distribution schemes since doing this does not affect the residual ($\vec{\lambda}\, \vec{\nabla} u^n = \vec{\lambda}_m\, \vec{\nabla} u^n$) and still corresponds to the linear wave solution (18) rewritten as

$$u(\vec{x}, t) = u_0 + |\vec{\nabla} u^n|(\vec{x}\, \vec{m} - \lambda_m\, t) \tag{34}$$

However, using the frontal speed for the definition of the k_i makes the scheme nonlinear since these coefficients will now depend on $\vec{\nabla} u^n$. Moreover, any of the schemes based on the frontal speed becomes linearity preserving since the frontal speed λ_m and hence all the k_i vanish for a triangle which is in equilibrium. Hence, the positive N-scheme discussed in the previous section becomes k-exact for linear polynomials when using the frontal speeds.

When implementing this idea it was found that the nonlinear scheme may change the targets in the distribution. In fact, contributions can be sent even to the upstream vertices in the case that there is only one downstream vertex according to the standard speed $\vec{\lambda}$. This hampers the convergence of the solution to a steady state and seems in conflict with the exact wave solution. In such a case however there is not even the need to use a nonlinear scheme since sending the entire fluctuation to one vertex also preserves linear polynomials. In conclusion, introducing the nonlinear scheme is only necessary if there is more than one downstream vertex for the *linear* scheme.

Hence, consider in more detail the 2D case with more than one downstream vertex, e.g. a triangle with two downstream vertices numbered 1 and 2. The updating (29) from this triangle using the N-scheme becomes

$$
\begin{aligned}
u_1^{n+1} &:= u_1^n - \tfrac{\Delta t}{S_1} k_1(u_1^n - u_3^n) \\
u_2^{n+1} &:= u_2^n - \tfrac{\Delta t}{S_2} k_2(u_2^n - u_3^n)
\end{aligned}
\tag{35}
$$

This update is positive for Δt sufficiently small, but it is not linearity preserving as mentioned before : for this to be true, both right hand sides should vanish for an exact linear steady state solution, while in reality only the sum of the two does vanish on a general triangulation.

If the use of the frontal speed leads to a two target scheme with the same downstream vertices 1 and 2 as in fig. 1 (left), the problem is cured by using this speed. However, if the frontal speed changes the targets in the distribution, fig. 1 (right), (35) would not longer be a positive update.

In such a case one can use a single target distribution to the common downstream node. This corresponds to selecting an advection speed vector $\vec{\lambda}^*$ along the edge pointing from the upstream vertex 3 to the common downstream vertex. Such a choice is still allowed as long as the residual is not affected, i.e. $\vec{\lambda} \, \vec{\nabla} u^n = \vec{\lambda}^* \, \vec{\nabla} u^n$. The locus of speeds satisfying this condition and the selected choice are plotted in fig. 1.

The resulting scheme is one of a number of nonlinear schemes which have very similar properties. They have been called "NN schemes", standing for Nonlinear N-scheme. Summarizing, the scheme in 2D is given by the following algorithm :

i. If the triangle has one inflow side according to $\vec{\lambda}$, send the residual to the unique downstream vertex.

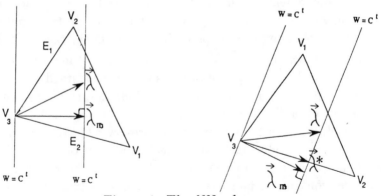

Figure 1: The NN scheme.

ii. If the triangle is two target according to $\vec{\lambda}$, say with nodes 1 and 2, compute $\vec{\lambda}_m$

1. if both $\vec{\lambda}_m \, \vec{n}_1 > 0$ and $\vec{\lambda}_m \, \vec{n}_2 > 0$ use the two target formula (35), with k_1 and k_2 based on the frontal speed.

2. else if $\vec{\lambda}_m \, \vec{n}_1 < 0$ and $\vec{\lambda}_m \, \vec{n}_2 > 0$ send the residual to node 2.

3. else send the residual to node 1

The extension of this nonlinear scheme towards 3D tetrahedra follows along the same lines.

Recently, Henri Paillère at the von Karman Institute has started a study of the capturing of steady discontinuities by the NN-scheme in 2D, revealing some interesting results. The conservation law considered is defined by :

$$\vec{F} = \frac{u^2}{2}\vec{1}_x + \vec{1}_y, \qquad (36)$$

leading to a linearized speed according to (23) :

$$\vec{\lambda} = \overline{u}\,\vec{1}_x + \vec{1}_y \qquad \text{with} \qquad \overline{u} = \frac{u_1 + u_2 + u_3}{3}. \qquad (37)$$

The computational domain is a square ($-0.0 \le x \le 1.0$, $-0.0 \le y \le 1.0$) with boundary conditions $u = 1$ at the left, $u = -1$ at the right and linear variation of u between 1 and -1 on the lower side. The exact steady state has a compression fan developing in a steady shock parallel with the y-axis and located at $x = 0.5$ with $u_{left} = 1$, $u_{right} = -1$. The grid is made of right isoceles triangles, using the configuration of fig. 3. In the vicinity of the shock, the NN scheme will reduce to a one target scheme for each triangle, sending the fluctuation of each triangle to the downstream node in the x-direction. Indeed, if the gradient is in the x-direction, the "frontal" speed will be given by :

$$\vec{\lambda}_m = \overline{u}\,\vec{1}_x \tag{38}$$

Unlike the linear schemes, the scheme is not sensitive to the presence of the y-component of the advection speed, thus propagating the information towards the shock in the direction *normal* to the shock.

It is clear that even with the NN-scheme, the discontinuity cannot be captured over one cell. Indeed, the fluctuation calculated over a cell whose nodes are at states related by the jump relation is not zero, since the linearized speed $\overline{u}$ is nonzero for a triangle with vertex values $(1, 1, -1)$ or $(-1, -1, 1)$. Therefore a fluctuation will be sent to the downstream node, destroying the exact solution (figure 2). This situation is totally different from the 1D Roe-scheme discussed before, where capturing over one cell is indeed possible.

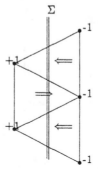

Figure 2: Distribution of the fluctuation using the NN scheme.

An analytical study reveals that a one parameter family of steady shock profiles is obtained, with two or at least one intermediate state. The conditions for equilibrium to exist are that no fluctuations are sent to states +1 and -1, and that the fluctuations sent to the intermediate states cancel each other.

The particular shock profile obtained in a calculation will depend on the position of the shock in the exact solution relative to the grid. For example, if the number of cells in the x-direction is odd, the shock is located on the grid and capturing in two cells with the intermediate state $u = 0$ is obtained. On the other hand, if the number of cells is even, capturing in three cells is obtained with two intermediate states $u_{ml} = \sqrt{\frac{3}{5}}$ and $u_{mr} = -\sqrt{\frac{3}{5}}$. Fig. 3 shows a number of shock profiles obtained by sliding the grid over the domain in the x-direction. The continuous evolution from a two-cell capturing to a three-cell and back to a two-cell capturing is illustrated for different values of ϵ, the subgrid position of the shock in the exact solution.

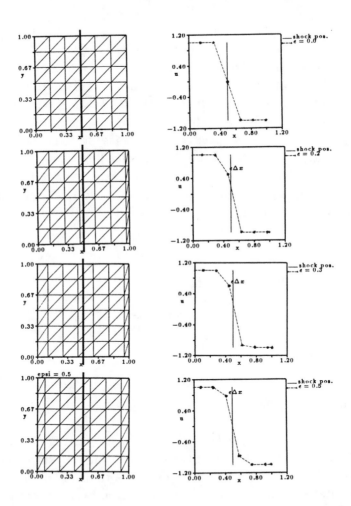

Figure 3: Shock profiles for different subgrid positions of the shock in the exact solution.

As a consequence of the fact that capturing over a single row of triangle is impossible, expansion shocks are eliminated as steady solutions and dissipated by the scheme. This was confirmed by numerical experiments and suggests that the multidimensional fluctuation splitting scheme using the linearization (23) satisfies an entropy condition.

In the early developments of this work, a conservative linearization different from (23) was proposed which is in fact a dimension by dimension extension of the 1D Roe linearization, given by

$$\overline{\lambda}_x = \frac{\int_{V_T} F_x dV}{\int_{V_T} u_x dV}, \quad \overline{\lambda}_y = \frac{\int_{V_T} G_y dV}{\int_{V_T} u_y dV}, \quad \overline{\lambda}_z = \frac{\int_{V_T} H_z dV}{\int_{V_T} u_z dV}, \quad (39)$$

where all gradients F_x, G_y, H_z, u_x, u_y, u_z are computed assuming linear variation of F, G, H, u over the cell. Apart from being inconsistent if the fluxes are nonlinear functions of u, this definition is difficult to extend to systems. Not surprisingly, it leads to shock capturing over one or maximum two cells and allows expansion shocks precisely as in 1D. Indeed, when applied to the above testcase, one obtains that $\overline{\lambda}_x$ using the above definition (39) vanishes for a triangle with vertex values $(1, 1, -1)$ or $(-1, -1, 1)$, and such triangles are in equilibrium, also for expansion shocks.

3.3 Comparison of scalar advection schemes

In this section we compare some numerical results for the Low Diffusion scheme A (LDA), the N-scheme and the NN-scheme introduced in the preceeding sections.

A first 2D testcase is an oblique shock version of the testcase introduced before. The conservation law is (36) over the unit square, discretized by a Delaunay-triangulated random point cloud. The boundary conditions are $u = 1.5$ at the left, $u = -0.5$ at the right, and linear interpolation between 1.5 and -0.5 on the lower boundary.

The mesh and the isolines of the steady state solution are given in figures 4 and 5

The improved accuracy of the solution for the linearity preserving schemes (LDA and NN) is apparent in the continuous part of the solution, while the positive schemes (N and NN) show a monotonic shock capturing. The NN-scheme combines both properties.

A second testcase is a comparison on 3D tetrahedra, made recently by Guy Bourgois. The equation considered is linear advection with advection speed $\vec{\lambda} = 0.75 \, \vec{1}_x + 0.875 \, \vec{1}_y + \vec{1}_z$. The basis of the

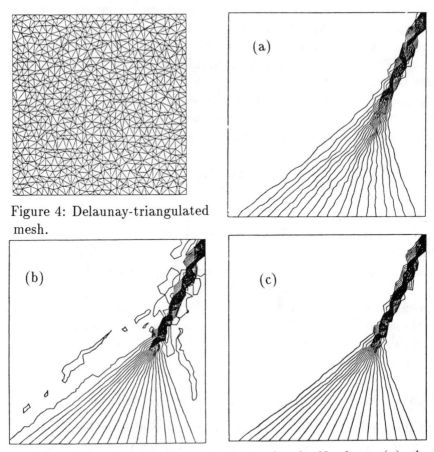

Figure 4: Delaunay-triangulated mesh.

Figure 5: Isolines of steady state solution for the N-scheme (a), the Low Diffusion A scheme (b) and the nonlinear NN-scheme (c).

testcase is a structured mesh on a cubic domain ($0 \leq x \leq 2$, $0 \leq y \leq 2$, $0 \leq z \leq 1$) with a mesh spacing of 0.1 (fig. 6).

Each cubic cell of the mesh is divided in 6 tetrahedra along the principal diagonal of the cube (the same decomposition is chosen for each cell). As boundary conditions, a step is imposed on the lower inlet surface of the domain along the line $x = 0.25 + 0.25y$, see fig. 6.

Figure 7 represents the steady state value of u at the upper surface of the cube for the different schemes discussed before, as well as some other schemes, namely : the first order upwind scheme on the structured mesh, a Lax-Wendroff fluctuation splitting scheme proposed in [13] and the optimal linear positive scheme on the structured mesh (the Roe-Sidilkover scheme, [14]).

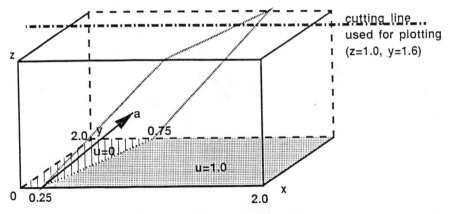

Figure 6: 3D shear on a cube.

The cut is performed at $y = 1.6$ in the outlet plane $z = 1$. We first remark that low-diffusion and Lax-Wendroff schemes lead to strong oscillations, explained by the non positivity of these schemes. The first order positive N-Scheme is much less dissipative than the standard dimensionally split first order upwind scheme and performs nearly as well as the optimal positive linear scheme [14].

The nonlinear NN-scheme combines the crisp capturing of the shear with a monotonic profile and is superior to any of the linear schemes.

3.4 Time integration and multigrid acceleration

In the previous section, simple Euler time integration with optimal local timestep has been used until a steady state is obtained. Excellent work has been done by former VKI student L.A Catalano (Politecnico di Bari) over the last two years to combine the fluctuation splitting discretization with optimally smoothing time marching schemes [1]. The positive compact stencil fluctuation splitting schemes combined with optimally smoothing multistage time integration turn out to be very well suited for multigrid acceleration [2].

Very recently, the approach has been extended towards the Euler equations, using the approach discussed in the next section [3].

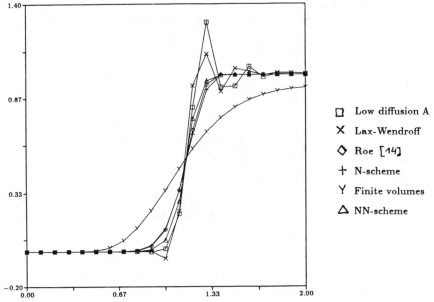

Figure 7: Comparison of scalar advection schemes for 3D shear over a cube. Profile of steady state solution at $y = 1.6$, $z = 1$.

4 The system of Euler equations in two space dimensions

Although the extensions discussed below are valid for 3D tetrahedra as well, we limit ourselves to a discussion of the 2D case. Following the steps of the development in section 2.3, we assume linear space variation of a particular set of variables, e.g. the conservative variable $\mathbf{u}$, and decompose the constant gradient at time level n in contributions corresponding to initial data for linear wave solutions :

$$\vec{\nabla}\mathbf{u}^n = \sum_k \alpha^k \, \bar{\mathbf{r}}^k \, \vec{m}^k. \tag{40}$$

Here, $\bar{\mathbf{r}}^k$ is an eigenvector of the Jacobian $(\mathbf{F_u} \, \vec{1}_x + \mathbf{G_u} \, \vec{1}_y) \, \vec{m}^k$ with eigenvalue $\bar{\lambda}_m^k$, corresponding to a wave propagating in the direction of the unit vector $\vec{m}^k$ with speed $\bar{\lambda}_m^k$. Suitable averages over the cell have been used such that the linearization is conservative, see section 4.1. Solving for the coefficients α^k and the direction angles θ^k of $\vec{m}^k$, the global solution for the linear initial data evolves in time according to

$$\mathbf{u}(\vec{x}, t) = \mathbf{u}_0^n + \sum_k \alpha^k \, \bar{\mathbf{r}}^k (\vec{x} \, \vec{m}^k - \bar{\lambda}_m^k t), \tag{41}$$

corresponding to the equation

$$\partial_t \mathbf{u} = -(\mathbf{F}_x + \mathbf{G}_y) = -\sum_k \alpha^k \, \overline{\lambda}_m^k \, \overline{\mathbf{r}}^k \qquad (42)$$

Note that the speeds $\overline{\lambda}_m^k \vec{m}^k$ which are obtained for the system are the *frontal speeds*, given for the scalar case by (32), and that solution (41) corresponds to the form (34). For the 2D Euler equations, the frontal speed vectors corresponding to entropy (e), shear (s) and acoustic (a) linear waves are given by

$$\vec{\lambda}_m^{e,s} = (\vec{u} \, \vec{m}^{e,s}) \, \vec{m}^{e,s}, \qquad \vec{\lambda}_m^{a} = (\vec{u} \, \vec{m}^{a} \pm c) \, \vec{m}^{a}, \qquad (43)$$

where $\vec{u}$ is the velocity vector and c the soundspeed. The corresponding *ray speeds*, are simply the velocity vector for shear and entropy waves ($\vec{\lambda}^{e,s} = \vec{u}$), and the bicharacteristic vector $\vec{\lambda}^a = \vec{u} \pm c \vec{m}^a$ for the acoustic waves. One easily verifies that $\alpha^k \lambda_m^k \overline{\mathbf{r}}^k = \alpha^k \lambda^k \overline{\mathbf{r}}^k$ precisely as in the scalar case. The linear schemes will make use of the ray speeds only, while a combination of the ray speeds and the frontal speeds will be used in the non-linear schemes like the NN-scheme.

According to the above decomposition, eq. (42), the fluctuation or residual of a cell T is splitted in parts

$$\Phi_T = \sum_k \Phi_T^k = -V_T \sum_k \overline{\lambda}^k \, \alpha^k \, \overline{\mathbf{r}}^k, \qquad (44)$$

which are distributed over the vertices of the cell according to the relevant speed vector $\vec{\lambda}^k$. Assembling the contributions from all surrounding cells one obtains for meshpoint i :

$$\mathbf{u}_i^{n+1} = \mathbf{u}_i^n + \frac{\Delta t}{V_i} \sum_T \sum_k (\beta_T^i)^k \Phi_T^k, \qquad (45)$$

The coefficient $(\beta_T^i)^k$ determines the fraction sent to vertex i due to linear wave k of cell T, following one of the scalar distribution schemes discussed in section 3.

Before we can apply the scalar distribution schemes for each linear wave we have to resolve two remaining issues : (1) the construction of the linearized quantities (speeds, eigenvectors etc.) such that discrete conservation is obtained and (2) a strategy for making the above decomposition in linear waves, which unlike in 1D is not trivial.

4.1 Discrete Conservation for the twodimensional Euler equations

Considering (44), the left hand side corresponds to the conservative flux balance over a cell (triangle in 2D), and the linearization has to be such that the right hand side reproduces this, for any value of the unknowns put at the vertices. The particular integration rule used to evaluate the flux integral is not important, as long as the integrals over common edges cancel when the fluctuations for two neighbour cells are summed. The Roe-linearization for the 1D Euler equations is easily generalized to satisfy these needs [15]. Using the results of section 2.3, we assume that the parameter vector $\mathbf{z} = \sqrt{\rho}(1, u, v, H)^\tau$ is the variable having linear change in space, then using the chain rule one has for the fluctuation :

$$\Phi_T = -\oint_{S_T} \mathbf{F}\vec{n}\, dS = -\int_{V_T} (\mathbf{F_z}\vec{1}_x + \mathbf{G_z}\vec{1}_y)\, dV \cdot \vec{\nabla}\mathbf{z}. \qquad (46)$$

Precisely as in 1D, this can be rewritten in terms of the conservative variable $\mathbf{u}$ using the relation $\vec{\nabla}\mathbf{u}(\mathbf{z}) = \mathbf{u_z}(\mathbf{z})\vec{\nabla}\mathbf{z}$. Since al the matrices involved ($\mathbf{F_z}$, $\mathbf{G_z}$ and $\mathbf{u_z}$) are linear in the components of $\mathbf{z}$, which itself is linear in space, the integrations are just arythmetic averages, and one obtains the conservative linearization :

$$\Phi_T = -V_T \left[\mathbf{F_u}(\overline{\mathbf{z}})\, \vec{1}_x + \mathbf{G_u}(\overline{\mathbf{z}})\, \vec{1}_y \right] \vec{\nabla}\mathbf{u}, \qquad (47)$$

where $\vec{\nabla}\mathbf{u}$ is constant over the cell, to be computed as

$$\vec{\nabla}\mathbf{u} = \mathbf{u_z}(\overline{\mathbf{z}})\, \vec{\nabla}\mathbf{z}, \qquad (48)$$

and $\overline{\mathbf{z}}$ is just the arythmetic average of the vertex values over the cell :

$$\overline{\mathbf{z}} = \frac{\mathbf{z}_1 + \mathbf{z}_2 + \mathbf{z}_3}{3} \qquad (49)$$

Hence, the averaged speeds $\overline{\lambda}_m^k$ and eigenvectors $\overline{\mathbf{r}}^k$ to be used in the previous section are simply the analytical expressions for the Jacobian matrix $(\mathbf{F_u}(\mathbf{z})\vec{1}_x + \mathbf{G_u}(\mathbf{z})\vec{1}_y)\vec{m}^k$, evaluated at the particular cell-averaged state $\mathbf{z} = \overline{\mathbf{z}}$, while the left hand side of (40) (used to compute the strengths α^k and the angles θ^k) has to be evaluated using (48).

4.2   Wave modelling for the 2D Euler equations

The crucial remaining issue to be discussed is the multidimensional wave decomposition used in (40) and (42). Quoting Phil Roe [16], the basic requirement is that it should have enough degrees of freedom to match an arbitrary linear variation of the initial data (the left hand side of (40)). Further, it should reduce to the 1D decomposition in the direction $\vec{m}$ whenever all gradients in the direction normal to $\vec{m}$ vanish and it should return the correct solution for initial data corresponding to an isolated plane wave.

For the Euler equations in two space dimensions this leads to 8 degrees of freedom corresponding to the 4 scalar gradients in two dimensions given by (40). The most successful models developed sofar [12] are based on 4 mutually orthogonal acoustic waves providing 5 degrees of freedom (4 acoustic strengths α^a and one direction angle θ^a), 1 entropy wave providing 2 degrees of freedom (1 strength α^e and one angle θ^e), and one shear wave with unknown strength α^s and imposed direction.

The choice of this imposed shear wave direction was a source of difficulties. In [12] the direction normal to the velocity was proposed, but this caused severe problems in the capturing of oblique shocks. Therefore, in [6] the pressure gradient direction was proposed. This solves the problem for shocks but leads to problems in isolated shears, where the pressure gradient vanishes and its direction becomes ill-defined.

A recent development by Phil Roe [16] is based on analysing the important directions in an isolated shock and shear : a shock is aligned with the principle axes of the velocity gradient tensor, while a shear is oriented at 45^0 with the principle directions as was already observed in [6]. Therefore, a choice of the imposed shear direction at 45^0 with the acoustic angle θ^a seems appropriate.

Substituting in (40) returns precisely the direction of the principle strain rate axes for the acoustic directions, confirming the validity of the choice for both isolated shears and shocks. On the other hand, solving for the entropy wave gives precisely the direction of entropy gradient as the direction selected by the model, with as strength the norm of the entropy gradient.

In conclusion, although only very limited experience is available with this new wave model it seems to be satisfactory from the point of theory, removing the ambiguity of the earlier developments. The numerical results for isolated shear and shocks confirm its ability to recognize contacts and shocks aligned with the mesh.

4.3 Numerical results

The results presented here are still to be seen as preliminar, as they have not been computed using the latest wave model discussed above. Also, it seems that the present treatment of the boundary conditions at solid walls is still unsatisfactory. This treatment is based on a characteristic variable decomposition in the direction normal to the wall.

Therefore, the numerical results given here should only be considered as a proof of feasibility of the wave modelling approach as a viable way to extend the scalar advection schemes to the system of Euler equations.

A first check has to be the ability to capture shocks and contacts aligned with the mesh. With respect to shock capturing, the scalar results of section 3.2 have been fully confirmed for the Euler equations on the same structured mesh, with a capturing in a layer of two or three cells depending on the subgrid position of the shock in the exact solution. Figure 8 shows a computation for a testcase with exact states connected by a shock imposed at the left and right boundary of the domain. A shear aligned with the mesh is preserved by the

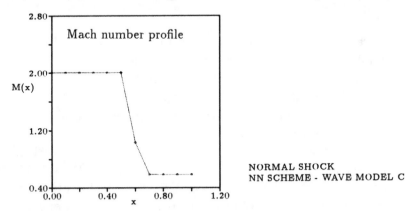

Figure 8: Capturing of grid-aligned shock for the Euler equations.

scheme if the new model is used [16]. Also for shocks and shears aligned with the *medians* of the triangles in the structured mesh, capturing is obtained in a layer of maximum two cells. This provides in total 6 directions in a structured mesh for which perfect capturing of shocks and shears is obtained. Deviations from these directions leads to a spreading of the shear as in the scalar case, while shocks are kept over a fixed layer of 2 or 3 cells.

Figure 9 shows the Mach lines for a more complex supersonic channel flow, with the unstructured mesh superposed. The excellent shock capturing for shocks not aligned with the mesh is confirmed.

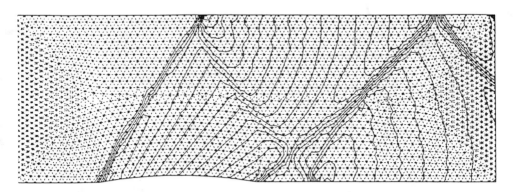

inlet Mach number of 1.4.

Figure 9: Iso-Mach lines for supersonic channel flow, superposed to the mesh.

The final testcase in figure 10 shows a computation for the transonic GAMM channel flow, made by former VKI students Catalano, De Palma and Pascazio (Politecnico di Bari) [3]. The computation is done on an isotropic mesh of 80X32 structured quadrilaterals subdivided by alternating left and right diagonals. As a consequence, the computational molecule is different for the odd and even points in the grid. The computation does not yet make use of the most recent wave model with respect to the shear wave direction. Also, large entropy errors are created along the wall boundary, indicating that the boundary conditions are not satisfactory. Nevertheless, although preliminar, this computation is converged to machine accuracy and demonstrates the ability to deal with subsonic flows.

Concluding remarks

It was the purpose of this contribution to show that some radically new ideas are coming up for the solution of the compressible flow

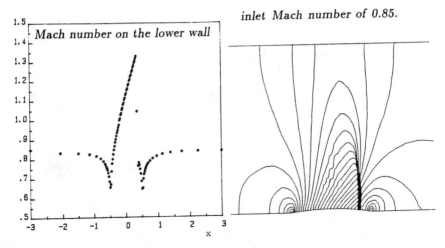

Figure 10: Iso-Mach lines for transonic GAMM channel flow.

equations. These ideas are conceptually not different from the suc-
cesfull upwind solvers developed in the eighties, as far as one dimensional flow is concerned.

The theory has been rebuilt only in the way the extension to two
and three space dimensions is handled, by curing the weakness of
the one dimensional physical modelling used in the state of the art
upwind methods. A satisfactory result in this respect is that the
results in two space dimensions carry over to the 3D case, both with
respect to scalar advection schemes and wave models.

The new approach has a strong potential, especially for codes
based on triangles and tetrahedra, bringing these unstructured grid
methods which suffer most from the one-dimensional modelling to the
same level of capability as structured grid solvers : perfect capturing
of shears and shocks aligned with one set of cell boundaries, and high
resolution on very compact stencils. It is hoped that the nineties will
see the fulfilment of these promises.

Acknowledgement

The work at VKI and Bari was supported by the Commission of
the European Community under Contract AERO-0003-C in Area 5
(Aeronautics) of the BRITE/EURAM Programme (1989-1993).

References

[1] L.A. Catalano, M. Napolitano and H. Deconinck. Optimal multi-stage schemes for multigrid smoothing of two-dimensional advection operators. Submitted to Comm. in Appl. Num. Meth., 1991.

[2] L.A. Catalano, P. De Palma and M. Napolitano. Explicit multi-grid smoothing for multidimensional upwinding of the Euler equations. Proc. GAMM conference, sept. 1991, Lausanne.

[3] L.A. Catalano, P.De Palma and G. Pascazio. A multi-dimensional solution adaptive multigrid solver for the Euler equations. Submitted for the 13th ICNMF conference, Rome, 1992.

[4] A. Dadone and B. Grossman. A rotated upwind scheme for the Euler equations. AIAA *Journal*, to appear.

[5] H. Deconinck, P.L. Roe and R. Struijs. A multidimensional generalization of Roe's flux difference splitter for the Euler equations. *4th ISCFD Conference*, U.C. Davis, 1991.

[6] P. De Palma, H. Deconinck and R. Struijs. Investigation of Roe's 2D wave models for the Euler equations. VKI TN 172, 1990.

[7] D.W. Levy. *Use of a rotated Riemann solver for the two-dimensional Euler equations*. PhD thesis, University of Michigan, 1990.

[8] I. Parpia. A planar oblique wave model for the Euler equations. AIAA paper 91-1545, 1991.

[9] P.L. Roe. Fluctuations and signals - A framework for numerical evolution problems. In K.W. Morton and M.J. Baines, editors, *Numerical Methods for Fluid Dynamics* , Academic Press, 1982.

[10] P.L.Roe. The use of the Riemann problem in finite differences. In W.C. Reynolds and R.W. MacCormac, editors, *7th Int. Conf. Numerical Meth. in Fluid Dynamics*, Springer, 1981.

[11] P.L.Roe. Approximate Riemann solvers, parameter vectors, and difference schemes. *J. Comp. Phys. 43, No 2, pp. 357*, 1981.

[12] P.L.Roe. Discrete models for the numerical analysis of time-dependent multidimensional gasdynamics. *J. Comp. Phys.*, 63:458-476, 1986

[13] P.L. Roe. Linear advection schemes on triangular meshes. CoA Report no 8720, Cranfield, november, 1987

[14] P.L. Roe, D. Sidilkover. Optimum positive linear schemes for advection in two and three dimensions. To be published, 1991

[15] P.L. Roe, R. Struijs and H. Deconinck. A conservative linearisation of the multidimensional Euler equations. Submitted to J. Comp. Phys 1991

[16] P.L. Roe. Beyond the Riemann problem I. Companion paper in this proceeding, 1991.

[17] C. Rumsey, B. van Leer and P.L. Roe. A grid-independent approximate Riemann solver with applications to the Euler and Navier-Stokes equations. AIAA paper 91-1530, 1991.

[18] R. Struijs, H. Deconinck, P.L. Roe. Fluctuation splitting schemes for the 2D Euler Equations. VKI lecture series 1991-01

[19] R. Struijs, H. Deconinck, P. de Palma, P.L. Roe, K.G. Powell. Progress on multidimensiomal upwind Euler solvers for unstructured grids. AIAA 91-1550

[20] D. Sidilkover. *Numerical solution to steady-state problems with discontinuities.* PhD thesis, Weizmann Institute, Israel, 1990.

[21] Y. Tamara and K. Fuji. A multi-dimensional upwind scheme for the Euler equations on structured grids. In M.M. Hafez, editor, *4th ISCFD Conference*, U.C. Davis, 1991.

[22] P. Van Ransbeeck C. Lacor and Ch. Hirsch. A multidimensional cell-centered upwind algorithm based on a diagonalization of the Euler equations. Proc. 12th Int. Conf. on Num. Meth. in Fluid Dynamics, Oxford,1990.

THREE DIMENSIONAL COVOLUME ALGORITHMS
FOR VISCOUS FLOWS

*R.A.Nicolaides**

Department of Mathematics
Carnegie Mellon University, Pittsburgh, PA 15213

ABSTRACT

The covolume technique is applied to the discretization of three dimensional flow problems. The two problems considered are the compressible Navier-Stokes equations and the primitive variable incompressible flow equations when the vorticity form of the convection terms is used. It is shown that the discrete covolume equations do not generate spurious vorticity in a sense made precise in the paper.

1 Introduction.

Staggered meshes are one of the reliable ways to avoid spurious mode problems in both compressible and incompressible flow computations. The type of staggering that we have in mind segregates the components of vectors such as velocities, from scalars such as pressure and energy and assigns them to different types of interlaced mesh points. The classical example of this approach is the original MAC scheme of Harlow and Welch (1965) and while there have been numerous evolutionary developments, the basic idea is widely used today.

One of the problems with the classical approach is its limitation to rectangular meshes. In order to solve problems in more complex geometries one usually resorts to some form of structured mesh generation. A recent report by Segal et al (1992) shows how to do this in an invariant way which preserves physical relationships similarly

---

*This work was supported by the United States Air Force under grant AFOSR 89-0359 et. seq.

to the cartesian scheme. This approach still does not permit the use of fully unstructured meshes.

An extension of the basic staggered mesh approach to unstructured meshes—the *complementary volume* or *covolume* approach—was given in Nicolaides (1989) and further developed and analysed in Nicolaides (1991) (1992a-c), and in Nicolaides and Wu (1992a-c). It turns out that there is a fairly satisfactory discrete vector field theory associated with these extensions of staggered mesh schemes. This theory is developed rigorously in Nicolaides (1992a) for the two dimensional environment. It contains analogs of results of vector field theory such as the identities curl(grad)$\equiv$ 0 and div(curl)$\equiv$ 0, as well as results on the existence of discrete potentials and an analog of the Helmholtz decomposition. In this paper we will present three dimensional versions of such results. As a first application we will discretize the compressible Navier-Stokes equations.

One advantage of the covolume approach is that it provides a uniform way to discretize different kinds of flow models including compressible, incompressible, potential and vortical flows. In addition, it is often possible to establish a relationship between the discrete models by taking various limits or making assumptions about the flow or using mathematical transformations as one does at the continuous level. To illustrate this, we will discretize the incompressible Navier-Stokes equations with the convection terms written in the vorticity format. Then we will derive by discrete transformations the conservative form of the discrete vorticity transport equation satisfied by the solution of the discrete Navier-Stokes equations. The discrete vorticity transport equation will turn out to be one which could have been derived directly from the continuous vorticity transport equation by the covolume method. More briefly, we may say that covolume discretization and vorticity transformation are commuting operations. One reasonable interpretation of this result is that the Navier-Stokes discretization does not generate spurious vorticity. This can be important in some situations.

In contrast to the two dimensional case, most of the three dimensional discretizations proposed here have not yet been implemented (for two dimensional results see Nicolaides and Wu (1992a), Hall et al (1991) and Hall et al(1992)). This lack of data reflects the relative newness of the approach. On the other hand, since covolume algorithms reduce to the tried and true staggered mesh schemes in simple enough geometries, there are reasons to expect good results.

2. Mesh Systems

The starting point for the unstructured mesh system is a set of nodes $\mathcal{N}$ and a set of tetrahedra $\mathcal{T}$ which have the nodes as vertices. $\mathcal{T}$ is assumed to cover the flow domain Ω in the usual way of a triangulation so that any two of its elements which intersect do so only on a common face, common edge or common node. The set of faces of the tetrahedra (or *elements*) is denoted by $\mathcal{F}$ and the set of edges by $\mathcal{E}$. This mesh is called the *primal* mesh.

In covolume methods generally, a dual mesh plays an essential part. The edges of this dual are obtained by joining the circumcenters of tetrahedra which share a common face (adjacent tetrahedra). The key property of the mesh system is that the edges of either mesh are orthogonal to the associated faces of the other mesh. For example each edge of the dual mesh is perpendicular to the common face of the adjacent tetrahedra which it links. This follows from elementary geometry, since the common face is a chordal plane of the circumscribing spheres of both of the tetrahedra. To visualize the dual property, note that from above, the elements of $\mathcal{F}$ which share a common element of $\mathcal{E}$ each orthogonally intersect an element of $\mathcal{E}'$ (the set of dual edges), and that these elements of $\mathcal{E}'$ must be coplanar. Such planes contain the set of faces $\mathcal{F}'$ of the dual mesh and these faces have as their edges the same dual edges. It follows that the common edge of the primal faces is perpendicular to the dual face. This establishes the mutual orthogonality property of the dual mesh system.

Figure 1 shows one tetrahedral face ABC and the dual face $PQRSTU$ associated with its edge AB. P, Q, R, S, T are the circumcenters of the various tetrahedra which share the edge AC. The dual face is polygonal in general. O denotes the point where the dual edge penetrates the triangular face. This point is the circumcenter of the triangle. O' denotes the point where the tetrahedral edge penetrates the dual face. In general this does not coincide with any standard geometrical position in the polygon. These interlacing loops of bounding edges will appear below in the discretization.

Some small adjustments must be made to the above picture for edges of $\mathcal{E}$ which lie in the boundary Γ of Ω since the dual edges of boundary faces have no finite node outside Ω to connect with. The situation is illustrated in Figure 2 where we make use of the circumcenters O and O'' of the two boundary triangles sharing the

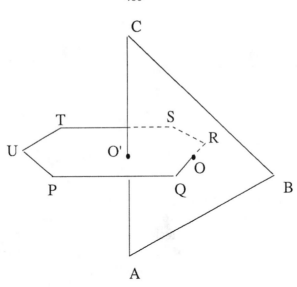

Figure 1: Two interlacing circulation loops made from primal and dual edges.

edge AC and the midpoint O' of the edge to complete the dual face. Q and P are the circumcenters of the tetrahedra standing on the faces ABC and ACD. These faces are noncoplanar in general. In this case the normal edge actually passes through the boundary of the dual face instead of through an interior point of it.

The polygonal faces of the dual mesh, including the boundary faces, enclose certain polyhedra which contain the mesh nodes of $\mathcal{N}$. There is exactly one node for each polyhedron. The boundary polyhedra may have their associated mesh nodes on their boundaries. The set of dual nodes consisting of circumcenters is denoted by $\mathcal{N}'$ and the sets of dual polyhedra and their faces are denoted respectively by $\mathcal{T}'$ and $\mathcal{F}'$.

Where it is necessary to distinguish this circumcenter based dual (from centroid based duals for example) we will refer to the *normal* dual.

3. Covolume discretizations

In this section we will discretize the vector field operations 'div', 'curl' and 'grad' and their combinations 'grad div' and 'curl curl' and show certain relationships between them.

First we must specify the mesh orientations. The nodes of the primal mesh are assumed to be numbered in some fixed way. The

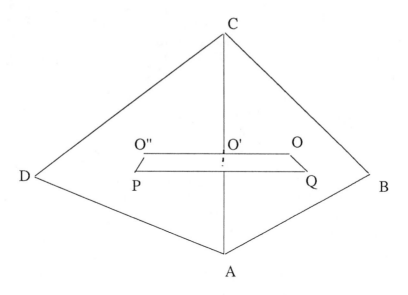

Figure 2: Situation at the boundary.

positive direction on a primal edge is from low to high node number. The normal to a primal face $\phi \in \mathcal{F}$ is directed by a right hand screw rule relative to the increasing order of the face's node numbers. The boundary of a primal face is oriented so that the positive direction of traverse follows increasing node numbers. The nodes of the dual mesh are also assumed to be numbered in a fixed way. The positive direction on a dual edge is that of the normal just defined. The boundary of a dual face is oriented by a right hand screw rule relative to the positive direction on the primal edge which constitutes its normal.

The volumes of the tetrahedra and their face areas are denoted as usual by $|\tau_i|$ and $|\phi_j|$ with a similar notation for the dual quantities. $|h_i|$ and $|h'_j|$ denote edge lengths in the two meshes.

A characteristic feature of covolume discretizations is that (at least to begin with) only single components of vector fields are introduced. These components are taken to be along the dual edges at the circumcenters of the primal faces. For example if $\mathbf{u}$ is a vector field defined on Ω we could work not with $\mathbf{u}$ itself, but with the collection of components $u'_i = \mathbf{u}.\mathbf{n}_i$ where $\mathbf{n}_i$ denotes the directed normal to the ith face, and $\mathbf{u}$ is evaluated at the face circumcenter. More generally we will work with collections of components $\{u'_i\}, \{v'_i\}$ etc. associated with the normal directions and with the circumcenters of the primal faces without being specific about their source.

Sets of dual components directed by the edges in $\mathcal{E}$ and associated with the point where the edge penetrates its dual face (O' in Figure 1) are similarly denoted by $\{u_k\}, \{v_k\}$ etc. the prime being absent in this case.

Now we will define the basic discretizations. First of all, the approximate flux out of a tetrahedron τ_j is defined by

$$f_j(u') := \sum_{i \in \partial \tau_j} u'_i A_i \qquad \tau_j \in \mathcal{T}$$

where the sum is over the four faces of the tetrahedron, and A_i denotes $|\phi_i|$ suitably signed. An approximate divergence is obtained by normalizing the flux by the volume $|\tau_j|$ of the tetrahedron

$$d_j(u') := f_j(u')/|\tau_j| \qquad \tau_j \in \mathcal{T}.$$

The approximate flux and divergence are most naturally assigned to the circumcenter of the tetrahedron since the four face normals meet there.

The dual quantities are the flux and divergence associated with an element τ'_k. The definitions are

$$f'_k(u) := \sum_{j \in \partial \tau'_k} u_j A'_j \qquad \tau'_k \in \mathcal{T}'$$

and

$$d'_k(u) := f'_k(u)/|\tau'_k| \qquad \tau'_k \in \mathcal{T}'.$$

These quantities are considered to be associated with the primal node in τ'_k.

Next we define a pair of circulations and their associated curl components. The approximate circulation around $\partial \phi'_l$ is

$$r_l(u') := \sum_{i \in \partial \phi'_l} u'_i h'_i \qquad \phi'_l \in \mathcal{F}'$$

where h'_i is signed positively if the direction of the the dual edge is directed parallel to the orientation of the dual boundary (circulation path) and negatively otherwise. The corresponding approximate component of the curl along the normal $\mathbf{n}_l$ is (Stokes's theorem)

$$c_l(u') := r_l(u')/|\phi'_l| \qquad \phi'_l \in \mathcal{F}'.$$

The duals are (with the corresponding sign convention)

$$r'_m(u) := \sum_{i \in \partial \phi_m} u_i h_i \qquad \phi_m \in \mathcal{F}$$

and

$$c'_m(u) := r'_m(u)/|\phi_m| \qquad \phi_m \in \mathcal{F}.$$

We also use a gradient approximation

$$g'_i(p) := (p_{i,+} - p_{i,-})/h'_i \qquad \epsilon'_i \in \mathcal{E}' \tag{1}$$

where p denotes a scalar field defined at the interior dual nodes, and the signs refer to the direction along the dual edge, with positive denoting the forward direction.

The dual definition of the gradient is omitted as it is not needed in this paper.

Notice that each operator maps a field from one of the mesh systems into a field in the other mesh system.

Second order operators are defined in terms of the first operators. The four basic second order expressions are grad div, curl curl, curl grad, and div curl of which the last two are identically zero. An analog of $\mathrm{curl}(\mathrm{grad}) \equiv 0$ is

$$c(g'(p)) \equiv 0$$

where we have supressed the subscripts for clarity. This follows by a simple calculation taking into account the sign conventions on the dual edge lengths h'_i. There is a corresponding dual result. Dual analogs to $\mathrm{div}(\mathrm{curl}) \equiv 0$ are

$$d(c'(u)) \equiv 0$$

and

$$d'(c(u')) \equiv 0. \tag{2}$$

These may be also proved by very simple direct calculations taking account of orientations. The appearance of the inner circulations in oppositely signed pairs is the essential point in both calculations. We often omit the outer parentheses in these second order expressions.

The remaining pair of operators are not discretized directly: only their normal components are discretized. For grad(div) the approximation is

$$\mathbf{n} \cdot \nabla \operatorname{div} \mathbf{u} \approx g'd(u'). \tag{3}$$

Here, $\mathbf{n}$ denotes normals to interior faces in $\mathcal{F}$ and $u'_i := \mathbf{n}_i \cdot \mathbf{u}$. The corresponding dual pair for curl(curl) are

$$c'c(u') \qquad \text{and} \qquad cc'(u). \qquad (4)$$

The first uses the normals to $\mathcal{F}'$ and the second uses the normals to $\mathcal{F}$.

When integrating over control volumes, the undivided operators f, f' and r, r' would be used to acheive the corresponding discretizations.

It is worth noting that the first order operators can be iterated as often as desired, alternating primed and unprimed operators, to produce higher derivative approximations.

4. Field extensions

For efficiency, we usually try to discretize equations using the variables of just one mesh system. Within a single system however, one often cannot avoid using (say) velocity components different from the particular normal components which appear as basic variables. If we wish to avoid carrying these components as additional variables, with the increased complexity that this implies, they have to be obtained from the already available normal components. This aspect is the subject of this section.

There are two field types which must be considered. The first type is a set of normal components relative to $\mathcal{F}$. The second type is a set of normal components relative to $\mathcal{F}'$. In either case the initial problem is to find the simplest vector field defined within a tetrahedron which has the given normal components on its faces or on its edges. This is a problem of extending the given set of components into a full vector field in each element. In general, such fields will be of less than full linear variation and in fact, constant in some cases. To obtain linear variation within the elements, we average the fields on both sides of the element faces at their circumcenters and define the interior field by standard linear interpolation from the four face values.

Consider first a tetrahedron τ_j with prescribed components u'_i in the directions normal to its four faces. Then there is a unique vector field defined in τ_j of the form

$$\mathbf{q}' := \mathbf{v}'_0 + \alpha\mathbf{x} \qquad (5)$$

where $\mathbf{v}_0'$ denotes a constant vector, α is constant and $\mathbf{x} := (x, y, z)$, such that

$$\mathbf{q}' \cdot \mathbf{n}_i = u_i' \qquad i = 1, 2, 3, 4.$$

This general result is due to Raviart and Thomas (1975). The special case when the the flux of the components is zero gives $\alpha = 0$ and was also given in Nicolaides 1989. A nice property of (5) is that the quantities $\mathbf{q}' \cdot \mathbf{n}_i$ have the same value no matter where they are evaluated on a face, since $\mathbf{n}_i \cdot \mathbf{x}$ is constant on the face. (5) gives the 'sublinear' extension of the normal components to the tetrahedron.

It remains to combine the $\mathbf{q}'$'s from the two sides of a face. Here we will use simple averaging, taking $1/2$ of each $\mathbf{q}'$ and denoting the result by $\mathbf{e}_k(u')$ for ϕ_k. Other weightings can be used with advantage. For example, for a two dimensional incompressible flow $(f(u') = 0)$ it is possible to choose the weighting to give zero mass flux out of the covolumes, i.e. $f'(u) = 0$ where u denotes the computed tangential components (Nicolaides 1992c): it is not yet known whether this can be done in three dimensions. Regardless of the weighting chosen, we denote the computed vectors by $\mathbf{e}_k(u')$. Note that $\mathbf{e}(u') := \{\mathbf{e}_k(u')\}$ has continous normal components on passing through faces. We denote the linear interpolant of $\mathbf{e}(u')$ in τ_j by $\hat{\mathbf{e}}_j(u')$. This interpolant is constructed using the four face values of $\mathbf{e}'(u')$ associated with τ_j.

For dual fields, defined on $\mathcal{E}$ and denoted by $\{v\}$ the extension into a tetrahedron is done differently. We will denote by $\mathbf{t}_j$ the various directions along the edges of the typical tetrahedron τ_k, and by ϕ_m its four faces with normals $\mathbf{n}_m$. The edge lengths are denoted as usual by h_j. We will have to use the matrix S of rank two defined by

$$S := \begin{pmatrix} 0 & z & -y \\ -z & 0 & x \\ y & -x & 0 \end{pmatrix}.$$

Let $\mathbf{t} \equiv (t_1, t_2, t_3)$ denote any unit vector and consider

$$\begin{aligned} \mathbf{t}^T S_{.,1} &= 0t_1 - zt_2 + yt_3 \\ &= (x, y, z) \cdot (0, t_3, -t_2) \\ &=: \mathbf{x} \cdot \mathbf{t}^{(1)}. \end{aligned}$$

It follows that $\mathbf{t}^T S_{.,1}$ is constant on planes normal to $\mathbf{t}^{(1)}$. Similarly, $\mathbf{t}^T S_{.,2}$ is constant on planes normal to $\mathbf{t}^{(2)} := (-t_3, 0, t_1)$. From $\mathbf{t}^{(j)} \cdot \mathbf{t} = 0$, $j = 1, 2$ it follows that $\mathbf{t}^T S_{.,j}$, $j = 1, 2$ is constant on every line parallel to $\mathbf{t}$.

We will seek $\mathbf{q}$ such that

$$\mathbf{q} \cdot \mathbf{t}_j = v'_j \qquad j = 1, 2, \ldots, 6$$

and assume that it can be found in the form

$$\mathbf{q} := \mathbf{v}_0 + S\alpha$$

where $\mathbf{v}_0$ and α are constant vectors in R^3. Although $\mathbf{q}$ depends on $\mathbf{x}$, the previous paragraph shows that $\mathbf{t}_j \cdot \mathbf{q}$ is independent of $\mathbf{x}$ on the edge ϵ'_j.

Since $\mathbf{q}$ is linear in $\mathbf{x}$, curl $\mathbf{v}$ is constant and Stokes theorem gives

$$\int_{\phi_m} \mathbf{n}_m \cdot \operatorname{curl} \mathbf{q}\, dA = \int_{\partial\phi_m} \mathbf{q} \cdot \mathbf{t}\, dL.$$

The left side is

$$-2\mathbf{n}_m |\phi_m| (\alpha_1 \mathbf{i}_1 + \alpha_2 \mathbf{i}_2 + \alpha_3 \mathbf{i}_3)$$

and the right side is

$$\sum_{i \in \partial\phi_m} v'_i h_i$$

using the midpoint rule to evaluate the integral. Hence we obtain the equations

$$\mathbf{n}_m \cdot \alpha = c'_m v'$$

for determining α. Only three of the possible four equations can be independent since the four face circulations necessarily sum to zero. Once α has been found, $\mathbf{v}_0$ is obtained by solving the equations

$$\mathbf{t}_j \cdot \mathbf{v}_0 = v_j - \mathbf{t}_j \cdot S\alpha.$$

Here we note that $\mathbf{v} - S\alpha$ has zero for its (discrete) face circulations and so it follows that only three of these six equations are independent.

As above, it remains to combine the $\mathbf{q}$'s from the two sides of a face. We will use simple averaging at the circumcenters taking $1/2$ of each $\mathbf{q}$ and denote the result by $\mathbf{e}_k(v)$ for ϕ_k. In this way we have a vector field at each face circumcenter and a linear field can be defined inside each element using linear interpolation of the four face values. We denote the linear interpolant of $\mathbf{e}(v)$ in τ_j by $\hat{\mathbf{e}}_j(v)$.

These linear interpolations can be viewed as being made (by a roundabout process) from the component data of the four neighboring tetrahedra.

A final point concerns the definition of control volumes. In addition to τ_m and τ'_k these may be of two types, one for the primal and one for the dual mesh. The primal control volumes are associated with faces of the tetrahedra. For the face ϕ_k the associated control volume is obtained by joining the circumcenters of the adjacent tetrahedra (tetrahedron in the case of a boundary face) to the vertices of ϕ_k. This control volume, denoted by V_k, is the union of two tetrahedra whose vertices are the three vertices of ϕ_k and the two adjacent circumcenters. Note that these control volumes have disjoint interiors and that their union provides a tesselation of Ω. The dual mesh covolumes V'_j are built on the common faces ϕ'_j just as v_k is built on ϕ_k by joining the adjacent vertices to the circumcenters surrounding a primal edge. They are no longer tetrahedral pairs, having more numerous faces. As before they have disjoint interiors and cover Ω.

5. Momentum equation

In the next two sections we will apply the operators of sections 3 and 4 to discretize the compressible Navier-Stokes equations. Our idea is not to provide a definitive discretization. Rather, we wish to illustrate the use of the special properties of the covolume approach as they concern these equations. In particular, we will not deal here with artificial viscosity or other methods for computing shocks, despite their obvious importance.

The compressible momentum equation is (Hirsch 1990),

$$\frac{\partial(\rho\mathbf{u})}{\partial t} + \operatorname{div}(\rho\mathbf{u} \otimes \mathbf{u}) = -\nabla p + \operatorname{div}(2\mu\hat{\nabla}\mathbf{u}) + \nabla(\lambda\operatorname{div}\mathbf{u})$$

where $(\hat{\nabla}\mathbf{u})_{pq} := (u_{p,q} + u_{q,p})/2$.

It is convenient to write the first viscous term in an equivalent form. For this, define the skew tensor W by $W_{pq} := u_{p,q} - u_{q,p}$ (here and in the previous paragraph we are temporarily using $\mathbf{u} := u_p$). Then

$$2\mu\hat{\nabla}\mathbf{u} = 2\mu(\nabla\mathbf{u})^T + \mu W. \tag{6}$$

It is easy to check that

$$\operatorname{div}(\nabla\mathbf{u})^T = \nabla(\operatorname{div}\mathbf{u}) \tag{7}$$

and by a longer calculation that

$$\mathrm{div}W = -\mathrm{curl\,curl}\,\mathbf{u}. \qquad (8)$$

These will be used below.

We will use the variables $\mathbf{u}$, ρ and e (internal energy per unit mass) and assume that p is given by an equation of state $p = p(\rho, e)$. Scalar quantities such as ρ, e, p etc. are defined at element circumcenters and we will make use of the normal components $\{u'\}$ relative to $\mathcal{F}$.

Let $\phi_i \in \mathcal{F}$ denote an interior face with oriented normal $\mathbf{n}_i$. We will discretize the $\mathbf{n}_i$ component of the momentum equation using the control volume V_i which was defined in the previous section.

Multiplying the momentum equation by $\mathbf{n}_i$, using (6) to replace the first viscous term and integrating over V_i gives for the left side

$$\frac{\partial}{\partial t} \int_{V_i} \rho \mathbf{u} \cdot \mathbf{n}_i \, dV + \int_{\partial V_i} \rho(\mathbf{u} \cdot \mathbf{n}_i)(\mathbf{u} \cdot \mathbf{m}) \, dA,$$

where $\mathbf{m}$ denotes the outer unit normal to V_i, and for the right side

$$\int_{V_i} \frac{\partial p}{\partial n_i} \, dV \;+\; \mathbf{n}_i \cdot \int_{V_i} \mathrm{div}(2\mu \nabla \mathbf{u})^T \, dV$$

$$+\; \mathbf{n}_i \cdot \int_{V_i} \mathrm{div}(\mu W) \, dV + \int_{V_i} \frac{\partial(\lambda \mathrm{div}\mathbf{u})}{\partial n_i} \, dV.$$

For simplicity we will assume that λ and μ vary sufficiently slowly that they may be considered to be constant in V_i and that their values there are given by the averages $\bar{\mu}_i$ and $\bar{\lambda}_i$ of their values at the circumcenters of $\tau_{i,L}$ and $\tau_{i,R}$ the tetrahedra from $\mathcal{T}$ which share ϕ_i. In this case, using (7) and (8) we can write the viscous terms as

$$(2\bar{\mu}_i + \bar{\lambda}_i) \int_{V_i} \frac{\partial(\mathrm{div}\mathbf{u})}{\partial n_i} \, dV - \bar{\mu}_i \mathbf{n}_i \cdot \int_{V_i} \mathrm{curl\,curl}\,\mathbf{u} \, dV.$$

Using (3) and (4) to approximate these viscous terms gives

$$|V_i|[(2\bar{\mu}_i + \bar{\lambda}_i)g'd(u') - \bar{\mu}_i c'c(u')].$$

If higher order variations in λ and μ have to be accounted for, it will be necessary to use interpolations of the second derivative terms at the circumcenters of $\tau_{i,L}$ and $\tau_{i,R}$, but we will not consider this here.

The term $\frac{\partial p}{\partial n}$ is approximated by

$$\frac{\partial p}{\partial n} \approx g'(p)$$

after p is computed from the equation of state.

The convection term is approximated by the midpoint rule applied at the centroid of each face of ∂V_i. The approximation may be expressed as

$$\sum_{k \in \partial V_i} (\hat{\mathbf{e}}(u') \cdot \mathbf{n}_i)(\hat{\mathbf{e}}(u') \cdot \mathbf{m}_k)|\partial V_{i,k}|$$

where $\mathbf{m}_k$ denotes the outer normal to the kth face of ∂V_i and $\hat{\mathbf{e}}(\cdot)$ denotes the linear interpolation defined in section 4. The evaluations of $\hat{\mathbf{e}}(u')$ are at the centroids of V_i's faces.

6. Energy equation

Defining $e_{tot} := \rho(|\mathbf{u}|^2/2 + e)$, the energy equation is

$$\frac{\partial e_{tot}}{\partial t} + \operatorname{div} G = 0$$

where

$$G := e_{tot}\mathbf{u} + p\mathbf{u} - 2\mu(\hat{\nabla}\mathbf{u})\mathbf{u} - \lambda\mathbf{u}\operatorname{div}\mathbf{u} - \kappa \nabla e.$$

The control volumes for discretization are the tetrahedra τ_k. Use of the divergence theorem and the approximations of section 3 give for the various terms the results listed below, in which $\mathbf{m}$ denotes the unit outer normal to the control volume, A_i denotes the areas of the faces of a control volume, and barred quantities denote averages of circumcenter values adjacent to tetrahedral faces.

$$\int_{\partial \tau_k} a(\mathbf{u} \cdot \mathbf{m})\, dA \approx \sum_{i \in \partial \tau_k} \bar{a}_i u_i' A_i \qquad a \equiv e_{tot} \text{ or } p$$

$$\int_{\partial \tau_k} \kappa\frac{\partial e}{\partial m}\, dA \approx \sum_{i \in \partial \tau_k} \bar{\kappa}_i g_i'(e) A_i$$

$$\int_{\partial \tau_k} \lambda(\mathbf{u} \cdot \mathbf{m})\operatorname{div}\mathbf{u}\, dA \approx \sum_{i \in \partial \tau_k} \bar{\lambda}_i \bar{d}_i(u') u_i' A_i$$

$$\int_{\partial \tau_k} 2\mu(\hat{\nabla}\mathbf{u})\mathbf{u}) \cdot \mathbf{m}\, dA = \int_{\partial \tau_k} ([2\mu \nabla \mathbf{u}^T + \mu W]\mathbf{u}) \cdot \mathbf{m}\, dA$$

$$\approx \sum_{i \in \partial \tau_k} 2\bar{\mu}_i g_i'(\mathbf{u}) \cdot \bar{\mathbf{u}}_i A_i \; + \; \sum_{i \in \partial \tau_k} \bar{\mu}_i \mathbf{m}_i \cdot (\mathbf{e}_i(\omega) \times \mathbf{e}_i(u')) A_i.$$

where $\omega := c(u')$.

The mass conservation equation is treated by similar techniques and involves nothing new, so we will omit the details.

7. Vorticity conservation

In this section will exhibit a connection between a covolume discretization of the incompressible Navier-Stokes equations and a conservative covolume discretization of the vorticity transport equation. In essence, the discrete Navier-Stokes equations convect vorticity in accordance with the latter.

We start with the momentum equation written with the convection term expressed in terms of vorticity as

$$\frac{\partial \mathbf{u}}{\partial t} + \omega \times \mathbf{u} = -\nabla q + \nu \triangle \mathbf{u}. \tag{9}$$

where ν is constant.

Taking the curl and rearranging in the standard way gives the vorticity transport equation

$$\frac{\partial \omega}{\partial t} + (\nabla \omega)\mathbf{u} - (\nabla \mathbf{u})\omega = \nu \triangle \omega$$

where the third term on the left is the vorticity stretching term. Introducing the tensor $M_{ij} := \omega_i u_j - \omega_j u_i$ allows this to be written in the conservative form

$$\frac{\partial \omega}{\partial t} + \operatorname{div} M = \nu \triangle \omega.$$

This may be easily checked using the definition $\operatorname{div} M := M_{ij,j}$ and $\operatorname{div} \mathbf{u} = \operatorname{div} \omega = 0$. Below, we will work with

$$\frac{\partial \omega}{\partial t} + \operatorname{div} M = -\nu \operatorname{curl} \operatorname{curl} \omega. \tag{10}$$

and return to the more general case at the end of the section.

Starting from a covolume discretization of (9) we will obtain a covolume discretization of (10) by manipulations paralleling those of the continuous case. Denote by $\mathbf{n}$ an arbitrary unit vector and by $\mathbf{t}$ an arbitrary vector orthogonal to $\mathbf{n}$. We will use the identity

$$|\mathbf{t}|^2 \mathbf{n} \cdot (\omega \times \mathbf{u}) = (\mathbf{n} \times \mathbf{t}) \cdot [(\mathbf{t} \cdot \mathbf{u})\omega - (\mathbf{t} \cdot \omega)\mathbf{u}]. \tag{11}$$

A proof of this is

$$(\mathbf{n} \times \mathbf{t}) \cdot [(\mathbf{t} \cdot \mathbf{u})\omega] - (\mathbf{t} \cdot \omega)\mathbf{u}] \;=\; \mathbf{n} \times [(\mathbf{t} \cdot \mathbf{u})\omega - (\mathbf{t} \cdot \omega)\mathbf{u}] \cdot \mathbf{t}$$
$$=\; \mathbf{n} \times [\mathbf{t} \times (\omega \times \mathbf{u})] \cdot \mathbf{t}$$
$$=\; [\mathbf{t}(\mathbf{n} \cdot (\omega \times \mathbf{u})) - (\omega \times \mathbf{u})(\mathbf{n} \cdot \mathbf{t})] \cdot \mathbf{t}$$
$$=\; |\mathbf{t}|^2 \mathbf{n} \cdot (\omega \times \mathbf{u}).$$

Now taking the inner product of (9) with the normal $\mathbf{n}_k$ to ϕ_k, keeping time continuous and applying the standard approximations gives

$$\frac{\partial u_k'}{\partial t} + \mathbf{n}_k \cdot (\mathbf{e}_k(\omega) \times \mathbf{e}_k(u')) = -g_k'(q) + \nu(g_k' d(u') - c_k' c(u')).$$

Using equation (11), split the convection term into two parts resulting in

$$\frac{\partial u_k'}{\partial t} \;+\; \frac{(\mathbf{n}_k \times \mathbf{t}_k)}{|\mathbf{t}_k|^2}[(\mathbf{t}_k \cdot \mathbf{e}_k(u'))\mathbf{e}_k(\omega) - (\mathbf{t}_k \cdot \mathbf{e}_k(\omega))\mathbf{e}_k(u')]$$
$$=\; -g_k'(q) + \nu(g' d(u') - c' c(u')).$$

where $\mathbf{t}_k$ is orthogonal to $\mathbf{n}_k$ but otherwise arbitrary.

Taking the circulation $c_l(\cdot)$ around the path with normal l where this direction is one of the edges of ϕ_k, eliminates the g' terms and gives

$$\frac{\partial \omega_l}{\partial t} \;+\; c_l(\frac{(\mathbf{n}. \times \mathbf{t}.)}{|\mathbf{t}.|}[(\mathbf{t}. \cdot \mathbf{e}.(u'))\mathbf{e}.(\omega) - (\mathbf{t}. \cdot \mathbf{e}.(\omega))\mathbf{e}.(u')]) \quad (12)$$
$$=\; -\nu c_l c'(\omega). \quad (13)$$

Let V_l' denote the dual control volume with normal $\mathbf{n}_l$ (normal to the symmetry plane) and denote the pair of faces containing $\mathbf{n}_k$ by $\partial V_{l,k}'$. We will choose $\mathbf{t}_k$ to lie in the plane of circulation and point outwards on the circulation path and to have magnitude

$$|\mathbf{t}_k|^2 = \frac{|\partial V_{l,k}'|}{|V_l'|}. \quad (14)$$

$\hat{\mathbf{t}}_k$ denotes the corresponding normalized vector, so that $\mathbf{n}_k \times \hat{\mathbf{t}}_k = -1$. (13) can now be written as

$$\frac{\partial \omega_l}{\partial t} \;+\; c_l[(|\mathbf{t}.|^{-2}[(\mathbf{t}. \cdot \mathbf{e}.(u'))(\mathbf{l}. \cdot \mathbf{e}.(\omega)) - (\mathbf{t}. \cdot \mathbf{e}.(\omega))(\mathbf{l}. \cdot \mathbf{e}.(u'))]$$
$$=\; -\nu c_l c'(\omega). \quad (15)$$

Equation (15) is the discrete vorticity transport equation which is automatically solved when we solve the discrete primitive variable equations. Now we will show that this equation can be rationally derived directly from (10) by the covolume method. This will imply that the discrete primitive variable equations conserve vorticity.

Th discretize (10), we take as basic variables the vorticity components ω_l directed along the primal edges ϵ_l. The velocity field is obtained by solving the div-curl system

$$d_j(u') = 0 \qquad \tau_j \in \mathcal{T} \qquad\qquad (16)$$
$$c_l(u') = \omega_l \qquad \epsilon_l \in \mathcal{E} \qquad\qquad (17)$$
$$u'|_\Gamma = \mathbf{u} \cdot \mathbf{n}|_\Gamma.$$

Existence and uniqueness for this system is dealt with in Nicolaides and Wu (1992c). We will also introduce the extensions $\mathbf{e}(u')$ and $\mathbf{e}(\omega)$ defined in section 4.

Multiplying (10) by 1 and integrating over V_l' we obtain

$$\frac{\partial}{\partial t} \int_{V_l'} 1 \cdot \omega \, dV' + \int_{\partial V_l'} 1^T M \mathbf{m} \, dV' = -\nu \int_{V_l'} \mathbf{m} \cdot \text{curl curl}\, \omega \, dV' \quad (18)$$

where $\mathbf{m}$ denotes the unit outer normal to $\partial V_l'$. Note that the convection term can be written

$$\int_{\partial V_l'} 1^T M \mathbf{m} \, dV' = \int_{\partial V_l'} (\omega_i u_j - \omega_j u_i) m_j l_i \, dV'$$
$$= \int_{\partial V_l'} (\omega \cdot 1)(\mathbf{u} \cdot \mathbf{m}) - (\omega \cdot \mathbf{m})(\mathbf{u} \cdot 1) \, dV'. \quad (19)$$

Recall that V_l' has mirror symmetry about the circulation plane ϕ_l'. Then it makes good sense to approximate (19) using the integrand's values at the midpoints of the edges in $\partial \phi_l'$ together with the mean normal, which lies in the plane ϕ_l' by symmetry, and is orthogonal to $\partial \phi_l'$. This normal is $\hat{\mathbf{t}}_k$ for the edge ϵ_k which is normal to ϕ_k. Putting these approximations into (18)-(19) gives

$$\frac{\partial \omega_l}{\partial t} + c_l \left[\frac{|\partial V'(l, \cdot)|}{|V_l|} ((\mathbf{t} \cdot \mathbf{e}(u'))(1 \cdot \mathbf{e}(\omega)) - (\mathbf{t} \cdot \mathbf{e}(\omega))(1 \cdot \mathbf{e}(u'))) \right.$$
$$= -\nu c_l c'(\omega). \quad (20)$$

This equivalent to (13) when we take into account the definition (14). It follows that the equation system (9) and (16) is equivalent to the equation system (20), (16) and (17) which is what we wanted to show.

To obtain the same equivalence if the viscous term in (10) is taken to be $\nu \triangle \mathbf{u}$, simply add ν times the dual of (3) $\nu\mathrm{gd}'(\omega)$ (which is necessarily zero by (17) and (2)) to the right side of (15) and (20). The full equivalence is therefore established.

An implication from this result is that one may choose to solve either the primitive variable or vorticity transport system and be sure of obtaining the same results. This property is unusual in incompressible flow algorithms, and it would be interesting to understand how vorticity is transported by other approaches including finite elements, but there appears to be little known about this.

References

C. A. Hall, J. C. Cavendish and W. H. Frey 1991: The Dual Variable Method for Solving Fluid Flow Difference Equations on Delaunay Triangulations. (Submitted to Computers and Fluids)

C. A. Hall, T. A. Porsching 1992: On a Network Method for Unsteady Incompressible Fluid Flow on Triangular Grids. (Submitted to Intnl. Jnl. Num. Meth. in Fluids)

F. H. Harlow and J. E. Welch 1965: Numerical Calculation of Time Dependent Viscous Flow of Fluid with Free Surface. Phys. Fluids. 8, p2182.

C. Hirsch 1990: Numerical Computation of Internal and External Flows, Vol 2. John Wiley and Sons.

R. A. Nicolaides 1989: Flow Discretization by Complementary Volume Techniques. AIAA paper 89-1978. Proceedings of the 9th AIAA CFD Meeting, Buffalo, New York, 1989

R. A. Nicolaides 1991: Covolume Algorithms. Proc. 4th Intnl. Symp. Num. CFD., Davis, Calif. Ed. H. A. Dwyer.

R. A. Nicolaides 1992a: Direct Discretization of Planar div-curl Problems. (To appear SIAM Jnl. Num. An.)

R. A. Nicolaides 1992b: Analysis and Convergence of the MAC Scheme. 1. The Linear Problem. SIAM Jnl. Num. An. (Submitted)

R. A. Nicolaides 1992c: The Covolume Approach To Computing Incompressible Flows. (To appear in Incompressible CFD – Trends and Advances, Eds. M. D. Gunzburger, R. A. Nicolaides. Cambridge University Press.)

R. A. Nicolaides and X. Wu 1992a: Numerical Solution of the Hamel

414

Problem by a Covolume Method. (To appear in Advances in CFD, ed. W. G. Habashi and M. Hafez.)

R. A. Nicolaides and X. Wu 1992b: Analysis and Convergence of the MAC Scheme. 2. The Navier-Stokes Equations. (Submitted to Math. Comp.)

R. A. Nicolaides and X. Wu 1992c: Covolume Solutions of Three Dimensional Div-Curl Equations. (In preparation.)

P. A. Raviart and J. M. Thomas 1975: A mixed finite element method for second order elliptic problems. Lecture Notes in Mathematics, 606. Springer-Verlag.

A. Segal, P. Wesseling, J. van Kan, C. W. Osterlee, and K. Kassels 1992: Invariant Discretization of the Incompressible Navier-Stokes Equations in Boundary Fitted Coordinates. (Submitted to Intnl. Jnl. Num. Meth. in Fluids.)

COMMENTS ON SESSION – INHERENTLY MULTIDIMENSIONAL SCHEMES

K. W. Morton

Oxford University Computing Laboratory
11 Keble Road
Oxford, OX1 3QD
ENGLAND

1 General observations

Session chairmen were asked to comment in the Panel Discussions on the talks in their session, to make general observations on the whole Workshop (especially in regard to their session topic), and to add their personal view of prospects for the nineties. It is impossible to make useful comments in any detail on the work of others without seeing their write-ups, so my comments on my session will be brief and combined with general comments. Then I will add some remarks on two specific classes of algorithms.

In thinking about what marks out "inherently multidimensional schemes" from the rest, one can only conclude that it consists of exploiting a direction defined by the problem and independent of the mesh. When one has dealt with the region boundary in a suitable way, one is left only with the "wind" direction; and the key question seems to be

"Is upwinding necessary or desirable? And if so, how should it affect the design of the algorithm?"

I will take this as my main theme and suggest that there are at least three rather distinct ways in which upwinding can occur, and it is virtually impossible to escape all of them if you want a really good algorithm. I shall call them:

(a) balancing convection and diffusion in steady problems;

(b) design of iteration schemes in steady problems;

(c) use of characteristics in unsteady problems.

So I find myself in general agreement with *Phil Roe* who in his talk observed casually "upwinding is inevitable"; and he then went on to outline exciting developments of his very influential 1D flux-difference splitting schemes on a triangular mesh in 2D. The work is being done in collaboration with *Herman Deconinck* and his group at VKI, and in his talk Herman described many more of the details. A key element in the algorithms is the decomposition of pressure waves into waves in a small number of directions. Another is the way in which information is transferred across a triangle depending on its orientation relative to the wave direction. The initial results that were presented at the workshop were very encouraging, and this is clearly going to be a line of development that will be pursued vigorously in the nineties, by at least two groups.

However, I have always been concerned over whether this level of sophistication in the implementation of upwinding is needed for steady or quasi-steady problems, which make up the majority in the aerospace CFD field. For truly unsteady problems, in which waves, shocks and other significant phenomena propagate at near characteristic speeds, I certainly think such techniques are needed. It is also clear that more attention will be paid to unsteady problems in the nineties than hitherto. But most of these will be quasi-steady problems, where changes are slow relative to the characteristic speeds; and then it would seem desirable to pay more attention to distinguishing and exploiting the differing timescales, e.g., as in flutter problems where such work has already been done. Several speakers mentioned this important and neglected class of problems.

The third and final talk in my session was by *Roy Nicolaides* who surveyed his recent work on covolume algorithms for the incompressible Navier-Stokes equations. This contains many attractive ideas which are truly multidimensional, with their emphasis on associating fluxes with the faces and edges of finite volumes, and on exploiting div-curl relationships. However, as with their staggered mesh antecedents, it is difficult to see how one can reconcile the need to separate the variables into two classes held at separate sets of points from the need to model local physical processes, such as at shocks and with turbulence, that one has in high speed compressible flows.

This talk was, though, one of the very few at the Workshop which was based on finite element methods. In any consideration of inherently multidimensional methods, it should be natural to think early on of what finite element methods can offer. With their geometric flexibility, well structured algorithms for dealing with unstructured meshes and their well developed theory, at least when ellipticity dominates, they should have much to offer. It is true that in the last decade the development of finite volume methods, using structured body-fitted meshes in a multi-block framework, and with their special attraction for conservation law modelling, have stolen much of the FEM thunder. Yet I still believe that the body of knowledge gained in the development of finite element methods has much more to offer to the CFD community than it has so far taken account of. Thus it was disappointing there was so little mention of the methods in the Workshop, and I will give a few examples where it could help.

(i) Upwinding for steady convection-diffusion. This was one of the first and is one of the most widespread occurrences of upwinding to improve accuracy and robustness. And it is now one of the best understood, at least in finite element terms. Yet there are still many comments made in discussion that imply the doubt that, by using "first order" differencing, the upwind methods must somehow be sacrificing accuracy. This issue should have been settled by the well developed theory of Petrov-Galerkin methods for these problems. It shows that, for a given choice of mesh and type of approximation

(e.g. piecewise multilinear), the best approximation in any given sense is obtained by using test functions (or weighting functions in the weighted residual terminology) that are upwinded relative to the basis functions for the approximation type. The decree of upwinding depends on the mesh Péclet or Reynolds number. Thus as $h \to 0$, central differencing (with the test functions equal to the approximation basis functions) is restored. I will return to this topic again in connection with finite volume methods.

(ii) Treatment of source terms. Several speakers referred to the difficulty of dealing with source terms in finite difference methods. Indeed, despite the work that Collatz published on Hermitian methods (generalisations of Numerov's method) forty years ago, difference methods are often guilty of losing accuracy and robustness by treating source terms in a cavalier manner. The topic is of course linked to upwinding. And, again, in the finite element framework the source term is treated in a natural logical way without causing any problem. For steady problems it is the choice of test function that determines the sampling of the source term: and it was Claes Johnson's application of this to Tom Hughes' streamline upwind method which turned it into a SUPG method and transformed its effectiveness. I will return to the unsteady case in the context of evolutionary-Galerkin methods below.

(iii) Development of finite volume theory. These methods still have problems which I believe will only be properly resolved when there is a better theory for them. It is becoming increasingly clear that this will result from the development of a non-conforming Petrov-Galerkin theory. I will return to this below.

2   Prospects for cell vertex methods

Originally developed by Ni, Denton, Hall and others for the Euler equations, the cell vertex methods have recently been extended to the Navier-Stokes equations [Crumpton et al 1990, Mackenzie 1991, Mackenzie and Morton 1991, Crumpton et al 1991, Morton 1991]. Their advantages over cell centre methods for the Euler equations, on quadrilateral conservation cells, include better accuracy on non-uniform meshes, more direct imposition of boundary conditions, and fewer spurious solution modes so that there is less dependence on artificial viscosity. These advantages have long been recognised in a wide variety of other fields where the same scheme on a rectangular mesh is called the box method and associated with the names of Thomas, Keller and Preissman among others.

When the viscous fluxes are added in a consistent way, gradients averaged along each cell edge in 2D or cell face in 3D need to be computed; and this leads to a less compact scheme than the corresponding cell centre scheme. Thus in 2D one typically needs to use 12 points instead of 9 points. But it turns out that there are several advantages to using such a scheme which is very compact for the inviscid terms and sacrifices this property for the viscous terms. The robustness to mesh distortion is retained from the inviscid case, as well as the advantages in the treatment of boundary conditions and the reduced dependence on artificial viscosity — despite the fact that the treatment of the viscous terms has now introduced more rather than fewer spurious modes. However, there are now added advantages which relate to the upwinding of convection-diffusion problems already discussed above.

In those methods, which are essentially cell centre methods, the upwinding parameters have to be chosen very carefully, with the optimum choice in 1D corresponding to exponential fitting. Of course, the upwinding can also be interpreted as addition of artificial viscosity, or enhanced diffusion, so this corresponds to the strong dependence of cell centre methods on the choice of artificial viscosity. But for convection-diffusion problems the cell vertex method needs no parameters at all. It gives accurate monotone solutions, including very sharp boundary layers, just by setting the unmodified residuals to zero — see [Mackenzie 1991, Mackenzie and Morton 1991, Morton 1991].

As always, however, there is a snag. We define the residual for the cell Ω_α in 2D as

$$R_\alpha(W) := \frac{1}{V_\alpha} \int_{\partial\Omega_\alpha} f\,\mathrm{d}y - g\,\mathrm{d}x, \qquad (2.1)$$

where $f = f^I + f^V$, $g = g^I + g^V$ are the combined inviscid and viscous fluxes, and V_α is the cell measure. What we then wish to do is to find the set of nodal unknowns $\{W_j\}$ for which

$$R_\alpha(W) = 0 \quad \forall \Omega_\alpha \in \Omega_h \qquad (2.2)$$

for some sets of cells. It is when this is done appropriately that one obtains the best results. But since there is no natural correspondence between the unknowns W_j at the cell vertices, x_j, and the cells, Ω_α, it is non trivial even to get the correct number of equations, let alone to solve them efficiently.

In the work reported in [Crumpton et al 1991, Crumpton et al 1990], we have resorted to using a generalised Lax-Wendroff iteration with multigrid acceleration. And the results are still very competitive. But the iteration takes the form

$$\delta W_j = \Delta t_j \sum_\alpha D_{\alpha,j} R_\alpha(W) \qquad (2.3)$$

with distribution matrices $D_{\alpha,j}$ mapping each residual to the nodes of its cell. Thus at convergence only a sum of residuals can be guaranteed set to zero. For the Euler equations we usually find that imposition of the boundary conditions is sufficient to ensure that individual residuals are set to zero; but for the Navier-Stokes equations we have not managed to drive the individual residuals to zero.

Thus the challenge of the nineties for those of us working in cell vertex methods is to find better ways of setting up and solving the system (2.2). Closely linked to this problem, I believe, is that of constructing a better theory, of finite volume methods and a sharper, more general error analysis. A start has been made on both these problems in [Morton 1991, Morton and Stynes 1991]. For the convection-diffusion problem the idea is to use a mapping between cells and vertices based on the convective velocity direction. This enables a unique mapping to be set up between the unknowns $\{W_j\}$ and either a single cell or a modified cell, obtained by splitting cells where the flow diverges or combining them where it converges. Thus, cells on an outflow boundary where Dirichlet boundary conditions are imposed are not used in the system (2.2). This turns out to be the key to the good accuracy in boundary layers.

Hence we see the second way in which upwinding should be used. The cell vertex equations for a steady problem, considered for just a single cell, would seem to be properly centred — a genuine central difference scheme with all terms differenced about the same point. Yet the upwinding comes in the assembly of the whole system, the imposition of boundary conditions (extra conditions have to be applied at inflow boundaries) and in the solution methods for the system. For fast iterative solvers can be based on this upwind mapping, at least in 1D — see [Morton et al 1991]. In 2D it would seem prudent to use upwinding in combination with relaxation in some way. Thus the errors corresponding to entropy and vorticity waves might be dealt with by an upwind iteration; while a pressure correction type relaxation might be used for the pressure waves. A similar point was made in discussion by Alain Lerat; and this viewpoint contrasts with that of Roe and Deconinck described earlier.

Moreover, for convection-diffusion this mapping plays a similar rôle to the symmetrising operator used in the Petrov-Galerkin theory in [Barrett and Morton 1984] and [Morton et al 1990]. It does not symmetrize the associated bilinear form but it does introduce positive definiteness. Thus if M is the mapping, S_0^h is the approximation space for the homogeneous problem, and $B(\cdot,\cdot)$ the bilinear form, we get for some $\gamma > 0$

$$B(U, MU) \geq \gamma^2 \|U\|_h^2 \quad \forall U \epsilon S_0^h \tag{2.4}$$

in a discrete norm $\| \cdot \|_h$. This is the basis of an error analysis developed in [Morton and Sobey 1991].

I hope for substantial progress in these two areas in the next few years. With success, the cell vertex method should become a highly effective technique for the steady Navier-Stokes equations at all Mach numbers. Furthermore, these developments would lead the way to its use for quasi-steady problems.

3 Exploiting the evolution-Galerkin approach

For truly unsteady problems, with hyperbolic dominance, the characteristics must surely be used in some way. This is the third rôle of upwinding, and probably its most irreplaceable.

Many authors, working in various areas of CFD, have combined the use of characteristics and the Galerkin projection — see [Morton 1991] for a partial review. This is the most effective way to use finite element methods for hyperbolic equations. More generally one should use an approximation to the evolution operation: hence *evolution-Galerkin* methods. For example, in [Morton and Sobey 1991] the convolution integral for unsteady convection-diffusion in 1D is used to generate generalisations of Leonards QUICKEST scheme; indeed QUICKEST can be generated this way using global polynomial approximation instead of finite elements. On the other hand, Donea's Taylor Galerkin methods can be regarded as lying in the evolution-Galerkin class with the evolution approximated by a Taylor expansion. Many other possibilities await our exploration.

However, it is the characteristic Galerkin or Lagrange Galerkin methods, in which evolution is approximated by constancy along approximate characteristic paths, which are the most generally useful. They seem to be so little known by attendees of the Workshop it seems worthwhile to present the basic idea here in a form that can be immediately related to the more familiar difference schemes.

Consider the scalar hyperbolic problem

$$u_t + f_x = 0, \quad \partial f / \partial u = a, \tag{3.1}$$

to be approximated at time level n by a finite element expansion

$$U^n(x) = \sum_{(j)} U_j^n \phi_j(x) \tag{3.2}$$

in terms of the basis functions $\{\phi_j\}$. The direct formulation of the method is written

$$\langle U^{n+1}, \phi_i \rangle = \int U^n(x) \phi_i(y) \mathrm{d}y, \tag{3.3a}$$

where

$$y = x + a(U^n(x))\Delta t; \tag{3.3b}$$

this reflects the fact that the value $U^n(x)$ is carried along the characteristic to the point y where it is integrated against the basis function to give the Galerkin projection. Note that this ensures that these methods are unconditionally stable.

However, the ECG formulation used in [Childs and Morton 1990] and earlier papers expresses the change in one time step in terms of the flux function as

$$\langle U^{n+1} - U^n, \phi_i \rangle + \Delta t \langle f_x(U^n), \Phi_i^n \rangle = 0, \tag{3.4a}$$

where $\Phi_i^n(x)$ is the average of the basis function over the distance that the characteristic travels in one time step

$$\Phi_i^n(x) = \frac{1}{a(U^n(x))\Delta t} \int_x^y \phi_i(s)\mathrm{d}s. \tag{3.4b}$$

This can be written in terms of flux differences. So, introducing the mass matrix M formed from the inner products $\langle \phi_i, \phi_j \rangle$ of the basis functions, we have in vector form with $\boldsymbol{U}^n = (U_i^n, i = \dots - 1, 0, +1, \dots)^T$

$$M(\boldsymbol{U}^{n+1} - \boldsymbol{U}^n) + \Delta t(\boldsymbol{F}_R^n - \boldsymbol{F}_L^n) = 0 \qquad (3.5a)$$

where

$$F_{R,i}^n = \int f(U^n) \left[-\frac{d\Phi_i^n}{dx} \right]^+ dx, \quad F_{L,i}^n = \int f(U^n) \left[\frac{d\Phi_i^n}{dx} \right]^+ dx. \qquad (3.5b)$$

Here $[\cdot]^+$ denotes positive values only. Thus $F_{L,i}^n$ picks up the values of $f(U^n)$ where Φ_i^n is rising and $F_{R,i}^n$ the values where it is falling. For example, if piecewise constant basis functions are used, the mass matrix is diagonal and (3.5a) becomes

$$U_i^{n+1} - U_i^n + \frac{\Delta t}{h_i}(F_{R,i}^n - F_{L,i}^n) = 0. \qquad (3.6)$$

This gives important generalisations of the Engquist-Osher scheme. One should note that, because

$$\sum_{(i)} \phi_i(x) \equiv 1 \Rightarrow \sum_{(i)} \Phi_i^n(x) \equiv 1, \qquad (3.7)$$

we have exact modelling of the conservation law, with $\sum h_i U_i^{n+1}$ differing from $\sum h_i U^n$ by the difference of the fluxes (3.5b) at the ends of the range of summation.

This is of course just the bare bones of the method. Source terms or other terms can be added to the equation (3.1); and it is clear that these should be averaged along the characteristics, which may themselves be changed as a result. The mesh may be adapted at each time step by predicting where the ϕ_i should be positioned at the new time level before projection. Systems of equations may be treated, the best schemes seeming to be based on Roe flux difference splitting, which occurs naturally in (3.4a) and (3.5a). But the most important general modification is the use of *recovery procedures* to give enhanced accuracy. The basic idea is that the projection at the end of the time step conceals information that may be of relevance to the next step. For example, calculating flux values from projected values u^n may give poor approximations or even physically unacceptable values. Thus at the recovery stage, all information $\{U_i^n\}$ together with any a priori information about smoothness, positivity, monotonicity etc. should be combined to give a recovered approximation $\tilde{u}^n$: the only constraint placed on this process, in order to preserve conservation, is that U^n should be the projection of $\tilde{u}^n$,

$$\langle U^n - \tilde{u}^n, \phi_i \rangle = 0 \quad \forall \phi_i. \qquad (3.8)$$

Within the ECG framework we then replace (3.4a) by

$$\langle U^{n+1} - U^n, \phi_i \rangle + \Delta t \langle f_x(\tilde{u}^n), \tilde{\Phi}_i^n \rangle = 0, \tag{3.9}$$

where $\tilde{u}^n$ replaces U^n in (3.4b) to give $\tilde{\Phi}_i^n$. These ideas were first presented at the Aachen conference [Morton 1982] and have been widely developed since, but have far greater potential than has yet been realised.

With recovery added we arrive at a class of schemes which we call PERU schemes, written operationally as

$$U^{n+1} = \text{PERU}^n. \tag{3.10}$$

That is, on the data U^n we first perform a recovery step R, then the evolution E, and finally the projection P.

To conclude with a return to the theme of this session, an important contribution that this class of methods can make is to the development of inherently multidimensional schemes for unsteady hyperbolic problems. Charles Hirsch was asking in his initial survey for a truly 2D convection scheme which showed which points to use at the old time level: the answer is given here by the position of the basis function shifted down the characteristic from its node point at the new time level. Indeed the most widespread use of characteristic Galerkin methods has been with the Lagrange Galerkin form in 2D and 3D, just dealing with the convective terms by using a Lagrangian derivative. Most recently Yang [Yang 1991] has shown impressive results with the Euler equations using the full characteristic Galerkin ideas but with Petrov-Galerkin projection.

References

J.W. Barrett and K.W. Morton. Approximate symmetrization and Petrov-Galerkin methods for diffusion-convection problems. *Computer Methods in Applied Mechanics and Engineering*, 45:97–122, 1984.

P.N. Childs and K.W. Morton. Characteristic Galerkin methods for scalar conservation laws in one dimension. *SIAM Journal of Numerical Analysis*, 27:553–594, 1990.

P.I. Crumpton, J.A. Mackenzie and K.W. Morton. Cell vertex algorithms for the compressible Navier-Stokes. Technical Report NA91/12, Oxford University Computing Laboratory, 11 Keble Road, Oxford, OX1 3QD, 1991.

P.I. Crumpton, J.A. Mackenzie, K.W. Morton, M.A. Rudgyard and G.J. Shaw. Cell vertex multigrid methods for the compressible Navier-Stokes equations. In K.W. Morton, editor, *Twelfth International Conference on Numerical Methods in Fluid*

Dynamics. Proceedings of the Conference held at the University of Oxford, England July 1990, volume 371 of *Lecture Notes in Physics*. Springer-Verlag, 1990.

J.A. Mackenzie. *Cell vertex finite volume methods for the solution of the compressible Navier-Stokes equations*. PhD thesis, Oxford University Computing Laboratory, 11 Keble Road, Oxford, OX1 3QD, 1991.

J.A. Mackenzie and K.W. Morton. Finite volume solutions of convection-diffusion test problems. Submitted for publication, 1991.

K.W. Morton. Shock capturing, fitting and recovery. In E. Krause, editor, *Proceedings of the Eighth International Conference on Numerical Methods in Fluid Dynamics, Aachen*, volume 170 of *Lecture Notes in Physics*, pages 77–93. Springer-Verlag, 1982.

K.W. Morton. Finite volume methods and their analysis. In J.R. Whiteman, editor, *The Mathematics of Finite Elements and Applications VII MAFELAP 1990*, pages 189–214. Academic Press, 1991.

K.W. Morton. Lagrange–Galerkin and Characteristic–Galerkin methods and their applications. In Björn Engquist and Bertil Gustafsson, editors, *Third International Conference on Hyperbolic Problems. Theory, Numerical Methods and Applications*, volume II, pages 742–755. Studentlitteratur, 1991. Proceedings, Uppsala, Sweden, June 11–15, 1990.

K.W. Morton, T. Murdoch and E. Süli. Optimal error estimation for Petrov-Galerkin methods in two dimensions. Submitted for publication, 1990.

K.W. Morton, M.A. Rudgyard and G.J. Shaw. Upwind iteration methods for the cell vertex scheme in one dimension. Technical Report NA91/09, Oxford University Computing Laboratory, 11 Keble Road, Oxford, OX1 3QD, 1991.

K.W. Morton and I.J. Sobey. Discretisation of a convection-diffusion equation. Technical Report NA91/4, Oxford University Computing Laboratory, 11 Keble Road, Oxford, OX1 3QD, 1991. Submitted for publication.

K.W. Morton and M. Stynes. An analysis of the cell vertex method. Technical Report NA91/07, Oxford University Computing Laboratory, 11 Keble Road, Oxford, OX1 3QD, 1991. In preparation.

S. Yang. Characteristic Petrov-Galerkin schemes for 2-D conservation laws. To appear in IMPACT, 1991.